Pharmaceutical Operations Management

Manufacturing for Competitive Advantage

Pankaj Mohan, Ph.D.
Eli Lilly and Company
Indianapolis, Indiana, United States

Jarka Glassey, Ph.D.
University of Newcastle upon Tyne
United Kingdom

Gary A. Montague, Ph.D.
University of Newcastle upon Tyne
United Kingdom

McGraw-Hill

New York Chicago San Francisco Lisbon London Madrid
Mexico City Milan New Delhi San Juan Seoul
Singapore Sydney Toronto

The McGraw-Hill Companies

CIP Data is on file with the Library of Congress

1 2 3 4 5 6 7 8 9 0 DOC/DOC 0 1 3 2 1 0 9 8 7 6

ISBN 0-07-147249-5

The sponsoring editor for this book was Kenneth P. McCombs and the production supervisor was Richard C. Ruzycka. It was set in Century Schoolbook by International Typesetting and Composition. The art director for the cover was Brian Boucher.

Printed and bound by RR Donnelley.

This book was printed on recycled, acid-free paper containing a minimum of 50% recycled, de-inked fiber.

McGraw-Hill books are available at special quantity discounts to use as premiums and sales promotions, or for use in corporate training programs. For more information, please write to the Director of Special Sales, McGraw-Hill Professional, Two Penn Plaza, New York, NY 10121-2298. Or contact your local bookstore.

Respectfully dedicated to my wife Piky (Swati)—my source of strength and character

Pankaj Mohan

I would like to thank Kevin, Mark and Rachael for their patience and support without which this task would have been much less enjoyable.

Jarka Glassey

I would like to thank Jill, Laura and Jamie for their understanding during the evenings and weekends that I spent in front of the computer.

Gary Montague

Contents

Foreword

It's all about productivity. Or is it? Well, that all depends on what you hope to accomplish. Is productivity simply mass production or is it the repetition of a specific task? You may suspect that there has got to be more. Allow me to introduce a reference that may help guide you: a reference that speaks to the evernearing paradigm of internal aspects of quality, cost, speed, and productivity vs. the external factors of the public, the market, and regulatory controls. Yes, I am speaking about the pharmaceutical manufacturing industry. And if your ailment matches that of the confused individual above, then reader, you have found your antidote.

Pharmaceutical Operations Management is not a typical business process book filled with general and archaic manufacturing processes. Rather, it is a book on the cutting edge of pharmaceutical manufacturing. It is not about cutting costs but rather, generating revenue. In the pharmaceutical and biotechnology markets, this translates into improved quality, reduced costs, and increased efficiency, all of which improve product time to market. This book offers a systematic framework of concepts and practices to get you there.

Written by an experienced practitioner Pankaj Mohan, PhD, from Eli Lilly and Company, United States, along with experienced academics, Jarka Glassey, PhD, and Gary Montague, PhD, of the University of Newcastle upon Tyne, United Kingdom, together they have compiled a key lever of manufacturing excellence along with the vision for a new paradigm. This book ultimately provides management, engineers, and scientists with an opportunity to work together to form a world-class manufacturing organization. The authors refer to it as "system thinking," but I refer to it as nothing less than a complete, absolute, and dead-on strategy and execution.

Pharmaceutical Operations Management offers diagrams, illustrations, and nine case studies to deliver a clear message. In this book, productivity and product quality are central themes, stressing that both must be optimized throughout the manufacturing system. The book explains a need for a common understanding from all aspects of a company, including its engineers, scientists, academics, and managers. Its four sections comprehensively cover the manufacturing life cycle, commercialization, process capability, variability reduction, and finally, the productivity of the manufacturing life cycle. The book also covers such topics as techniques for identifying suitable scales for manufacturing operations, tips for laboratory and pilot scale-up, simplified process integration, ways to capture critical process information, as well as efficient risk management and validation plans. The authors provide a thorough overview of process analytical technology (PAT), critical process management, data management, knowledge management, and knowledge management strategies. In addition to introducing the concept, tools, and design behind Six Sigma, the authors present root cause analysis and its role in identifying special cause variations.

The book is designed for leaders, from top management to today's engineers and scientists. The book helps you to learn to develop, implement, and sustain a robust manufacturing process.

Congratulations! You are about to embark on a journey, one that will show you the way to compete in the everchanging world of pharmaceutical and biotechnological manufacturing.

JOHN CLIFTON
Group Publisher
Pharmaceutical Online and Corry Publishing

Preface

Concept

The new pharmaceutical and biotechnology industry paradigm is recognizing manufacturing as a critical aspect of the value chain. The evolving paradigm is a fine balance between internal and external dynamics. The internal dynamics include enhancement of quality and safety of products; reduction in cost of manufactured products; enhancement of speed to market new products; and a total rework on taking productivity to a new level. The external dynamics, which is one of the major drivers of the internal improvements, are public perception, market, and regulatory expectations.

Manufacturing is facing various opportunities in the changing paradigm:

- Market projections and forecasts are variable at best, leading to pressures for a reliable supply of medicine on one hand and better asset utilization on the other. The manufacturing organization has to be flexible and adaptable to changing capacity needs.
- A series of warning letters and 483s from FDA and other regulatory agencies have encouraged the pharmaceutical sector to further focus on quality.
- Cost of the manufactured products needs a closer scrutiny for cost reduction opportunities.
- Quality by design, by enhancing scientific and engineering rigor during design of the manufacturing processes.
- Six Sigma methodology or its variation to instill a more rigorous productivity mindset.
- Transformational leadership that can lead the pharmaceutical manufacturing operations into the new paradigm.

Thus, the new paradigm is encouraging manufacturing excellence and the realization that manufacturing is not just about cost but more importantly it is about generating revenue. This book presents key levers of manufacturing excellence and the vision for the new paradigm, which should help to fill a real gap in pharmaceutical manufacturing operations.

This book offers modern manufacturing methodologies along with quantitative concepts for improving quality, reducing costs, increasing efficiency, and improving time to market. This book sets the systematic framework for building a world-class manufacturing organization with focus on quality by design, scientific-risk-based approach, productivity, and transformational leadership. This book will underpin the systems approach to process design including the value of scale-up and scale-down, understanding of key elements of manufacturability, control strategy, variability reduction, as well as of Six Sigma as an improvement methodology required to unleash the manufacturing potential.

The unique aspects of the book include

1. An attempt to integrate various stages of a manufacturing life cycle using "systems thinking" as the construct-integrating tools, techniques, and strategy to engineer metamorphosis in manufacturing performance.
2. A focus on uncovering engineering and management opportunities encountered in the evolution of world-class manufacturing organizations so as to transform a state of flux into an organized manufacturing system.
3. A systems approach to manufacturing life cycle. This involves about 127 diagrams and illustrations conveying to readers the manufacturing picture with clarity.
4. A team of authors ideally composed for this compilation—an international partnership of a practitioner from a leading pharmaceutical industry (lead author—Pankaj Mohan, PhD, MBA, AMP, CEng, FIChemE from Eli Lilly and Company, United States) and academics from a leading center of academic excellence (coauthors—Jarka Glassey, PhD and Gary A. Montague, PhD from University of Newcastle, UK). The team's work also includes combined research in manufacturing for over a decade and refers to the work of various jointly supervised graduate students.
5. A focus on the pharmaceutical and biotechnology sector, with relevance to other process industries. It includes over nine industrial case studies designed to assist readers with the application of concepts.
6. A book design planned to bring engineers, scientists, academics, management, and executives to a common understanding of manufacturing, which is critical for integration of cross-functional teams to achieve metamorphosis in manufacturing performance.
7. Central themes of productivity and product quality across the entire manufacturing life cycle. Six Sigma is presented as the overarching

productivity improvement methodology in the context of manufacturing life cycle. This aspect is especially relevant to the current business climate and competitive landscape.

8. The transformational leadership model, which is truly unique as it details the leadership attribute required to develop and lead world-class manufacturing organizations.

In summary, this book is an attempt to integrate various stages of a manufacturing life cycle using "systems thinking" as the construct—integrating tools, techniques, and strategy to engineer a metamorphosis in manufacturing performance. A manufacturing system is a group of interacting, interrelated, or interdependent system components that form a complex and unified whole. Optimizing productivity and designing product quality into manufacturing are the central themes of this book. It is designed for management and technical leadership, including new engineers and scientists at the final stages of their academic learning as it covers systematic approach to manufacturing strategy backed by quantitative methodology with industrial application examples.

Case study	Descriptor
A	Stirred tank application
B	Crystallization process
C	Process control
D	Knowledge management
E	Product size variability
F	Distillation advanced control
G	Bioprocess advanced control
H	Filling—Six Sigma
I	Productivity

The case studies, listed in the accompanying table, will demonstrate a systematic application of quantitative techniques in the industry. These concepts could be extended to other manufacturing sectors not covered in the case studies.

Organization of the Book

This book will uncover engineering opportunities encountered in the evolution of world-class manufacturing organizations so as to transform a state of flux into an organized manufacturing system. A manufacturing process is designed to produce a product within certain quality specifications. The primary intent of any manufacturing business is to develop, operate, and sustain a robust process producing a high-quality product (within specifications) reproducibly at optimum productivity. The book is divided into four parts covering the manufacturing life cycle and operations.

Part 1—Commercialization

Decreasing time to market is an important aspect for pharmaceutical and biotechnology sectors. The first product to enter the market often gains a competitive advantage. This calls for an aggressive commercialization of new products and a robust process design phase that would set the foundation of manufacturing success. Although there are various aspects of design, that is, facility design, utility design, and so forth, Part 1 focuses on the engineering aspects of the late process development phase and manufacturability as key aspects of the commercialization process for new products.

Chapter 1—Process design. The aim of Chap. 1 is to demonstrate a systematic approach to process design with the focus on delivering a robust manufacturing process which is "right the first time."

Learning outcomes of this chapter will include

- Techniques for identification of a suitable scale for manufacturing operation
- Scale-down methodology for applying engineering principles
- Tips for productive laboratory and pilot-scale development
- Scale-up philosophies

The learning is then applied to a comprehensive case study to gain practical understanding of the application of the concepts.

Chapter 2—Manufacturability. Manufacturability implies that the process design is capable of delivering a robust process. This is the final phase of late stage commercialization. The focus of this chapter is to systematically introduce various aspects of manufacturability.

Learning outcomes of this chapter will include

- Introduction to process integration and process simplification
- Introduction to process flow description as an established way to capture information critical to manufacturability
- Detailed discussion on risk management and FMEA
- Introduction to modeling and tips for process optimization
- Introduction to commissioning, qualification, and validation plans

A few examples are discussed to illustrate the practical application of these concepts.

Part 2—Process capability

A manufacturing process must be capable of producing a high-quality product within specification, reproducibly. A successful in-process control strategy is key to a capable manufacturing process. In-process control strategy implies that adequate quality control is built in the operation to deliver a capable process and product within specification. Manufacturing should not only test quality but should also build quality into the process. This part focuses on in-process control to proactively minimize deviations in the manufacturing process. Another aspect of a capable manufacturing operation is its ability to manage knowledge and to put knowledge in action. Process capability has two elements: (1) critical control strategy and (2) knowledge management.

Chapter 3—Critical control strategy. The aim of this chapter is to introduce a systematic approach to develop a critical control strategy for a robust manufacturing platform, with the objective of building quality control in the design (rather than testing for it).

Learning outcomes of this chapter will include

- The control hierarchy
- *Process analytical technology* (PAT) and critical process parameter measurement
- Alarm strategy
- Data management
- Control of critical process parameters

The learning is then applied to a comprehensive case study to gain practical understanding of the application of the concepts.

Chapter 4—Knowledge management. The aim of this chapter is to demonstrate a systematic approach to knowledge management with the focus on building a learning and innovative organization.

Learning outcomes of this chapter will include

- An appreciation of the critical need for building a learning and innovative organization
- Intellectual property, a key business differentiator
- Knowledge-management strategies
- A novel knowledge-extraction technique
- Turning intellectual property into action

The learning is then applied to a comprehensive case study to gain practical understanding of the application of the concepts.

Part 3—Variability reduction

Variability could render a product out of specification leading to rejection and rework. When variability is out of control there can be a significant threat to manufacturing. If variability is controlled so that the mean productivity of the process improves, this will have significant business impact too. Besides the cost issue, variability makes process improvement efforts difficult as any small improvement may be masked by variability. Therefore reducing variability should be an important priority for a manufacturing operation. This part discusses industrial examples of variability reduction.

Chapter 5—Fundamental strategies for variability reduction. The previous chapters (Chaps. 1 to 4) have dealt with process design and implementation of a manufacturing process. A key aspect of implementing a robust process and its subsequent sustenance is the ability to reduce variation.

Learning outcomes of this chapter will include

- Historic perspective on variation
- Understanding types of variation and process capability quadrants
- Statistical process control concepts
- Root cause analysis and its role in identifying special cause variation

The learning is then applied to a case study to gain practical understanding of the application of the concepts.

Chapter 6—Emerging monitoring and control strategies. This chapter introduces new and emerging monitoring and control strategies aimed at further reducing variability and enhancing process performance.

Learning outcomes of this chapter will include

- Techniques for advanced process control including artificial neural networks and model predictive control
- Introduction to *multivariate statistical process control* (MSPC)
- Introduction to artificial neural network pattern recognition
- Introduction to knowledge-based systems

The learning is then applied to a comprehensive case study to gain practical understanding of the application of the concepts.

Part 4—Productivity

In a competitive environment, organizations focus on productivity to generate a winning proposition for their customers. The aim of productivity improvement is to continuously or radically enhance manufacturing metrics. At the organizational level, productivity has to be envisioned at a broader level encompassing all aspects of the business. In a complex, ambiguous, and dynamic business environment, productivity is the key differentiator of a successful business. Knowledge, leadership, and synergy are three key pillars of optimal productivity. This part focuses on the methodology and philosophy for optimal productivity.

Chapter 7—Productivity improvement methodologies. The aim of this chapter is to present an integrated picture of productivity improvement methodologies under the overarching improvement philosophy of Six Sigma.

Learning outcomes of this chapter will include

- Introduction to Six Sigma
- The lean Six Sigma concept
- Overview of Six Sigma tools
- Introduction to design for Six Sigma

The learning is then applied to a case study to gain practical understanding of the application of the concepts.

Chapter 8—Productivity improvement philosophy. Productivity is the underlying message in the book applied systematically to "process life cycle management." This final chapter focuses on putting the concepts discussed in the previous chapters in tune with an overarching philosophy for productivity improvement. The overarching philosophy focuses on key elements such as knowledge, leadership, and siloless synergy.

Learning outcomes of this chapter will include

- An understanding of productivity improvement in the larger organizational context. While Six Sigma methodology is critically important for productivity, it is a subset of a larger design that includes a well-balanced cocktail of knowledge, leadership, and siloless synergy.
- Other critical aspects of leadership. This book so far has focused on the analytical intelligence aspect of leadership, however, other aspects of leadership are equally critical.
- Logistical strategy to engage the organization in the productivity drive.

The learning is then applied to a comprehensive case study to gain practical understanding of the application of the concepts.

Part

1

Commercialization

Decreasing time to market is an important aspect for pharmaceutical and biotechnology sector. The first product to enter the market often gains a competitive advantage. This calls for an aggressive commercialization of new products and a robust process-design phase that would set the foundation of manufacturing success. Although, there are various aspects of design, i.e., facility design, utility design, and so forth, Part 1 focuses on the engineering aspects of late process-development phase and manufacturability as key aspects of the commercialization process for new products.

Chapter

1

Process Design

You never have enough information, and you never have enough time. WOODY FLOWERS

Objective

The aim of this chapter is to demonstrate a systematic approach to process design with the focus on delivering a robust manufacturing process, which is "right the first time."

Learning outcomes of this chapter will include:

- Techniques for identification of suitable scale for manufacturing operation
- Scale-down methodology for applying engineering principles
- Tips for productive laboratory and pilot scale development
- Scale-up philosophies

The learning is then applied to a comprehensive case study to gain practical understanding of the application of the concepts.

1.1 Introduction

Commercialization of new products is the key competitive lever for manufacturing industries. Decreasing time to market is important in virtually all manufacturing sectors. The first product to enter the market often gains a competitive advantage. Launch of a new product ahead of the competition is important for the financial success of a manufacturing business.

At a simplistic level, commercialization has four phases (Fig. 1.1). The first phase is discovery of new products (product design), where a new business opportunity is born. This is followed by early process development phase in small-scale laboratories; the focus of this phase is to screen the opportunity by manufacturing small quantities of material for trials and investigation. The successful products are then developed further in the late development phase to a scalable manufacturing process with optimal productivity, which is safe and environment friendly. The final phase of the commercialization process is the assurance of manufacturability. Manufacturability implies that the process design is capable of delivering a robust process.

Process development includes both early and late development phases. The focus of the early process development stage is to rapidly produce and screen as many compounds (received from the discovery phase) as possible. Thus, the methods and reagents are chosen to assist rapid laboratory synthesis and purification. Little consideration is given to the cost of the reagents, since only milligram quantities are typically being synthesized. Scalability of the process is also not a consideration at this point. Although this will require the synthesis pathway to be redeveloped before designing a production facility, it is appropriate because less than one-tenth of the molecules screened will have any commercial potential. Rapid screening is essential to find likely candidate molecules.

For pharmaceutical and bioprocess business the manufacturing process is classified into two broad categories—drug substance and drug

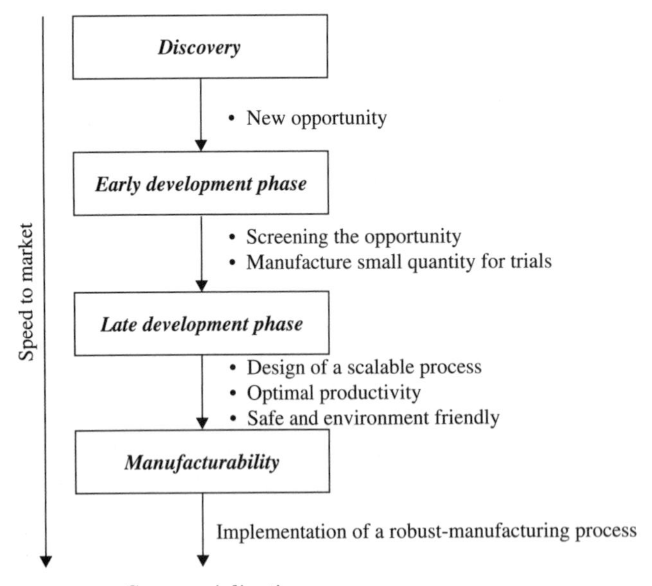

Figure 1.1 Commercialization process.

product. Drug substance incorporates the manufacturing process that leads to the formation of the active pharmaceutical ingredient. Drug product incorporates the process that transforms the active pharmaceutical ingredient into a formulated product ready to be delivered to the patient.

1.1.1 Drug substance—early development phase

For the drug substance process development, during the early development phase, scientists (chemists/microbiologists) begin looking at various synthetic routes that can be used to create the desired molecule. The focus is to produce enough material of acceptable purity and potency for clinical trials. The chemists begin by brainstorming various possible routes based on their general knowledge of chemical reactivity. The most promising of these routes are then selected for exploratory process research, where the scientists determine whether the chemistry will actually work and what yields can be achieved. This lays the groundwork for the process design that will eventually be used in the production facility. After sufficient exploratory process research has been completed, a team meets for route selection. All the data for the various routes under consideration are brought together for detailed analysis. One route is selected for more detailed research and will eventually be passed to process development. The selected route is the one that appears to have a high certainty of success. Further experimentation will attempt to enhance profitability associated with the route, most often by increasing product yields, and reducing cycle time and cost of product.

Once a potentially viable compound is identified, development moves to an exploratory process chemistry group. Concurrently, the molecule will be undergoing further testing (e.g., toxicology, potency, environmental impact) that could cause the project to be terminated. With such a huge incentive to decrease time to market, companies have embraced the ideas of life cycle design and concurrent engineering. A fundamental requirement is to improve development and manufacturing processes by integrating the development process across traditionally distinct functions. One way this integration can be implemented is by concurrently performing various activities that would typically be segmented into different departments. For example, while continuing the development process, a new product may be in the midst of field trials and obtaining regulatory approval. Another way is to consider the downstream ramifications of early design decisions in detail. Such an approach helps to maximize the effective patent life by making development of the manufacturing process faster, thus reducing the likelihood that market entry will be delayed because of process design considerations.

The key outputs of the early stage process development are the screening of new opportunities identified during the discovery phase and the manufacture of small quantities for testing and trials. At the same time, speed is of essence—failing fast has significant business implications for rapid commercialization. Failing fast is important, as only a few molecules screened actually become a viable product. The competing priorities of failing fast and of process development for clinical trial material of acceptable quality create interesting dynamics. Detailed discussion on product development and early stage process development is beyond the scope of this book.

1.1.2 Drug product—early development phase

The active ingredient called drug substance is formulated to produce final drug product for the patients. The final dosage form could be oral or parenteral. The oral dosage form can be solid or liquid while the parenteral dosage form is administered intravenously or intramuscularly. This section provides a brief overview of the early process development stage of drug product formulation.

The early process development stage for drug product focuses on developing the right formulation and dosage form. Bioavailability is the North Star for drug formulation. Bioavailability stands for the availability of drug substance to the patient. A simplistic bioavailability kinetic model could include absorption of the drug substance in the blood and subsequent decay (assuming first order) of the drug concentration in the blood either by metabolism or by excretion. One of the initial challenges pharmaceutical scientists face is the selection of a dosage form. Primary dosage forms include liquid or solid. The liquid dosage form that is administered by *intravenous* (IV) injection has the maximum bioavailability.

Oral dosage form. The oral solid dosage form undergoes a longer absorption process than does the IV dosage form. This is because the IV dosage form is directly injected into the bloodstream whereas the oral solid dosage form has to undergo absorption in the intestine or other parts of the GI system. That is why liquid administered through an IV has a better bioavailability as compared with oral solid or liquid dosage form. This comparison between solid and liquid dosage forms is very important for formulation and is quantified by a factor called *formulation efficiency*. Formulation efficiency for solid dosage is the ratio of absorption of solid dosage form to liquid form. One of the major advantages of the solid dosage form is product stability as compared with the liquid form.

Traditionally, the dissolution test has been used as an indicator with the assumption that the faster the dissolution, the more readily will the drug substance be absorbed in blood. This may not necessarily be true in all cases; however, this provides a good portal for relative comparison of various dosage forms. Typically, a plate method of dissolution is used by making a tablet in a die and rotating it at a given speed (rpm) in a dissolution medium contained in a small tank. The dissolution medium can be water, HCl, or buffer. The dissolution temperature and speed of rotation are controlled and specified. Engineers should be engaged at this early stage of formulation development to assist scientists in applying engineering principles in developing pharmacokinetic models describing the absorption mechanism and relating it to the dissolution characterization.

In the event of solid dosage form selection, the pharmaceutical development team has to figure out the type of solid dosage form. In order of increasing complexity, solid formulation has various forms, which includes pure drug substance, powders, hard-shell capsules, uncoated tablets, film-coated tablets, and sugarcoated tablets. Selection of excipients for dosage form (other than pure drug substance) is the next significant exercise in the drug formulation. Excipients work as fillers, diluents, and in some cases, stability enhancers for pure drug substance; and depending on the dosage concentration, appropriate amount of excipients are added. Excipients exhibiting physical and chemical compatibility are subsequently recommended for formulation development.

Engineering partnership gains importance during the solid dosage form selection as solids' properties and related unit operation such as milling, wet granulation, coating, drying, and blending become significant in ensuring product uniformity and dissolution. For example, selection of unit operations to ensure content uniformity during tablet formulation requires an engineering analysis of powder flow, blending, and compressibility, based on which unit operations such as direct compaction, roller compaction, or wet granulation can be selected in addition to mixer/kneader and drying technology selection. Mixer/kneader selection requires technology selection of suitable agitation systems—Hobart mixers, Sigma-type kneaders, high-intensity mixers, and the like. Another key aspect of technology selection is drying technology such as microwaves, vacuum dryers, tray dryers, and fluid-bed granulators. The drying aspect poses a challenge for engineers and scientists in the determination of end point. Typically, for a fluid-bed dryer, change in exit temperature is an indication of the end point, while for other technologies establishing drying characteristic and a mass balance can help determine the end point (Carstensen, 1998).

Parenteral formulation. An important aspect of dosage form, which is gaining importance in pharmaceutical formulations, is *sustained release.* Sustained release formulations release active drug substance in a controlled manner over a length of time. The formulators have to overcome the risk of dumping, that is, a large amount of drug substance released earlier than expected. There are various ways of achieving sustained release—erosion, coated beadlets, insoluble matrix, osmotic pump, hydrogels, and microspheres (Avis et al., 1993). The details of formulation are beyond the scope of this book.

Another important aspect of drug product is *packaging development.* Packaging is designed to ensure:

- Product stability and protection
- Presentation to patient
- Identification of drug and dosage
- Information including any warning
- Convenience of usage by the patient
- Containment and compliance

The details of packaging development are also beyond the scope of this book, but can be found in comprehensive literature, such as Dean et al. (2000).

The earlier section gave a brief overview of some of the important aspects of drug formulation in the early development stage. This section will focus on the late process development stage, highlighting the systematic application of engineering for rapid commercialization. The objective of late-process development phase is to develop a robust process in the laboratory that is scalable to commercial production. The focus of this phase is to develop a manufacturing process with optimal productivity (and profitability) that is safe and environment friendly. A systematic approach to late process development phase is proposed in Fig. 1.2. According to this approach the first step is to identify the manufacturing scale.

Economies-of-scale analysis helps to identify the manufacturing scale. Profitability projections are generally based on results obtained by the process at the laboratory and pilot scale. Any losses of efficiency, quality, or productivity caused by the increase in scale will adversely affect the technoeconomic feasibility of the project. Thus, accurate prediction of process parameters at the manufacturing scale is important for the industrial application of the new process. Section 1.2 describes a methodology for understanding economies of scale, which can assist in finalizing the optimal manufacturing scale.

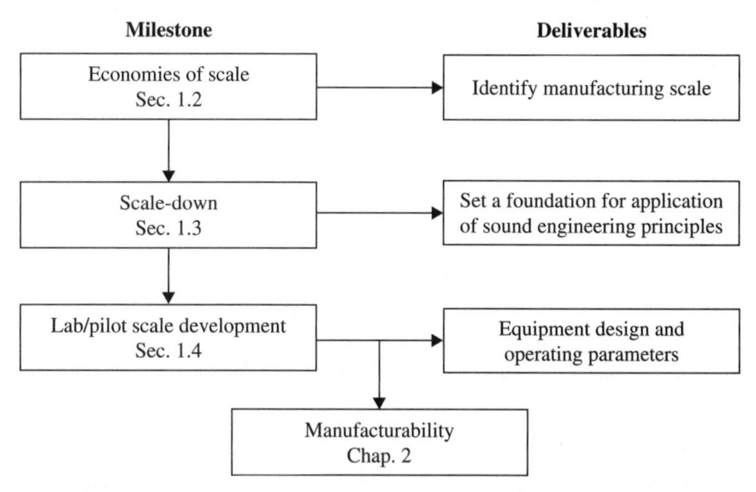

Figure 1.2 Systematic approach to late-stage process development.

It is equally important to understand how the process should be designed to minimize the effects of scale-up on production. Therefore, once the manufacturing scale is determined, simulation of the process conditions in a small-scale laboratory (called scale-down) can be undertaken. This is key to the success of scale-up. Scale-down ensures a sense of realism in process development leading to a process design with a high probability of success upon commercialization. If performed correctly, the scale-up becomes a mere verification stage and a very competent organization may skip the pilot scale verification stage to enhance speed to market. Scale-down methodology is described in Sec. 1.3 and the laboratory/pilot scale development is discussed in Sec. 1.4. The validity of the systematic approach to late process development phase is demonstrated for an industrial case study discussed in Sec. 1.5.

1.2 Economies of Scale

Identification of manufacturing scale is an important step in the late stage of commercialization process. It sets the stage for the scale-down operation with "end in mind." An appreciation of the economies of scale by engineers and scientists is an important factor for an optimal design. The application of a rigorous economy-of-scale methodology is important to finalize the scale of operation. This section describes factors that must be considered in finalizing the scale of operation, along with scale-up accounting techniques introducing some basic financial concepts.

The economies of scale could be represented as a cyclical process as indicated in Fig. 1.3. The initiation point of the cycle is demand projection to understand the business deliverables; this is followed by business

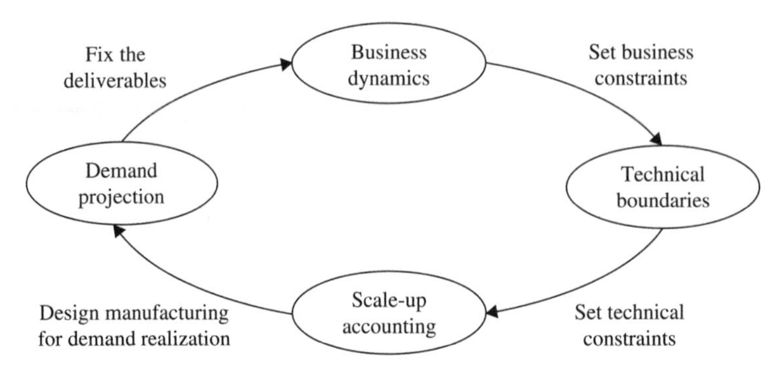

Figure 1.3 Cycle of economies of scale.

dynamics that sets the business constraints. Demand projection and business constraints set the business expectation for the manufacturing facilities, which gets translated into the technical boundaries. Once the technical boundaries are set, then scale-up accounting methodology helps achieve the optimal utilization of capital with regard to the demand. Even after a product is commercialized any changes in demand would lead to repeating this cycle.

This section focuses on technical boundaries and scale-up accounting methodologies of the economies of scale in detail.

1.2.1 Demand projection

For estimating future demand, the company may use several methods—buyers' survey, composite of sales force opinion, expert opinion, market tests, time series analysis/dynamic modeling, and statistical demand analysis. These methodologies could help define total market potential. This is typically defined as:

$$Q = nqp \tag{1.1}$$

where Q = total market potential
n = number of buyers in the specific product/market under the given assumptions
q = quantity purchased by an average buyer
p = price of an average unit

The number that is most relevant to the economies of scale for manufacturing is

$$M = nq \tag{1.2}$$

where M is the manufacturing target.

Manufacturing target sets the plant capacity for producing a particular product. Plant capacity is typically designed in light of future projections and market potential. Various assumptions are used in collecting data for market potential. These assumptions may range from conservative to optimistic estimates. The dilemma is—should the company create manufacturing facilities based on optimistic estimates and suffer idle capacity, or should it take a conservative course and suffer lost opportunity. Each company has its own philosophy for risk taking.

1.2.2 Business dynamics

Demand projection gives management enough information to decide whether to launch the new product or not. Commercialization is a significant cost in the product life cycle; therefore, careful financial consideration is key to the economic viability. If the economic forecast is favorable, the next step is the timing. In commercializing a new product, market-entry timing can be critical. Speed of commercialization is often stressed by companies as a key competitive advantage and is especially important for the pharmaceutical industry because of patent protection aspects. Location of the manufacturing site is another important issue facing business leaders. Considerations, such as taxes and locations, are probably significant factors to consider. Certain locations (e.g., Ireland, Singapore, and Puerto Rico) currently have tax advantages for manufacturing in the global economy. Diversifications of risk or supply chain consideration are also critical in deciding whether to develop multiple sites or a single manufacturing site.

1.2.3 Technical boundaries

The demand projection and business dynamics help formulate the boundaries around the economies of scale. Another vital factor in the economies of scale is operability of the manufacturing scale under technical constraints that may be process or facility specific. The process constraints are determined during the development stage. The scalability aspect of the process developed (to be described later in this chapter) is especially relevant. For financial analysis, that precedes detailed process development work, typically a very approximate estimation of technical boundaries is made with the help of process experts.

1.2.4 Scale-up accounting

After setting approximate boundary conditions—demand, business dynamics, and technical boundaries—the challenge is to specify the scale of operation. This is not a trivial task and there is no unique answer. Scale-up in industry has been achieved by and large by trial and error. For example, while some biotech companies have successfully scaled up bioreactors and benefited from the economies of scale, others have successfully used numerous batteries of small-scale equipment to achieve the manufacturing target. A few small-scale companies have gone out of business either because they had an incorrect scale of operation or their competitors had a more efficient scale of operation. Thus, a rigorous financial analysis of scale of operation is critical for the survival of the company. An application of scale-up accounting methods is illustrated in case study A in Sec. 1.5.

In scale-up accounting, manufacturing or production costs are critical. They can be classified into three basic categories:

1. Direct materials
2. Direct labor
3. Factory overhead

Direct materials. All materials that can be identified with the production of a finished product can be easily traced to the product, and represent a major material cost of producing the product, for example, water as a raw material used in manufacturing.

Direct labor. All labor directly involved in production that can be easily traced to the product, and that represents a major labor cost of production, for example, production operators directly involved in manufacturing.

Factory overhead. This all-inclusive collection is used to accumulate the cost of indirect material, indirect labor, depreciation of capital assets, and the like. Factory overhead includes costs that may be classified as variable, fixed, or semivariable. Fixed factory overhead costs are those that remain constant with the volume of production, while the fixed cost per unit varies with the volume of output. Depreciation of capital assets is an example of a fixed cost.

Direct materials and labor are variable costs. Variable costs are those in which the total cost changes in direct proportion to changes in volume or output, while the unit cost remains constant.

These three categories of the manufacturing cost flow through the work in a process inventory account. The costs of direct materials, labor, and factory overhead used in production are charged to work in progress. When goods are completed, the total costs incurred to manufacture the goods are transferred from work in progress to the finished goods inventory account. While these accounting principles are adhered to in product costing, the scale-up aspect is more sensitive to the fixed and variable cost concepts. Variable costs are key levers in cost control of any manufacturing operation. The value of this concept is explained using a scale-up case study described in Sec. 1.5.

The scale-up accounting illustrated by case study A (Sec. 1.5) is based on historical data. However, in the industry the financial projection is made prior to the construction of the manufacturing facility, leading to a heavy reliance on cost estimation. There are several sources for cost estimation:

1. Equipment vendor's estimate
2. Data from previous experience
3. Published guides to capital cost estimation, for example, *A Guide to Capital Cost Estimating*, Institution of Chemical Engineers, 3d ed., 1988.

Above all, the experience of the estimator is most crucial as "fuzz factors" are generally used to ascertain more realistic financial figures.

The detailed accounting for "product costing" is complex and involves meticulous calculations. In this chapter, a simple aspect of the accounting is discussed focusing on the generic learning around the economies of scale. This basic knowledge of product costing is useful for both engineers and scientists, as it helps develop a better appreciation of financial realities. The systematic approach to scale-up accounting consists of four steps. The first step considers analysis of the total capital employed, followed by accounting for product cost, cost of product, and ratio analysis.

Step 1: Accounting for total capital employed. Total capital employed is the quantification of total cost in building the manufacturing facility and operating the plant for a year. This cost is very important for the economies of scale as it indicates the level of financial commitment and forms the basis of financial decision.

Some of the cost categories that are involved in this accounting include:

- Total equipment cost such as process equipment (fermenter, seed tank, and so forth); utilities equipment (steam, water, waste systems, and so forth); installation; and instrumentation.
- Engineering and supervision cost includes expenses incurred for architects, structural engineers, mechanical and electrical consultants, process consultants, and surveyors.
- Utilities costs include operating expenses for steam, cooling water, and electricity.
- Consumables include filter membranes, sample bottles, and the like, and rates include the interest rate on loans.

Raw material (medium or substrate) is an important item in the cost estimation. Utilities (electricity, water, and gas) to the main plant are also a major cost. This cost is dictated by the location of the site and the efficiency of the process.

Step 2: Accounting for product cost. It is important to account for the manufacturing cost needed to produce the desired product. Simplistically, manufacturing cost is the sum of operating cost, depreciation on equipment and buildings, and any loan payment. Additional cost may include hidden costs such as cost of rework and loss of product due to poor quality.

$$\text{Manufacturing cost} = \text{operating cost} + \text{depreciation} + \text{loan payment} + \text{hidden costs} \tag{1.3}$$

A manufacturing operation manages the manufacturing cost to ensure the viability of the business and to deliver value for money to the customers and shareholders.

Step 3: Cost of product. *Cost of product* (COP) is a fundamental financial measure for a manufacturing operation. The elements of COP include manufacturing volume, cycle time of a batch, cycles per year, productivity, and annual throughput.

$$\text{COP} = \text{manufacturing cost per unit volume or weight of product} \tag{1.4}$$

Step 4: Economies of scale—ratio analysis. The comparison of financial ratios of two scales is an important aspect of the economies of scale. The primary financial metric is COP (per unit volume), typically the unit cost of production is scale dependent with a lower COP for the larger scale (see Sec. 1.5). The ratio of COP of one scale to another gives a good measure of scale-related impact. The ratio of total capital employed to COP gives insight into the investment rationale of the economies of scale with regard to the cost of products. Scale-up case study A demonstrates the financial analysis involved in choosing the right scale for an economically viable operation.

The initial estimation of the commercial scale of operation sets the stage for the next phase of the late-stage process development. For successful commercialization it is important to undertake the process development (small-scale laboratory work) under simulated commercial process conditions. This critical development step is called *scale-down.*

1.3 Scale-Down

The foremost challenge for scientists and engineers is to develop a process capable of producing a desired product in a small-scale laboratory. A key aspect of success is to start with a sense of "realism" from

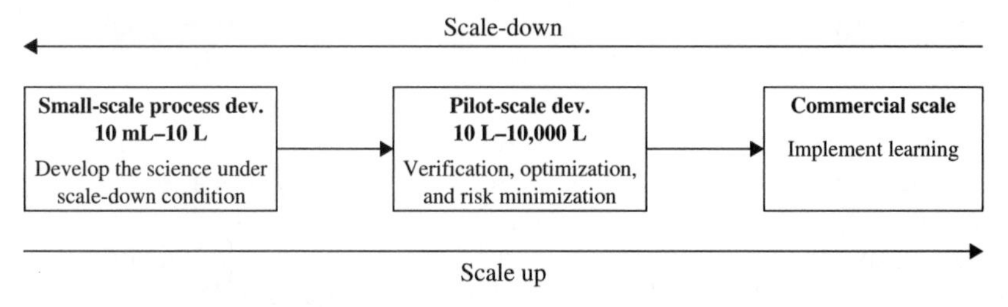

Figure 1.4 Process development scale-up/scale-down.

the outset by adopting scale-down philosophy. Scale-down can be defined as developing the small-scale process within the boundaries or constraints of a large-scale commercial process.

Figure 1.4 summarizes scale-up/scale-down with regard to scale of operation and the deliverables from each stage. The process is typically developed in small volumes, approximately 10-mL to 10-L scale, where the focus is developing the science under scale-down conditions. The next step in the scale-up protocol is to verify the process at a pilot scale, before developing the commercial scale, which is typically in the volume range of 10 to 10,000 L. The objective of the pilot-scale development is to verify the process in relation to the product attributes, optimize the process, and minimize risk to product, people, and environment. What is learned from small scale and pilot scale is implemented in the commercial scale of operation.

There are two basic approaches to scale-up/scale-down: the empirical approach is a data-driven approach, the essence of which is experimentation; and the mechanistic approach, which is based on physical understanding of the system, driven by universal laws. An optimal approach to scale-up/scale-down should include a combination of empirical and mechanistic approaches called the *hybrid approach*. Figure 1.5 summarizes the scale-up/scale-down methodologies related to the interaction of

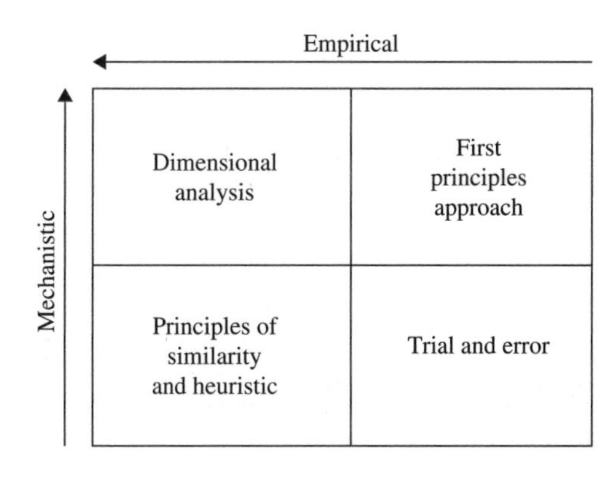

Figure 1.5 Approaches for scale-up/scale-down.

empirical and mechanistic approaches. Principle of similarity methodology is primarily an empirical approach, while first principles methodology is primarily a mechanistic approach. Dimensional analysis provides a good combination of both empirical and mechanistic approaches, while trial and error is neither empirical nor mechanistic. Another approach to scale-up/scale-down is heuristic, which is based on human knowledge and experience (rule of thumb), and is discussed in detail in Chap. 4. The scale-up/scale-down approaches will be discussed in detail in this section followed by a case study example in Sec. 1.5.

1.3.1 Principles of similarity

The scales differ in many ways, though the similarity in some aspects may lead to ease of scaling up or down. Johnstone and Thring (1957) refer to four types of similarity as aides to effective-process translation between scales, namely, chemical or biochemical similarity, geometric similarity, mechanical similarity, and thermal similarity.

Chemical or biochemical similarity is an important aspect of scale-up, it refers to the similarity in the underlying science—chemistry or biochemistry—at a molecular level. This includes stoichiometry and the reaction kinetics rate.

Geometric similarity implies similar geometrical aspects of the equipment. Look alike is the essence of geometric similarity. Typically, linear scale ratios are compared across scales. For example, for a mixing tank, tank diameter to impeller diameter is a widely used ratio for geometric similarity.

Mechanical similarity includes force applied to a stationary or moving system, which can be described in static, kinematic, or dynamic terms. *Static similarity* relates to the deformation under constant stress of one body or structure by that of another. It exists when geometric similarity is maintained, even as elastic or plastic deformation of stressed structural components occurs. *Kinematic similarity* encompasses the additional dimension of time, while *dynamic similarity* involves the forces that accelerate or retard moving masses in dynamic systems.

Thermal similarity includes the ratio of thermal fluxes for different modes of heat transfer—convective, conductive, and radiation. This also includes similarities related to the flow of fluids and fluid thermal capacity. It is also important to add mass transfer rate similarities as a scale-up guide in some cases.

1.3.2 Dimensional analysis and dimensionless numbers

Dimensional analysis is a hybrid approach to scale-up/scale-down, engaging both mechanistic and empirical philosophies. Dimensional analysis is a method of generalizing the physical problem using basic dimensions

TABLE 1.1 International System of Units

Fundamental parameters	Fundamental dimensions	Fundamentals units (SI)
Length	L	m (meter)
Mass	M	kg (kilogram)
Time	T	s (second)
Thermodynamic temperature	θ	K (kelvin)
Amount of substance	N	mol (mole)
Electric current	I	A (ampere)
Luminous intensity	I_v	cd (candela)

(length, mass, time, and so forth). It is based on the principle that a process equation has to be dimensionally homogenous, and that it could be generalized for any scale. This is an integrated approach between formal mathematics and empiricism. The dimensionless numbers generated from the analysis require the application of first principles for interpretation.

The currently used *International System of Units* (SI) is based upon seven basic dimensions. They are as shown in Table 1.1:

A scientific approach of using the dimensions to understand the interrelation of the physical phenomenon is called the *pi theorem*. The pi theorem (Boucher and Alves, 1959 and Catchpole and Fulford, 1966) is defined as:

> Physical relationship between n physical quantities can be reduced to a relationship between $m = n = r$ mutually independent dimensionless groups, where r is the rank of the dimensionless matrix, made up of the physical quantities in question and generally equal to (or in some case smaller than) the number of the base quantities contained in their secondary dimensions.

Scale-independent attributes of the pi theorem form the foundation of scale-up. This theorem could be interpreted as: two processes may be considered completely similar if they take place in a similar geometrical space and if all the dimensionless numbers necessary to describe them have the same numerical value.

To illustrate the pi theorem approach, let us apply this to a widely used piece of process equipment—the mechanically agitated tank. For the design of a baffled mechanically agitated tank, understanding the interaction of energy input (power) of the impeller (type used is Rushton) and the related impact of physical properties on process parameters is key for a successful scale-up. The following steps will aid understanding of the application of the pi theorem.

Step 1. From experience and knowledge (first principles), list the parameters influencing energy (power) input in the mechanically agitated tank.

$$\text{Power} = \text{fn}(d, \rho, v, N, Q, g) \tag{1.5}$$

TABLE 1.2 Pi Theorem Example

Parameter	Symbol	SI units	Dimensions
Impeller diameter	D	m	L
Impeller speed	N	1/s	T^{-1}
Density of fluid	ρ	kg/m^3	$M L^{-3}$
Kinematic viscosity	ν	m^2/s	$L^2 T^{-1}$
Gas flow rate	Q	m^3/s	$L^3 T^{-1}$
Acceleration due to gravity	G	m/s^2	$L T^{-2}$
Power	P	W	$M L^2 T^{-3}$

where the variables are described in Table 1.2 together with SI units and dimensions of each variable.

Step 2. Create dimensional matrix. Matrix rows are formed by the fundamental dimensions and represent the rank r of the matrix (see Table 1.3). The columns of the matrix consist of the parameters. The dimensional matrix consists of a square core matrix and a residual matrix. A residual matrix should consist of the target and other related parameters (for power it is kinematic viscosity, gas flow rate, and acceleration due to gravity). Core matrix consists of the essential variables that define the target variables in some relationship determined by the solution of the dimensional matrix (for power it is density, impeller speed, and diameter). A square matrix may appear in more than one dimensionless group.

Step 3. Solve dimensional matrix—linear transformation. Create a unity matrix by manipulating the core matrix. A unity matrix is characterized by the main diagonal consisting of numeric value of only one, and the remaining elements are zero. The dimensionless number is a fraction created by the elements of the residual matrix as a numerator and the core matrix as the denominator, with the exponent as in the residual matrix.

The manipulations to achieve the unity matrix are shown in Table 1.4.

By definition the number *pi* (Π) is given by the difference: $m = n - r = 7 - 3 = 4$.

TABLE 1.3 Dimensional Matrix Example

	Core matrix			Residual matrix			
	ρ	d	N	P	ν	Q	g
Mass, M	1	0	0	1	0	0	0
Length, L	−3	1	0	2	2	3	1
Time, T	0	0	−1	−3	−1	−1	−2

TABLE 1.4 Linear Transformations

	Unity matrix			Residual matrix			
	ρ	D_i	N	P	v	Q	G
Multiply row 2 by 3 times row 1							
M	1	0	0	1	0	0	0
$3M+L$	0	1	0	5	2	3	1
T	0	0	–1	–3	–1	–1	–2
Multiply row 3 by –1							
M	1	0	0	1	0	0	0
$3M+L$	0	1	0	5	2	3	1
$-T$	0	0	1	3	1	1	2

Applying this to the unity and residual matrices shown in Table 1.4 for each column of the residual matrix the Π number can be calculated by using the values shown as powers of variables in the unity matrix in that order.

In this case for the first column of the residual matrix (representing power requirement P) the following relationship can be written from Table 1.4

$$\Pi_1 = P/\rho^1 D_i^5 N^3 = N_P \text{ (power number or Newton number)} \quad (1.6)$$

Similarly for the remaining columns of the residual matrix, Π numbers can be written as follows:

$$\Pi_2 = v/\rho^0 D_i^2 N^1 = \text{Re}^{-1} \text{ (Reynolds number)} \quad (1.7)$$

$$\Pi_3 = Q/\rho^0 D_i^3 N^1 = \text{Fl (flow number)} \quad (1.8)$$

$$\Pi_4 = g/\rho^0 D_i^1 N^2 = \text{Fr}^{-1} \text{ (Froude number)} \quad (1.9)$$

Similar principles can be applied to heat and mass transfer and reaction engineering as shown in Table 1.5. This table summarizes the significance of some important dimensionless numbers in their respective areas of unit operation.

Dimensional analysis is a widely used methodology in scale-up/scale-down, the analysis holds true at all scales. Importantly, since the dimensionless numbers are independent of scale, they could be very effectively used in comparing various scales for the same process or even two different processes. While detailed discussion of dimensionless analysis is beyond the scope of this book, the importance of such an analysis will

TABLE 1.5 Dimensionless Numbers and Their Significance

Name	Symbol	Group	First principle interpretation
		Fluid flow	
Reynolds	Re	vl/v	Ratio of inertial and viscous forces
Power no.	N_P	$P/\rho\, d^5 n^3$	Impact of impeller type on fluid flow
Froude	Fr	V^2/lg	Fluid flow with free surface
Archemede	Ar	$\rho(\rho_s - \rho)\, gl^3/\mu^2$	Gravitational settling of particle in fluid
Weber	We	$\rho\, v^2\, l/\sigma$	Fluid flow with interfacial forces
Mach	Ma	v/v_w	Gas flow at high velocity
Flow no.	Fl	$q/n\, d^3$	Gas flow-dynamics in an agitated reactor
		Heat transfer	
Nusselt	Nu	hl/k	Ratio of convective to conductive heat transfer
Prandtl	Pr	$C_P\, \mu/k$	Heat transfer in flowing fluid
Stanton	St	$h/C_P\, \rho v$	Heat transfer in flowing fluid
		Mass transfer	
Schmidt	Sc	$\mu/\rho\, D$	Mass transfer in flowing fluid
Sherwood	Sh	$h_D\, l/D$	Mass transfer in fluid
Lewis	Le	Sc Pr^{-1}	Simultaneous heat and mass transfer
		Reaction engineering	
Arrhenius	Arr	E/RT	Activation energy of reaction
Hatta	Hat_1	$(k_1\, D)^{1/2}/k_L$	Kinetics of first-order reaction

be demonstrated in case study A in Sec. 1.5. Further details on dimensionless numbers and their significance can be found, for example, in Boucher and Alves (1959) and Catchpole and Fulford (1966).

1.3.3 First principles approach

The first principles approach is based on applying the fundamentals of engineering. This requires a high level of theoretical understanding of the subject matter. Figure 1.6 depicts a hierarchy of fundamentals and theory that form the building blocks of a manufacturing process.

The base of the first principles hierarchy is material properties. It is critically important to understand the properties of the raw materials, intermediates, and product. The material properties could be divided

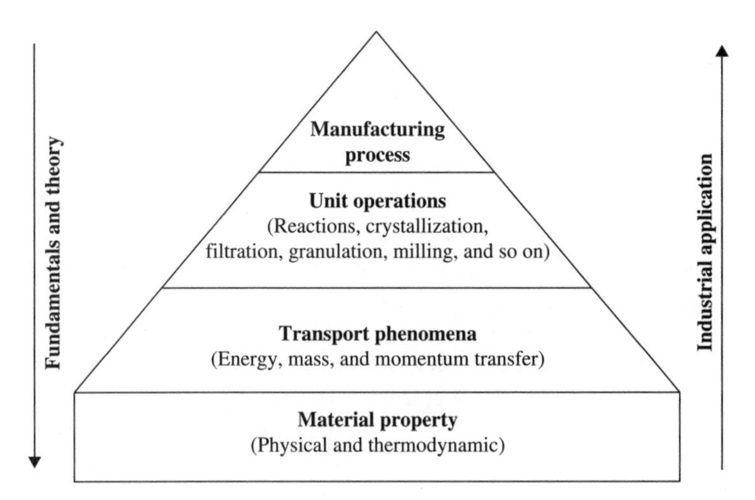

Figure 1.6 First principles approach.

into physical and thermodynamic properties. Physical properties define the physical state of the material, which would include dimensions, hardness, porosity, pressure, density, viscosity, friction coefficient, and the like. The thermodynamic properties could be subdivided into intrinsic and extrinsic properties. Intrinsic properties are independent of the amount of substance present; examples of such thermodynamic properties include temperature, pressure and stress. Extrinsic properties, on the other hand, are dependent on the amount of substance present; examples of such thermodynamic properties include volume, strain, charge and mass.

The knowledge of material properties feeds into transport phenomenon. Transport phenomena focus on the transport of the basic entities of physics—energy, mass, and momentum in any chemical process. There are various types of transport that aid the transfer of these basic entities. These various types of transport include (Bird, Stewart, and Lightfoot, 1960):

1. Molecular motion (viscosity, thermal conductivity, and diffusivity)
2. Laminar flow in one dimension (shell momentum, energy, and mass balances)
3. Laminar flow with two independent variables
4. Arbitrary continuum (equation of change)
5. Turbulent flow
6. Between two phases (interphase energy, mass, and momentum transfer)

7. Radiation
8. Large flow system (macroscopic mass, energy, and momentum balance)

Material properties and transport phenomenon are critical aspects of the fundamentals in chemical engineering, which are classified as unit operations. Unit operations are the basic building blocks of chemical engineering. McCabe et al. (2000) describe unit operation as "an economical method of organizing much of the subject matter of chemical engineering based on two facts: (1) systematic organization of individual processes, each one of which can be broken down into a series of steps, called operations, which in turn appear in process after process; (2) the individual operations have common techniques and are based on the same scientific principles." Because the unit operations are a branch of engineering, they are based on both science and experience.

A number of scientific principles and techniques are basic to the treatment of the unit operations. Some are elementary physical and chemical laws, such as the conservation of mass and energy, physical equilibria, kinetics, and certain properties of matter. Theory and practice must combine to yield designs for equipment that can be fabricated, assembled, operated, and maintained. A balanced discussion of each operation requires that theory and equipment be considered together. Examples of unit operations include reaction engineering, crystallization, granulation, milling, centrifugation, chromatography, and solids processing. The unit operation concept allows systematic study of these operations themselves that clearly cross industry and process lines, thus enabling the treatment of all processes in a unified and simplified form. Various unit operations then form a manufacturing process. Textbooks and published research literature provide a rich source of information for first principles knowledge. These aspects are applied in case study A (Sec. 1.5) to further clarify the concept. The manufacturing process is an integration of various unit operations.

The first principles approach focuses on the fundamental understanding of the manufacturing process. This approach is scientific; however, some of the process complexities may be very difficult to interpret by first principles, and a balanced combination of first principles and an empirical approach could be better suited.

1.3.4 Trial and error

Trial and error integrates a little knowledge and experimental data and extrapolates it with the help of experience and opinion to generate a scale-up design. In many cases, luck may support a favorable outcome; however, failures could be disastrous leading to delay in the launch of product. Failure may also create an external perception (to competitors or regulatory agencies) of weakness in the commercialization process. This aspect of scale-up will not be discussed in this book.

The challenge is to apply these principles of scale-up/scale-down appropriately to maximize the process-design value in an aggressive time frame. A good strategy would include a balanced approach incorporating both empirical and mechanistic aspects. It is also critically important to appreciate the fact that scale-up or scale-down requires juggling of unit operations. It is unrealistic to expect that all aspects of scale change similarly during scale-up/scale-down. The next section explains this point in detail.

1.3.5 Scale-down rules—understanding the rate-limiting step

A good scale-down approach relies on a balance between both empirical and mechanistic approaches. Typical approach includes the following rule of thumb:

1. Apply law of similarity—keep the scale as similar as possible especially around the key dimensions (i.e., tank diameter, tank height, and the like).
2. Apply first principles to gain insight into the process and the critical process parameters.
3. Identify critical dimensionless numbers and keep them constant.

An important reality of scale transformation is the selectivity of rules, so as to maximize the benefits to the process design. An example is discussed in this section to illustrate the point that the scale-up rules should be treated in the context of the process under study. For example, for a mechanically agitated tank, if we scale up with one rule, the

TABLE 1.6 Variation of Scale Factors

	Tip speed	Power	Power/volume	Mixing time	Reynolds number
Scale on constant*	Varies with scale as:				
Tip speed	Constant	s^2	s^{-1}	s	s
Power	$s^{-2/3}$	Constant	s^{-3}	$s^{5/3}$	$s^{1/3}$
Power/vol	$s^{1/3}$	s^3	Constant	$s^{2/3}$	$s^{4/3}$
Turbulent mixing time	s	s^5	s^2	Constant	s^2
Reynolds number	s^{-1}	s^{-1}	s^{-4}	s^{-2}	Constant

*S is the linear scaling factor, which is equal to tank diameter of large scale/tank diameter of small scale.

question is: how do the others vary? The relationships between potential scale parameters are illustrated in Table 1.6. The table indicates that during scale transition, if one scale parameter is kept constant while others vary, the challenge is to determine which scale parameter to focus on.

Literature offers some insight in solving this challenge by proposing an approach called *regime analysis*. Pace (1980) suggested that, for a more realistic approach to scale-up, the rate-limiting step has to be determined first. The laboratory scale process is then designed by optimizing the rate-limiting step. One of the methods to determine the rate-limiting step in a process is regime analysis. This analysis is based on a comparison of the characteristic times of various mechanisms in a process. The lowest characteristic time is indicative of the rate-limiting step. Kossen (1992) suggested four steps for scale-down using regime analysis.

- Regime analysis of process at production scale—rate-limiting step or running regime
- Simulation of the rate-limiting mechanism at laboratory scale
- Optimization and modeling of the process at laboratory scale
- Optimization of the process at production scale by translating the optimized laboratory conditions

The concept of regime analysis is based on small-scale simulations using various compartment models. With the parameter identified and literature reviewed, it is important to design small-scale experiments focused on the critical scale-up factors. The critical scale-up factors form the basis of constraints around the large-scale commercial process. In the mechanically agitated tank, for example, shear is a rate-limiting aspect for scale-up and tip speed is identified as the scale parameter most critical to shear. Tip speed is scaled down in laboratory scale to simulate production scale shear environment for late scale process development activities. Besides regime analysis, it is also important to use dimensional analysis and the principles of similarity to devise small-scale experiments.

The next step is to apply these concepts to experimentally develop the process in the laboratory. The next section focuses on a systematic approach to laboratory-pilot-scale development, which includes tips for experimental setup and measurement strategy.

1.4 Laboratory-Pilot-Scale Development

Laboratory- and pilot-scale development aim to integrate the scientific and engineering efforts. Chemists, biochemists, and chemical engineers clearly have complementary capabilities, and a team effort is critical to the success of commercialization. Solving technical problems that arise during process development may require input from both chemists and chemical engineers. Chemical engineers will play a more active role once the project enters the pilot plant (scale-up) and moves into commercial start-up. This means that it is critical to identify and solve important chemistry problems before pilot production. A fundamental flaw in the chemistry that results in low yields, for example, may be framed and tackled as a chemical engineering problem. Changes in flow rates, reactor designs, and other engineering parameters may help, but they will not fix the root cause of the problem—the chemistry. Moreover, attacking the type of issues generally dealt with in pilot development, such as optimization and combining steps (process integration) can be impeded if there are deeper problems with the fundamental chemistry. For example, if steps are combined before all the impurities are identified, the source of the impurities will be difficult to trace. The goal is to speed development by bringing *process chemistry* closer to *engineering design*. This is a challenge for management as scientific and engineering communities are generally divided by common goals!

1.4.1 Laboratory scale development—experimental program

Literature analysis, combined with knowledge from the early process development phase, helps develop the laboratory-scale experimental strategy. The strategy for experimental programs focuses on the interaction of three components of the process (Fig. 1.7):

1. Science is the fundamental aspect of the design and is scale and equipment independent.

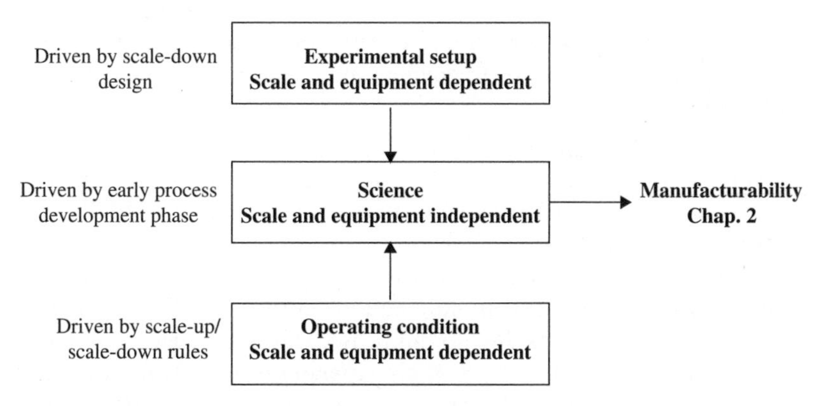

Figure 1.7 Elements of experimental design.

2. Experimental setup includes the type of equipment, its configuration and geometry at a particular scale, and the component integration with other equipment set for the experimentation.
3. Operating condition includes the specification of equipment-operating state.

The experimental design can now be formulated with all the information from the preceding sections of this chapter. A good experimental design is critical to the success of the process development phase and a rigorous experimental tool is required to assess the interaction between science and engineering. *Design of experiment* (DOE) is a statistical tool extensively used in industry to maximize information with a minimum number of experiments by using statistics. DOE is discussed in detail in Chap. 7.

The critical aspects that integrate science and engineering could be classed as critical product parameters, critical process parameters, critical operating parameters, and proven acceptable ranges.

1. *Critical product parameter.* Parameters that primarily characterize the product in all its quality attributes, for example, efficacy, dissolution, impurity level, and potency. For case study A (Sec. 1.5) the critical product parameters were potency and level of impurities.
2. *Critical process parameter.* Process parameters that impact the critical product parameters hence, impacting product quality and also safety. These parameters are scale, equipment, and engineering environment dependent. For case study A (Sec. 1.5), power input was the critical parameter as it controlled (given all things similar) K_La, cell suspension, dissolved carbon dioxide, and shear.
3. *Critical operating parameter.* The Critical operating parameter is the operating setpoint of the equipment required to maintain the critical process parameter. For example, impeller speed is a critical operating parameter that impacts power input, which could be a critical process parameter.
4. *Proven acceptable range (PAR).* In a typical operating environment it is impossible to hold a critical operating parameter constantly at one operating setting without any tolerance for deviation. Deviation from a constant setpoint may be caused by measurement uncertainties, process control, and process variations (these will be discussed in detail in subsequent chapters). Thus, a major deliverable from process development is to identify a range of operating parameters and prove through experimentation that variation within the range does not impact any critical process parameter.

Thus, PAR provides the necessary assurance that if the process drifts within the boundary of PAR, the product quality and safety is still acceptable.

There are three major objectives for the late process development phase (as illustrated in Fig. 1.1) that set the stage for manufacturability, namely, design of a scalable process, optimal productivity, and a safe and environment friendly process.

The interaction (Fig. 1.8) between critical product parameters and critical process and operating parameters are key to a successful design of a process and its optimization.

Impact of scale vs. operating space. The interaction of critical operating parameter with critical process parameter will change with scale, which needs to be studied for generating an appropriate setpoint for operation (operating space). The initial assessment of setpoints for critical operating parameter at various scales comes from the theoretical analysis.

Proven acceptable range. Once the operating space at a particular scale is fixed, then the next step is to understand how much variation around this setpoint the process can tolerate before deviating from outside quality and safety specifications.

Optimization. Within the boundaries of critical process parameter the process is optimized with regard to productivity, cost, capacity, waste minimization, and ease of operation. These attributes of optimization could be classed under process intensification and simplification and are discussed in detail in Chap. 2.

Laboratory-scale experiments set the stage for pilot-scale development to focus on scalability and robustness of the process.

1.4.2 Pilot-scale development

Small-scale laboratory development is followed by pilot-scale development. Pilot development entails moving the process to an intermediate

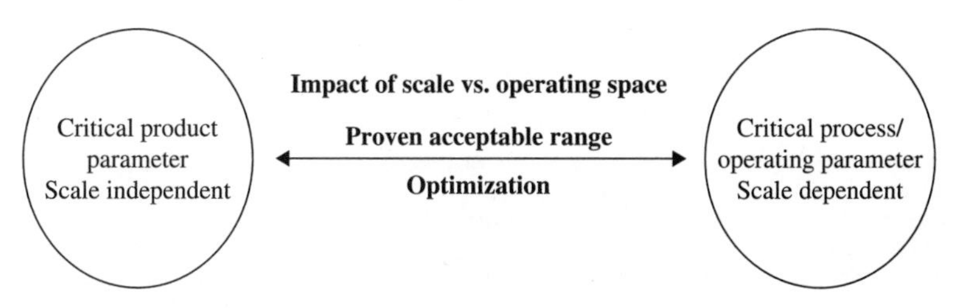

Figure 1.8 Objective of process development.

scale and selecting operating parameters (such as agitation rate, temperature, and airflow) that optimize efficiency. Pilot plants, unlike laboratories, are more representative of the final commercial environment and are larger in scale. Two opportunities may be discovered at this stage. First, a process that had high yields in the laboratory might have lower yields when run in the pilot plant. These scale-up effects may be the result of changes in the science (microbiology, chemistry) rather than of scale per se, and lower-than-anticipated process performance would trigger a search for how the science might have been altered. The second opportunity is operational in nature. Based on the understanding at the small-scale development stage, scale-up calculations could be undertaken to predict operating parameters and engineering boundaries at the pilot and manufacturing scale (see case study in Sec. 1.5).

Figure 1.9 illustrates the key deliverables from pilot-scale development. The critical deliverable is risk minimization due to—scalability assessment; equipment and process design robustness; and gaining operational experience of the integrated process. The scalability assessment projects, with more confidence (as compared with laboratory scale), the boundaries (engineering and operations) of manufacturing scale process. Pilot scale provides the avenue for evaluating engineering boundaries on the process. Planned experiments are then carried out to assure the "manufacturability" of the process, which will be discussed in detail in Chap. 2.

Before the experimental program is designed and executed, it is critical to develop the strategy for measurement of key parameters.

1.4.3 Measurement strategy

During the experimental program many measurements are typically made. Measurements serve several purposes. First, the function of the experimental program in process development is to quantify and understand

Scalability assessment - Verification of lab scale process - Technology assessment - Prediction of manufacturing scale operation	Equipment robustness - Material of construction assessment - Equipment robustness, pump, tank, and the like - Maintenance and reliability - Safety assessment
Operational experience - Operating parameters confirmation - Operator training - Debugging of SOP and recipes - DCS and software platform experience	Process design robustness - Process integration - Process simplification - Control strategy - Process optimization

Figure 1.9 Deliverables from pilot-scale evaluation—risk minimization.

behavior. From this understanding improvements are sought and indications of scale-up requirements are obtained. Measurements provide insight into behavior either directly (i.e., a high level of a component concentration may indicate excessive feed) or offer clues to the scientist (e.g., in a fermenter change in respiratory quotient indicates change in metabolic activity). In the latter case, the expertise of the scientist is called upon to a greater extent to diagnose problems or to point to opportunities. But this is not the only purpose of measurement. Standard measurements providing basic regulatory control are essential. In the case of a bioprocess these would be temperature, pH, substrate (glucose and the like), and possibly dissolved oxygen. Other measurements such as off-gas concentrations are typical and serve as important indicators. For example, it may be the case that off-gas measurements provide an indication of when to transfer broth to a larger vessel. Here they are providing indirect measures of organism state and broth concentrations. The development program would involve correlating performance to a feature of the off-gas profile and forming a procedural rule of the standard operating policy when the shift from development lab to production is made. Other measurements are made via sample analysis in the laboratory, with the assays undertaken being specific to the process considered.

Experiments may be undertaken with, say, variation in environmental control set points and other aspects of operational policy. In the early stages of the development program the purpose is to search for improvement, but in the later stages the objective is to assess the "robustness" of operation. A window of operation is that the PAR can be specified in which significant deviation is not observed in terms of product quality. It is necessary to define PAR in terms of the operational window as specified by measurement of condition. In this case, the measurements defining the PAR may turn out to be less comprehensive than those required for improvement purposes. PAR must be assessed on a routine basis when it comes to production and therefore cannot be defined by a complex and expensive series of measurements, unless absolutely justified.

Recent advances, both in technology and the growing acceptance of methods by regulatory authorities, are now leading companies to implement spectroscopic techniques such as *near infrared* (NIR) and *mid infrared* (MIR) spectroscopy to provide a "fingerprint" of a process condition. In the past, off-line use was reported but on-line implementation is growing. They can be used either to identify specific components or to simply assess an overall state and as such provide a vital additional source of information. Routine use in an on-line sense is not yet commonplace and technical problems still remain, but the prospects are encouraging. The measurement strategy is an important aspect of *process analytical technology* (PAT), which is further discussed in Chap. 3.

The scale-up/scale-down concepts discussed in the earlier sections will be applied to an industrial case study to demonstrate the value of applying the systematic approach to commercialization.

1.5 Scale-Down in the Biopharmaceutical Industry—Case Study A

Previous sections have considered the fundamental strategies employed in scale-up/scale-down. This section applies those concepts and techniques in a biopharmaceutical case study. A biopharmaceutical process has several unit operations. A typical bioprocess will include following unit operations:

1. *Bioreaction or fermentation.* Microorganism produces a product of interest
2. *Filtration.* Separation of waste
3. *Advanced filtration (tangential flow filtration).* Concentration of product
4. *Chromatography.* Purification of product

This case study focuses on bioreaction or fermentation unit operation to demonstrate the application of scale-up/scale-down concepts. In most cases, several scales of fermentation are required to develop a bioproduct. Usually, laboratory fermentation (1 to 20 L) is used to screen strains and optimize media and fermentation conditions; pilot-plant fermentation (10 to about 10,000 L) is applied to verify if the process conditions are optimal and to produce sufficient data for evaluation. After the laboratory identifies a novel or improved product candidate, much effort will be devoted to developing a commercial process for its production. The objective of process development is the production of sufficient new and modified product to meet market demand in the quickest time possible, to meet all safety requirements, and to have a cost-effective and reliable process. A vital technical part of this development is the translation of the fermentation from laboratory to production scale.

In this case study, late stage process development was conducted in a 7-L bioreactor, and the production scale was 450 L for the antibiotic production (Mohan, 1996). Pilot scale (100-L bioreactor) was used to assess and enhance the robustness of the process and to gain operational experience prior to implementing at the manufacturing scale. Pilot scale was also used to manufacture clinical trial material. This scale-up/scale-down approach followed the strategy discussed in the previous sections, which included understanding the economies of scale for the financial justification for the 450-L scale as compared to the 7-L scale and, applying scale-down techniques to develop appropriate laboratory-scale experiments. Scale-up was undertaken in the 100-L pilot scale to verify the operating conditions, and a model of the process was developed for scenario analysis.

1.5.1 Economies of scale

As discussed in Sec. 1.2, an important aspect of economies-of-scale analysis involves a step-by-step scale-up accounting approach leading to the selection of appropriate scale based on sound economic rationale. The step-by-step approach includes accounting for total capital employed, product cost, and cost of product and ratio analysis.

Step 1: Total capital employed. Total capital employed has two key cost elements (Table 1.7) namely: prime plant cost, which includes costs of total equipment, building, and engineering and supervision; and operating cost, which includes costs of equipment maintenance, building maintenance, operating labor, maintenance labor, raw material, utilities, consumables, laboratory cost, insurance, and rates. Total capital employed for 7 L and 450 L were £ 541,519 and £ 3,146,690, respectively.

TABLE 1.7 Accounting for Total Capital Employed

Type	Type	Description	7 L (£)	450 L (£)
		Prime plant cost		
Direct	Fixed	Total equipment cost	336,210	1,088,926
Direct	Fixed	Building cost	36,196	542,787
Indirect	Fixed	Engineering and supervision	72,790	1,091,551
		Operating cost		
Indirect	Fixed	Equipment maintenance	26,729	80,495
Indirect	Fixed	Building maintenance	181	27,139
Direct	Fixed	Operating labor	45,000	60,000
Overhead	Variable	Maintenance labor	10,000	20,000
Direct	Variable	Raw material	340	21,843
Direct	Variable	Utilities cost	507	32,642
Direct	Variable	Consumables	454	29,232
Overhead	Fixed	Laboratory cost	11,000	16,000
Overhead	Fixed	Insurance	704	45,358
Overhead	Fixed	Rates	1408	90,717
Total Capital Employed			541,519	3,146,690

*One pound (£) equals approximately 1.8 U.S. dollars ($) (October 2005).

Step 2: Accounting for product cost. Equation 1.3 indicates the elements of product cost, which are operating cost, depreciation, loan payment, and hidden cost. Assuming that the hidden costs are negligible, the product cost is detailed in Table 1.8.

Step 3: Cost of product. Cost of product is a fundamental financial metric of a manufacturing organization. It depends on the manufacturing cost and the annual throughput. The cost of product accounting is shown in Table 1.9.

Step 4: Ratio analysis. Table 1.10 summarizes key financial aspects of the economy of scale. The scale-up case demonstrates the financial analysis involved in choosing the right scale for an economically viable operation. The unit cost of production is significantly different, with a much lower value for the larger scale. The comparison on the basis of variable cost to fixed cost ratio is very low for 7-L scale, suggesting that the cost of equipment and related items is far too high to support an economically viable process for the 7-L scale. With scale-up, one observes an increase in this ratio. Perhaps one of the best indications of the economic efficiency of scale is *total capital employed/unit cost.* This indicates the units of product produced with the investment. There is a big difference between the scales with a relatively much higher value at the larger scale.

TABLE 1.8 Accounting for Product Cost

Type	Type	Description	7 L (£)	450 L (£)
		Depreciation/loan payment		
Overhead	Fixed	Equipment depreciation	33,621	108,893
Overhead	Fixed	Building depreciation	1810	27,139
Overhead	Fixed	Eng and sup–loan payment	7279	109,155
		Operating cost		
Indirect	Fixed	Equipment maintenance	26,729	80,495
Indirect	Fixed	Building maintenance	181	27,139
Direct	Fixed	Operating labor	45,000	60,000
Overhead	Variable	Maintenance labor	10,000	20,000
Direct	Variable	Raw material	340	21,843
Direct	Variable	Utilities cost	507	32,642
Direct	Variable	Consumables	454	29,232
Overhead	Fixed	Laboratory cost	11,000	16,000
Overhead	Fixed	Insurance	704	45,358
Overhead	Fixed	Rates	1408	90,717
Manufacturing Cost			139,033	668,613

TABLE 1.9 Cost of Product

Cost of product–variables	7 L	450 L
Manufacturing volume (L)	7	450
Cycle time of one batch, CT (days)	15	15
Cycles per year, CPY	24	24
Productivity, P (g/l)	20	18
Annual throughput, T = C*CPY*P(kg)	3.36	194
Manufacturing cost, M	139,033	668,613
Cost of product, COP = M/T [£/kg)]	41,379	3447

An analysis of the economy of scales enables the manufacturing organization to assess the economic viability of the scale of operation. This is key to manufacturing success and productivity. Having a wrong scale or a less efficient scale as compared with a competitor is fatal in the business environment. Identifying the scale of operation helps set the end goal for the late-stage process development. The next stage is to scale down the critical aspects of the process at the lab scale for process development activities.

1.5.2 Scale-down

Scale-down methodology is both a science and an art. Section 1.3 summarizes the key aspects of the scale-down methodology. This section on scale-down methodology generally recommends that an integrated approach be adopted for scale-up/scale-down. The integrated approach should include a well-balanced combination of empirical and mechanistic approaches. This would include the dimensionless number approach, the principle of similarity approach, and the first

TABLE 1.10 Economies of Scale—Ratio Analysis

Ratios (450 L/7 L)	7 L	450 L	Ratios 450 L/7 L
COP ratio	£/Kg 41,379	£/Kg 3447	12
Manufacturing cost	£ 139,033	£ 668,613	4.8
Annual throughput	3.36 Kg	194 Kg	58
Total capital employed	£ 541,519	£3,146,690	5.8
Fixed cost	£ 127,732	£ 564,896	4.4
Variable cost	£ 11,301	£ 103,717	9
Important ratios for economies of scale			
Variable cost/fixed cost	0.09	0.18	2
Man. cost/total capital emp.	0.26	0.21	0.8
Total capital employed/COP	13	913	70

principles approach. These various aspects of scale transformation are integrated to generate case-specific rules for scale-up. Such an integrated approach is systematically illustrated for the case study A in the following section.

Understand the science: Biochemical reaction similarity. Scale-up of bioreaction is a complex process. A number of parameters influence the microbial response, i.e., growth rates, nutrient and oxygen utilization rates, or cell viability and proliferation. The science must be similar at all scales. The similar aspects of bioscience could be categorized into biological and chemical factors.

The biological factors include:

Culture selection. The strain must be the same at different scales. Operating conditions are identified and optimized to increase productivity. The production process is developed with these optimal conditions; however, these conditions are scale dependent. For example, most of the strain selection is carried out in shake flasks. For a baffled shake flask, high oxygen transfer rates can be achieved when broth volume is kept low and high shaker speed is used. Because *dissolved oxygen* (DO) is less likely to be a limiting factor under these conditions, the optimal strain may have a high *oxygen uptake rate* (OUR).

Number of generations and selection pressure. The stability of plasmids must be considered. For a production 1,000 times larger than the laboratory pilot scale, 10 additional generations are needed to reach the final production stage. For genetically engineered strains, the stability of the plasmids harboring the production gene must withstand this larger generation number, which may increase mutation probability. In smaller scale fermentations, selection pressure, such as the use of antibiotics, can be applied to maintain reasonable plasmid copy numbers in the cultured cell. For large-scale fermentations, the cost of antibiotics may be prohibitive, so nonproducing variants are more likely to develop and compete with the producing strain at production levels.

Medium sterilization. Both the fermentation surface area and total initial microbial count in a medium before sterilization are functions of tank volume. To achieve the same degree of sterility as for a small tank, larger tanks require longer sterilization, which also affects medium quality. Due to the retention of condensate in the medium, quality of steam then becomes another variable in large-scale fermentation. There are parameters that are used to compare similarity in sterilization aspects, namely, R_0 and F_0; these parameters are discussed in details by Boeck et al. (1988) and Alford et al. (1989).

Biological variation may result from scale-dependent changes in cell morphology, which in turn may influence broth rheology. Clumping of cells or excessive wall growth in small fermenters introduces diffusional resistances for oxygen and nutrients, and hence alters the mesoenvironment of the cells. Such wall growth also hinders kinetic analysis since it can lead to cell retention in continuous or semibatch cultures.

The chemical factors include:

Medium quality. For economic reasons, industrial grades of raw material usually are used (complex medium instead of a defined one) and thus the composition and quality of raw material may vary batch to batch.

Water quality. In large-scale industrial fermentation, well or potable water is preferred to deionized water for the fermentation broth. The concentrations of metal ions and other compounds depend on the plant location and season of operation.

The outcome of this analysis leads to the development of the scientific boundary (see Fig. 1.10) for scale-down. The development of process in the 7-L scale used the same strain, raw material, water, and sterilization time (similar F_0) as projected in the 450-L scale. This ensured that the process would be developed based on similar scale-down biological/chemical conditions.

However, these biological/chemical conditions are scale dependent on the engineering aspects. The next step is to evaluate the engineering conditions that would maintain a similar biological environment at different scales (Lab scale—7 L, pilot scale—100 L, and commercial scale—450 L).

Identify engineering scale-dependent aspects: comprehensive literature search. One way of identifying engineering aspects involved in the process is to undertake a rigorous literature search in the subject area. A literature

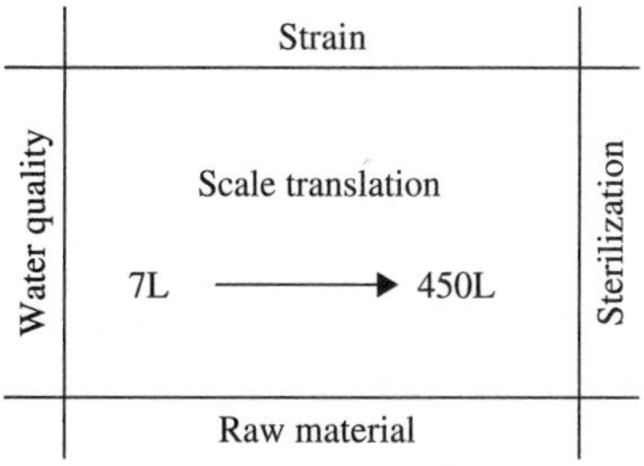

Strain

Water quality

Scale translation

7L → 450L

Sterilization

Raw material

Figure 1.10 Developing scientific boundary for 7-L scale.

search is very important for benefiting from past experience and building the much-required knowledge base (first principles) for successful scale-up. In this section, the literature search for case study A demonstrates the value of such an approach. The first principles aspects of scale-up relevance discussed are: mixing, mass transfer, rheology, shear, and heat transfer.

Unit operation and transport phenomenon

Oxygen mass transfer. Oxygen mass transfer has been a key aspect of study in the literature. Gas mass transfer at the gas-liquid interface (gas bubble) is limited by two factors, namely, liquid film and the interfacial area of the gas bubble. These resistances to mass transfer are lumped in a single mass transfer coefficient, namely, K_{La} (SI unit 1/s). Many articles in the literature focus on the criticality of maintaining K_{La} for successful aerobic bioprocessing (Manfredini et al., 1983; Larsson and Enfors, 1985; Furukawa et al., 1983; Harrison and Pirt, 1967). These studies show that K_{La} is dependent upon power per unit volume, gas flow rate, and also on the fluid property (ionic strength and coalescing property). A detailed understanding of mass transfer constraints is key to scale-up, as the oxygen supply capacity of a large-scale fermenter is usually lower than at a smaller scale. Oxygen enrichment could be achieved by various methodologies such as improved mixing, gas blending (enriched oxygen), and higher backpressure. But, even if a more standard backpressure is applied, the water head pressure at the bottom of a large tank can be twice the pressure found in a small tank.

Carbon-dioxide inhibition. The dissolved carbon dioxide (CO_2) concentration may be higher in the large tank than in the small tank. The dynamic equilibrium of CO_2 is affected by both pH and pressure. Pressure fluctuation, due to the hydrostatic head, increases the partial pressure of CO_2 in a large-scale bioreactor, which can adversely affect productivity. Kobayaski et al. (1992) stress the importance of CO_2 levels in scale up. Though CO_2 inhibition can be a problem, the degree of inhibition varies from one microorganism to another. In practice, in large-scale bioreactors, there will be an axial gradient of CO_2 concentration depending on the head pressure, and the cycling effect between low and high concentration of CO_2 can be important.

This knowledge helps develop an appreciation for the critical process design considerations regarding oxygen mass transfer and CO_2 inhibitions.

Cell suspension. Homogeneity in the vessel for substrate concentration gradient, pH, and dissolved oxygen is essential for microbial growth. Large-scale vessels increasingly show heterogeneous tendencies due to inadequate mixing, which adversely affects cell suspension. The extent of heterogeneity varies with the geometrical configuration of vessel and agitator, which influences the global and local hydrodynamics. Mixing in the bioreactor is vital for the effectiveness of the transport processes. Hansford and Humphrey (1966) suggest that the critical factor that determines the overall effectiveness of oxygen uptake by microorganisms is the frequency at which cells are circulated through the highly oxygenated impeller region. The circulation time should be less than the oxygen consumption time to prevent oxygen starvation of cells. They also discussed the relationship between mixing times and growth yields in continuous fermentations and noted that poor distribution of the substrate caused a decrease in the yield. Also, it was pointed out that pH gradient, due to poor mixing of acid or base additions, can influence the biological performance of the microorganism. Bioreactor models integrating a fluid description of the physical conditions of the reactor, and a biokinetic description of the response of the microorganisms to the physical conditions, are potentially very powerful tools for analyzing scale-dependent changes. Fluid dynamic models, for example, *computational fluid dynamic* (CFD) models, could be independent of the scale where the governing partial differential equations are in the form of the equations of motion, being derived from the first principles. These models rely on the numerical procedure being accurate enough (see “Simulation using computational fluid dynamics” in Sec. 1.5.2).

Heat transfer. The heat-transfer demand increases in volumetric ratio with scale. To accommodate the need for increased heat-transfer surface area, the aspect ratio (height to diameter) can be increased. Intensive bioreaction processes are increasingly found to be rate-limited by inadequate heat-transfer capability. Heat transfer is particularly important for large-scale production vessels because an increase of the vessel size leads to a reduction of surface area to volume ratio. Mohan et al. (1992) have shown that significant axial variations of heat-transfer coefficient exist, with the maxima near the impeller plane. Furthermore, the heat-transfer coefficient is influenced not only by aeration rate (at least below the flooding rate), impeller speed, and bulk flow, but also by the morphology of the filamentous microorganism (i.e., the fungal mycelium).

Shear damage. Little data are available on how scale-dependent stress conditions could influence product quality, either by a direct influence upon product formation of proteolytic stress proteins, the production of metabolites, or the release of intracellular contaminants. Mechanical shear stresses or gas-liquid interfacial forces (particularly foaming) leading either to molecular or cellular damage, must be avoided upon scale-up to ensure product integrity and avoid the release of intracellular DNA. Important interactions between the turbulence intensity at different scales and the morphology (and hence, the metabolic state) of certain organisms can be expected. These scales range from the largest eddies, on the scale of the height of a turbine blade of an agitator (about 0.1 m) to the smallest eddies, which are produced by the cascade of turbulent eddies. In agitated bioreactor systems, the smallest eddy size is in the order of 20 to 100 μm (Bailey and Ollis, 1980). In the literature, researchers have listed a number of factors affecting morphology and productivity of microorganisms. Makagiansar et al. (1993) suggest that hyphal breakup depends on the frequency of mycelial circulation through the zone of high-energy dissipation around the impeller. They showed good, quantitative agreement with circulation frequency ($1/t_c$) and power dissipation (P/d^3t_c). They proposed that the greater the frequency of circulation of mycelia through the impeller zone, the greater the damage and the lower the rate of penicillin synthesis by the culture. A study by Yim and Shamlou (2000) suggests that the hyphal breakage in case of *Penicillium chrysogenum* is likely to occur in the impeller zone as a result of the direct fluctuating stresses acting on the opposite of the hypha.

The above literature search sets the stage for selecting critical engineering areas of focus for the scale-down setup. In many ways, this reflects the regime analysis approach:

- Simulation of the rate-limiting mechanism at laboratory scale
- Optimization and modeling of the process at laboratory scale
- Optimizing the process at production scale by translating the optimized laboratory conditions

The listed analysis identifies important engineering aspects of scale-up/scale-down, which are captured in Fig. 1.11. This sets the stage for the experimental program at the small 7-L laboratory scale.

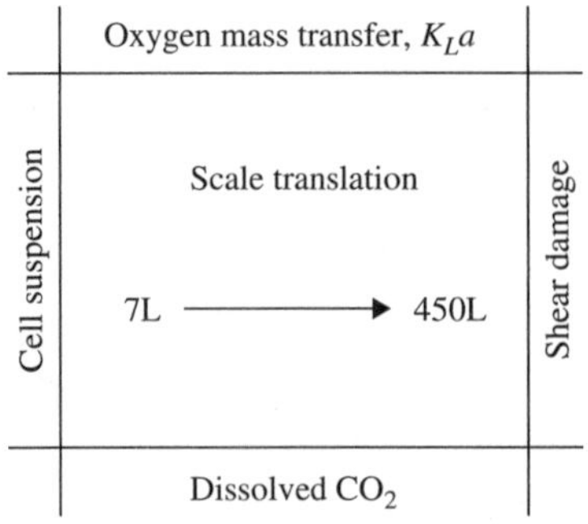

Oxygen mass transfer, K_La

Cell suspension

Scale translation

7L → 450L

Shear damage

Dissolved CO_2

Figure 1.11 Engineering boundary for scale-up/scale-down—case study A.

Experiments were conducted at the 7-L scale, within the scientific boundary of similar strain, sterilization conditions, water quality, and raw materials, to understand the impact of these engineering parameters on bioprocess performance, which was process-yield (antibiotic product). Oxygen mass transfer K_La was found to be the most critical parameter influencing process performance. Scale-up and scale-down is a juggling of unit operations. It is vitally important that the critical parameter for scale-up must be similar during scale transformations.

Scale transformation rules. The application of first principles in understanding the impact of unit operation (through literature search) and the subsequent experimentation to understand the critical engineering parameter(s) for scale transformation are key steps in building robust engineering principles in designing the processes. The next steps include systematic identification of scale transformation rules:

Step 1: Characterize the scales with end in mind (commercial scale). The starting point in the engineering scale-up/scale-down analysis includes the characterization of all scales including physical dimensions and physical properties. Physical dimension would include the tank, impeller, baffle, and air sparger dimension along with the motor capacity. The physical properties would include fluid properties, such as density and viscosity. Start the characterization with the commercial scale first (end in mind). Table 1.11 lists the numerical values of physical properties for all scales.

Step 2: Apply the law of similarity using the commercial scale as the benchmark. Table 1.11 shows similarities being applied to geometry, mechanical/operation, and dimensionless number and unit operations. Geometric similarity was based on some key geometrical conventions or rules of thumb:

Aspect ratio: the height to diameter ratio

Impeller to tank diameter ratio

Impeller placement: one-third of impeller diameter

Baffle width: one-twelfth of the tank diameter

With commercial scale as the benchmark, the pilot-scale and laboratory-scale dimensions are calculated following the conventions discussed. This generates design for the laboratory- (7-L) and pilot-scale (450-L) equipment that is similar in configuration to the commercial scale (450 L).

Step 3: Match most critical process parameter (K_{La}) at all scales. K_{La} is theoretically assessed and then experimentally verified. Table 1.11 lists K_{La} for various scales: For 450 L the K_{La} is 113 per hour, which is derived theoretically as the 450-L scale is still in the conceptual design phase. With this information the operating conditions (impeller speed, power

TABLE 1.11 Scale-Up Calculations

	Commercial scale 450 L	Pilot scale 100 L	Lab scale 7 L
	Physical properties		
Liquid density (kg/m^3)	1050	1050	1050
Cell density (kg/m^3)	1150	1150	1150
Viscosity (cP)	50	50	50
Surface tension (N/m^2)	2	2	2
Mean particle size (micron)	50	50	50
20% Cell volume (L)	90	20	1.4
80% Liquid volume (L)	360	80	5.6
	Geometric similarity		
	Tank similarity – tank aspect ratio – H/t = 2		
Operating tank volume (L):	450	100	7
Tank diameter (t, cm):	66.7	40.6	16.5
Operating height (H, cm)	133.4	81.2	33
	Impeller similarity – impeller/tank diameter = 1/3		
Type	Rushton	Rushton	Rushton
Number of impellers	1	1	1
Diameter of impeller (D, cm)	22.2	13.5	5.5
	Impeller placement – 1/3 height of the tank		
Impeller placement from the base (cm)	44.5	27	11
	Baffle similarity – width = t/12		
Number of baffles	4	4	4
Baffle width (t/12, cm)	5.6	3.4	1.4
	Mechanical/operational similarity		
Impeller speed (rpm)	100	295	800
Power/volume (kW/m^3)	0.5	0.58	0.96
Tip speed (m/s)	1.2	2.1	2.3
Mixing time (s)	8.5	5.7	2.7
Air-flow rate ratio (VVM)	2	2	2
Air-flow rate (l/m)	900	200	14
Minimum suspension speed (RPM)	85	130	230
	Dimensionless numbers and unit operation similarity		
Reynolds number	1760	1920	861
Gassed power number	3.85	3.85	3.85
Froude number	0.063	0.33	0.998
Mass transfer coefficient K_{la} (1/h)	113	109	109
Gas-flow number	0.82	0.28	0.1

per unit volume, airflow rate) were selected for the pilot and the laboratory scale.

Step 4: Check that other important aspects of scale transition are also met. Evaluate the differences in other important unit operations and their impact on process performance.

Step 5: Set criterion for success—experimentally verify. Criteria for success will include equivalence of the antibiotic productivity of the batch at various scales. Productivity for case study A is the weight of antibiotic produced per total volume of broth. The following productivity was noted for case study A: 20 g/L for 7 L, –20 g/L for 100 L, and 18 g/L for 450 L. This indicates that the scale-up was successful in delivering a capable process with equivalent productivity.

Simulation using computational fluid dynamics. The process stage model will lead to valuable knowledge capture, leading to better-informed decision making for scalability and process improvement. Process stage modeling has three aspects—mechanistic (first principles), empirical, and hybrid (mechanistic + empirical). Mechanistic models include "computational fluid dynamics" modeling, which involves defining finite volume grids and solving of Navier-Stokes equations with appropriate boundary conditions. For example, for case study A, it was important to understand and predict cell-suspension behavior at the manufacturing scale. A computational model was created and verified at the small scale with experimental data, and subsequently the model was scaled up to predict the cell-suspension behavior at the manufacturing scale. The model (Fig. 1.12) indicates that the cell-suspension behavior would be similar at the two scales.

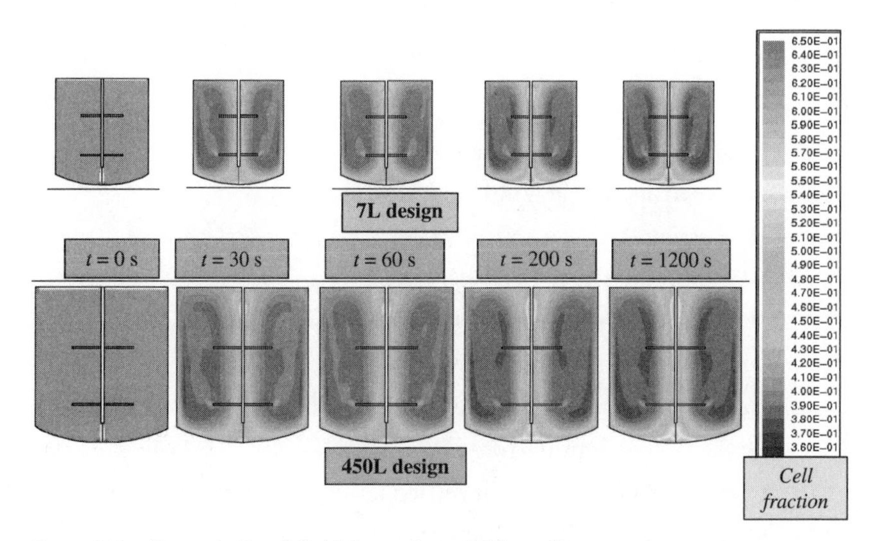

Figure 1.12 Computational fluid dynamic model for cell suspension.

This model also provides the opportunity to evaluate numerous scenarios and experiments to further minimize risk during scale-up. Such models allow virtual laboratory space for additional experimental programs.

The output from the laboratory- and pilot-scale development of the late commercialization phase could be summed up into manufacturability. A compilation of deliverables from the process development phase leading into a robust manufacturing process includes process integration, process simplification, critical-control strategy, risk analysis, process optimization, process qualification, and validation. Chapter 2 discusses this concept of manufacturability in detail.

Concluding Comments

Successful manufacturing has its roots in a well-executed development phase. In this chapter, the late development phase has been discussed with a case study. Some key conclusions from this chapter include:

1. Apply economies of scale to identify the manufacturing scale and to start with the end in mind (identification of the commercial scale).
2. Apply a structured approach to scale-up/scale-down to minimize risk and maximize time, investment, and the consequent knowledge base.
3. The late process development phase is to be performed within the constraints of the commercial scale manufacturing process to build realism and to achieve scalability successfully.
4. Modeling and simulation are important avenues to further minimize risk during scale-up.

Nomenclature

v = velocity, m/s

v_w = velocity of sound, m/s

l = characteristic length, m

P = power, W

d = diameter, m

n = impeller speed, rps

g = acceleration due to gravity, m/s^2

h = heat transfer coefficient, W/m$^2 \cdot$ K

k = thermal conductivity, W/m $\cdot$ K

C_p = specific heat capacity at constant pressure, J/kg $\cdot$ K

h_D = mass transfer coefficient, m/s

D = diffusion coefficient, molecular diffusivity, m^2/s

R = universal gas constant, 8314 J/kmol · K

E = activation energy, J

T = temperature, K

q = gas-flow rate, m^3/s

k_1, k_2, k_L = reaction kinetic constants

Greek symbols:

μ = dynamic viscosity, Pa

ν = kinematic viscosity, m^2/s

ρ_L = liquid density, kg/m^3

ρ_S = solid density, kg/m^3

σ = surface tension (interfacial), N/m

References

Alford, J. S. Jr., L. D. Boeck, R. L. Pieper, and F. M. Huber (1989), "Interaction of media components during bioreactor sterilization: definition and importance of Ro," *J. Ind. Microbiol.*, **4**: 247–252.

Avis, K. E., H. A. Lieberman, and L. Lachman (1993), *Pharmaceutical Dosage Forms: Parenteral Medication*, vol. 3, 2d ed., Marcel Dekker, New York.

Bailey, J. E., and D. F. Ollis (1980), *Biochemical Engineering Fundamentals*, 2d ed., International Edition, McGraw-Hill, New York.

Bird, R. B., W. E. Stewart, and E. N. Lightfoot (1960), *Transport Phenomena*, Wiley, New York.

Boeck, L. D., R. W. Wetzel, S. C. Burt, F. M. Huber, G. L. Fowler, and J. S. Alford, Jr. (1988), "Sterilization of bioreactor on the basis of computer calculated thermal input designated as Fo," *J. Ind. Microbiol.*, **3**: 305–310.

Boucher, D. F., and G. E. Alves (1959), "Dimensionless numbers," *Chemical Engineering Progress*, **55**(9)(September): 55–83.

Carstensen, J. T. (1998), *Pharmaceutical Preformulation*, CRC Press, Boca Raton, FL (July 7).

Catchpole, J. P., and G. Fulford (1966), "Dimensionless group," *Industrial and Engineering Chemistry*, **58**(3)(March): 46–78.

Dean, D. A, E. R. Evans, and I. H. Hall (2000), *Pharmaceutical Packaging Technology*, Taylor & Francis (August 1), New York, NY.

Furukawa, K., E. Heinzle, and I. J. Dunn (1983), "Influence of oxygen on the growth of Saccharomyces cerevisiae in continuous culture," *Biotech. Bioeng.*, **25**:2293–2317.

Hansford, G. S., and A. E. Humphrey (1966), "The effect of equipment scale and degree of mixing on continuous fermentation yield at low dilution rates," *Biotech. Bioeng.*, **8**:85–96.

Harrison, D. E. F., and S. J. Pirt (1967), "The influence of dissolved oxygen concentration on the respiration and glucose metabolism of Klebsiella aerogenes during growth," *J. Gen. Microbiol.*, **46**:193–211.

Johnstone, R. E., and M. W. Thring (1957), *Pilot Plants, Models, and Scale-up Methods in Chemical Engineering*, McGraw-Hill, New York, pp. 12–26.

Kobayaski, Y., M. Hara, H. Fukui, K. Sakamoto, and T. Koyano (1992), "Optimization of CO_2 level in a bioreactor—a key to successful scale up," *Abstracts of Papers of the American Chemical Society*, **10**:203.

Kossen, N. W. F. (1992), "Scale up in biotechnology," in F. Vardar-Sukan, S. S. Sukan (eds.), *Recent Advances in Biotechnology*, Kluwer Academic Publishers, Middleton, The Netherlands, pp. 147–182.
Larsson, G., and S. O. Enfors (1985), "Influence of oxygen starvation on the respiratory capacity of *Penicillium chrysogenum*," *Appl. Microbiol. Biotechnol.*, **21**:228–233.
Makagiansar, H. Y., P. A. Shamlou, C. R. Thomas, and M. D. Lilly (1993), "The influence of mechanical forces on the morphology and penicillin production of *Penicillium chrysogenum* fermentation," *Bioprocess. Eng.*, **9**:83–90.
Manfredini, R., V. Cavallera, L. Marini, and G. Donati(1983), "Mixing and oxygen transfer in conventional stirred fermenters," *Biotech. Bioeng.*, **25**:3115–3131.
McCabe, W., J. Smith, P. Harriott (2000), *Unit Operations of Chemical Engineering*, 6 ed., McGraw-Hill Science/Engineering/Math, New York, NY.
Mohan, P. (1996), "Financial Analysis of Scale-Up in Biotechnology," Master's dissertation, Middlesex University Business School, London.
Mohan, P., A. N. Emery, and A. W. Nienow (1992), "Heat transfer in a mechanically agitated bioreactor vessel," *Trans. IChemE*, **70** (December): Part C, 25–85.
Pace, S. C. (1980), "1980 Triple engineering–conference keynote," *Hydraulics and Pneumatics*, **33**(9):56–58.
Yim, S. S., and P. A. Shamlou (2000), "The engineering effects of fluid flow on freely suspended biological macro-materials and macromolecules," *Adv. Biochem. Eng. Biotechnol.*, **67**:83–122.

Chapter 2

Manufacturability

Manufacturability has three key elements: trained people, qualified equipment, and a validated process.
P. MOHAN

Objective

Manufacturability implies that the process design is capable of delivering a robust process. This is the final phase of late stage commercialization. The focus of this chapter is to systematically introduce various aspects of manufacturability.

The learning outcomes of this chapter will include:

- Introduction to process integration and process simplification
- Introduction to process flow description as an established way to capture information critical to manufacturability
- Detailed discussion on risk management and FMEA
- Introduction to modeling and tips for process optimization
- Introduction to commissioning, qualification, and validation plans

A few examples are discussed to illustrate the practical application of these concepts.

2.1 Introduction

The process development stage is followed by design and commissioning of manufacturing, the focus of which is manufacturability. Manufacturability implies that the manufacturing platform is robust with ease of operation, producing high-quality product reproducibly (with little variability), and is safe and environment friendly. The responsibility for the design and commissioning passes to engineering, where plant and equipment design become the focus of attention. The activities in plant design include selecting and sizing equipment, determining energy and material requirements, and optimizing recycle and energy streams. Plant and equipment design must be focused around successful process scale-up leading to a robust operating design that will form the basis of successful manufacturing. This design has to withstand rigorous manufacturing trials to prove its suitability as a robust operating platform (manufacturability). The elements of manufacturing that have to undertake this trial are people, equipment, and process.

Figure 2.1 illustrates the elements of manufacturability namely: people, equipment, and process. The people must be trained to perform manufacturing operations, equipment must be qualified to perform the set operation, and the process must be validated to meet the product specification.

People are at the heart of manufacturing operations and are by far the most important element of manufacturing. A highly trained and motivated workforce is a prerequisite of a successful manufacturing operation. When a new process is scaled up, the development of operational requirements and the consequent *individual training plans* (ITPs) of the people operating the plant are top priorities for manufacturing leadership. Development of an ITP is a challenge as it depends on the skill set and potential of the individual.

For example, the requirement for an operator conducting crystallization operation in a chemical/pharmaceutical manufacturing process could be: "Should be able to conduct crystallization operation consistently and reproducibly." A short job description could look like:

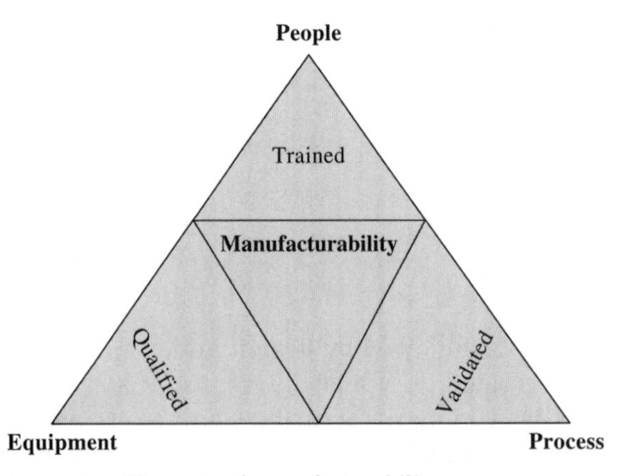

Figure 2.1 Elements of manufacturability.

- Essential qualification—associate degree preferable in engineering or related discipline
- Essential skill set—operational experience (two years or more) in a manufacturing facility, preferably in the crystallization aspect of the operation

In the interview process the current skill set should be mapped with the required skill set, and a development plan including the ITP should be identified. Operator training and its implementation is critical for the success of the operation (Fig. 2.2). It is generally observed that in many instances manufacturing deviations are caused by untrained operators. Trained operators are an asset to the organization. Given the turnover of operators, an organization faces an ongoing need to develop new talent. In Chap. 4 of this book a strategy for knowledge extraction from experienced operators is discussed along with its implementation to develop a robust operation. Knowledge extraction around key decision making in an operation can help new operators benefit from the knowledge base of trained and experienced operators. Another way to enhance operational robustness is to engage in process control and automation to enable operators to enhance their effectiveness. Chapter 3 presents new and existing techniques of process control and automation to minimize operator interface with the process by enabling them with improved process control and automation options.

Process and equipment are inseparable entities of any manufacturing operation. The equipment has to be suitable for appropriate processing and the process should be able to optimize equipment capability. The starting point for the design of commercial scale is sound development work (as discussed in Chap. 1). This scientific work is generally compiled in technical documents, that is, integrated process development reports, process flow documents, and the like, and is discussed in detail in Chap. 4. These form the foundation of manufacturability and ensure that the process works at small/pilot scale. The challenge then is to ensure that the development knowledge gets translated into design and operation at the manufacturing scale.

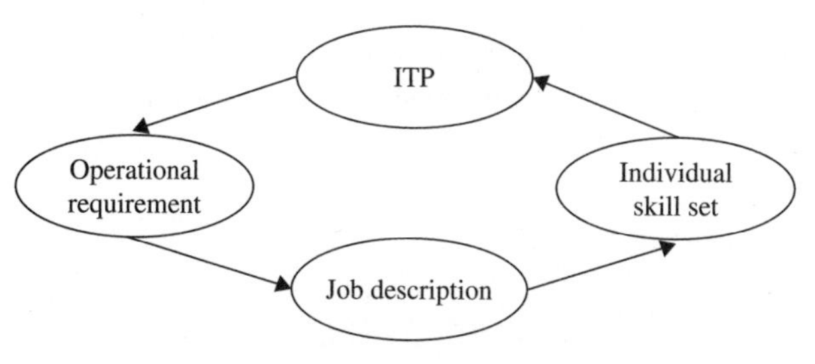

Figure 2.2 Individual training cycle.

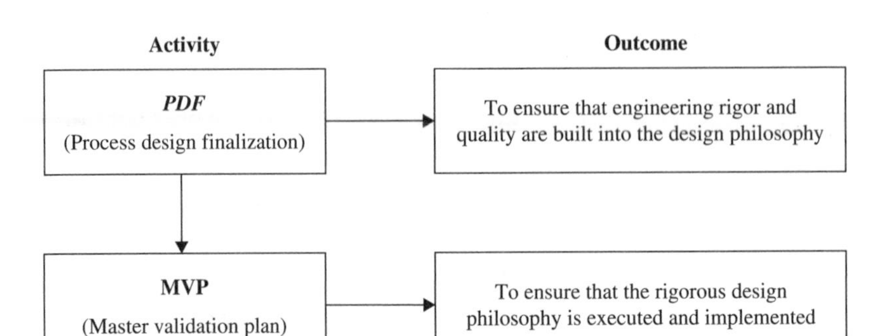

Figure 2.3 Structured approach to manufacturability.

The systematic approach to manufacturability (Fig. 2.3) includes:

Process design finalization (PDF). An organized effort to integrate various aspects of process development into a holistic final design ensuring that appropriate rigor and quality is built into the design philosophy

Master validation plan (MVP). A structured effort to execute and implement the design philosophy

2.2 Process Design Finalization

This is the final stage of late-stage process development and is a systematic approach to integrate various aspects of process development into a holistic design and to rigorously examine the design with the focus on the process-engineering aspects. One of the central aspects of the process design is to ensure that quality is built into the design. A systematic approach for *process design finalization* (PDF) is shown in Fig. 2.4. This approach is illustrated with the help of an industrial case study.

A case study (case study B) on crystallization for a solid form product is used to demonstrate the application of the structured approach. Crystallization processes are widely used in process industries. Typically, crystallization takes place in a stirred tank, followed by filtration and drying, thereafter the particle size is reduced using a mill, and finally the product with the right size particle and size distribution is dispensed in appropriate packaging. Figure 2.5 shows various unit operations involved in the process.

2.2.1 Process design finalization team

Formation of a process design finalization (PDF) team is an opportunity to integrate a diverse workforce (different departments) with a common objective of successfully launching the product (see Fig. 2.6).

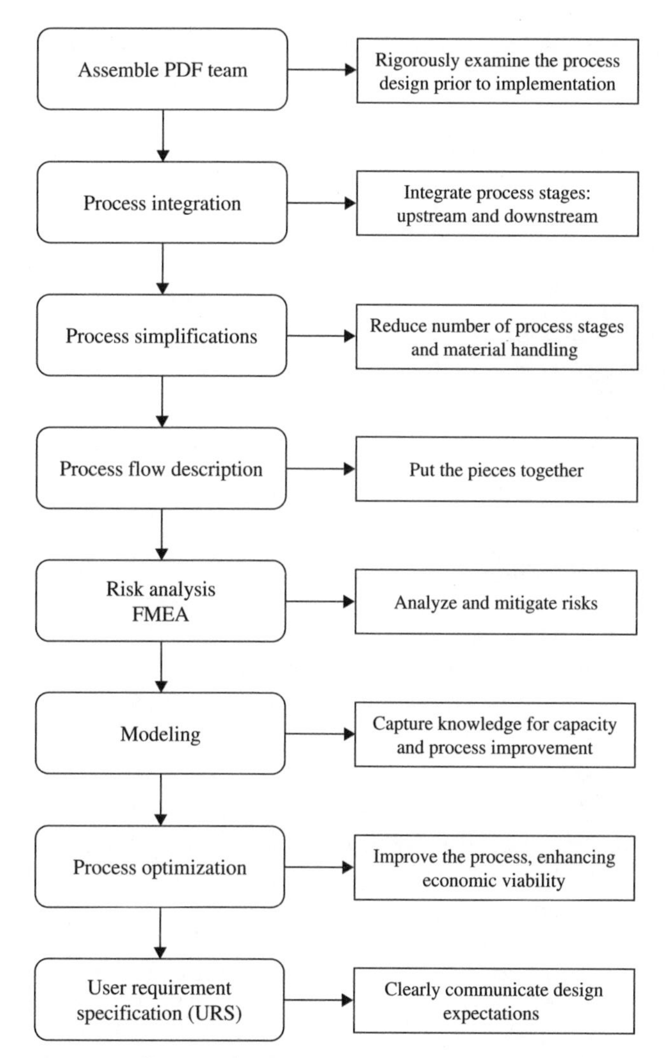

Figure 2.4 Systematic approach to process design finalization.

This team should consist of scientists and engineers involved in the process development including design engineers, operations associates, and quality representatives. The team should also include a design engineer from the engineering firm responsible for building the manufacturing facility. One of the first tasks of the team is to review the

Media makeup ⟹ Crystallizer ⟹ Filtration ⟹ Drying ⟹ Milling

Figure 2.5 Integration of unit operations into a manufacturing process.

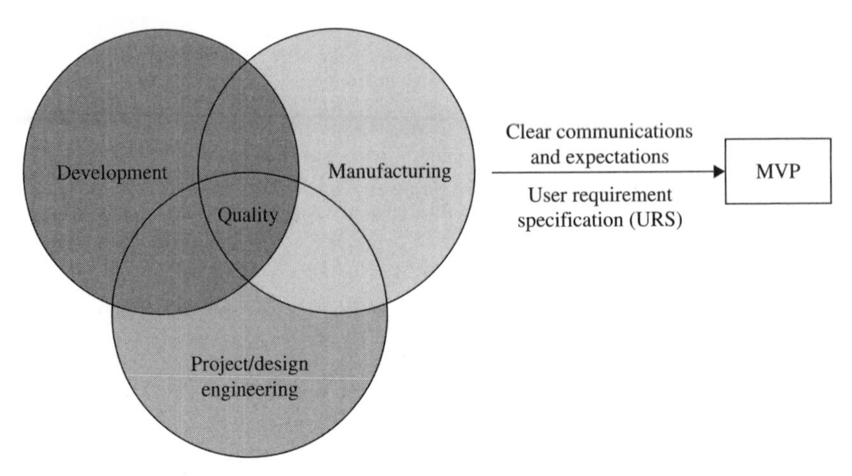

Figure 2.6 Process development review team.

knowledge base from development. It is important for everyone to have a similar level of understanding regarding the process development prior to embarking on a collaborative journey in finalizing the design.

The first stage of design finalization is the integration of various aspects of the design of the process. This will be discussed in the next section.

2.3 Process Integration

Process integration is the integration of various stages and the infrastructure into one manufacturing entity. The complete manufacturing process needed to transform raw material into products has various unit operations, as illustrated in case studies A and B. To increase the speed to commercialization, process development activities occur concurrently, often in different departments and different sites. It is critically important that these process stages be integrated. The key aspect is the integration of upstream and downstream parts of the process, with utilities and the plant infrastructure.

Traditionally in pharmaceutical and process industries, the development of upstream and downstream processes is independent leading to challenges in process integration. Such activities ensure the ease of forward processing by reviewing aspects such as physical properties between transfers and equipment capability to handle the variation in the input material from previous stages. The process integration activity should be initiated early on in late-stage development. At the process design review stage the process integration should culminate to produce a process flow description, finalizing the overall alignment of the process. This overall process integration should be intensely reviewed for completeness and rigor by the process design team. For case study B, the process integration could be simply depicted as in Fig. 2.5.

2.3.1 Mass and energy balance

Process integration involves rigorous analysis of each unit operation in light of the available data from development so as to integrate these unit operations into a manufacturing process. One aspect of process integration is establishing material and energy balances for each unit operation and also waste-stream assessment. This exercise helps integrate with the process utilities demand, for example, heating and cooling load. Mass and energy balance also helps calculate the components in the waste stream, which is important for developing an environment friendly process. The mass and energy balance is based on the universal law: In any unit operation, mass and energy are always conserved.

Mass balance. Figure 2.7 shows the mass balance for case study B for the crystallizer T1. For the reactive crystallization, reactants A and B react to produce product P and gaseous emission E.

The concept of mass balance could be illustrated by assuming that a component y is present in all streams and does not participate in any reaction. The mole fraction of y in streams A, B, E, and P are respectively, y_a, y_b, y_e, and y_p.

Thus, the mass balance for component y is:

$$\mathrm{P}y_p + \mathrm{E}y_e = \mathrm{A}y_a + \mathrm{B}y_b \tag{2.1}$$

The mass balance could be used for:

1. Estimation of product concentration
2. Estimation of waste-stream composition

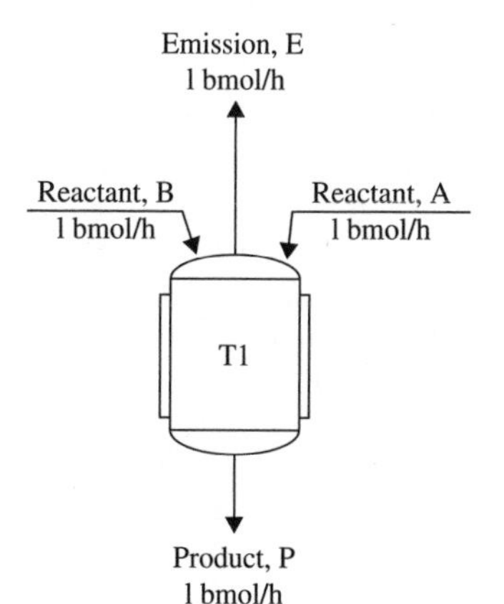

Figure 2.7 Mass balance.

3. Efficiency of the unit operation
4. Identification of opportunities to improve the unit operation

Energy balance. For the energy balance the rule of thumb is that as the scale increases, heat demand increases volumetrically, whereas the heat-transfer capability increases by surface area. For mechanically agitated tanks the heating/cooling is provided by a heat-transfer jacket or coil. As an example, for case studies A and B the energy balance for the mechanically agitated tank (T1) is shown in Fig. 2.8.

The arrows pointing out indicate the heat removed from the tank T1. The main heat removal Q_j comes from heat exchange through the cooling jacket or cooling coil. Cooling fluid at the appropriate flow rate removes the heat generated from the tank. This is also referred to as service-side heat transfer. Other avenues of heat removal include evaporation Q_e and heat loss (losses are assumed to be negligible in this case).

The arrows pointing in indicates heat added to the crystallizer tank T1. These include:

1. Heat generated due to exothermic reaction during crystallization (or microbial heat generation for case study A). This heat generation is typically the largest contributor.
2. Heat generated due to impeller motion Q_{im}.
3. The expansion of gas inside the tank also generates heat Q_s.
4. Any process stream that enters the tank at a different temperature to the fluid in the tank adds or removes sensible heat Q_{sen}.

Usually, the limiting resistance to heat transfer is from the service to the process side and is equal to $1/h$. Where h is the process side heat-transfer coefficient, which also depends on impeller type (Mohan, et al., 1992). This approach could be extended to other stirred tank operations.

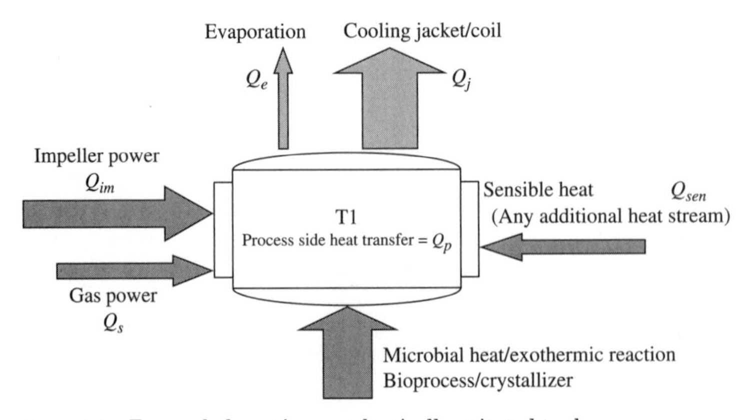

Figure 2.8 Energy balance in a mechanically agitated tank.

General formulae for calculating overall heat balance

Heat loss due to evaporation	$Q_e = mL$ (joules)
Heat input due to impeller power	Q_{im} = Shaft Power Draw (watts)
Energy due to expansion of gas (approximate calculation)	$Q_s = MV_s g$ (watt)
Service side (cooling jacket/coil)	$Q_j = mc_p \Delta T_j$ (joules)
Process side heat transfer	$Q_p = hA\Delta T_s$ (joules)

$$Q_p = (\text{exothermic reaction/microbial heat}) + Q_{\mathrm{im}} + Q_s + Q_g + Q_{\mathrm{sen}} - Q_e \tag{2.2}$$

This information is important to determine the size of the heat-transfer equipment (jacket, coil, or both), the type and quantity of heat-transfer fluid, and heat-transfer operating parameters, that is, flow rate, and so forth.

Once the process has been integrated aided by mass and heat transfer, the next step is to focus on process simplification, that is, identifying opportunities to make the process simpler and cheaper to operate.

2.4 Process Simplification

The drivers for process simplification are robustness, safety, cost, efficiency, capacity, and ease of operation. Levers of process simplification include: process, equipment, recipe, material handling, level of automation, and ergonomics. One key aspect of simplification is minimizing process hazards, which can be identified using techniques like HAZOP (hazard and operability studies discussed in Sec. 2.6.2). For example in case study A, an ammonia source is required as a continuous feed material. In the original design gaseous ammonia was used and was blended with the sparged air to provide the required nitrogen source. However, there are numerous hazards associated with storage and maintenance of gaseous ammonia. The process was redesigned to use aqueous ammonium phosphate. This not only eliminated the need for gaseous ammonia, but also replaced a phosphorus source in the raw material eliminating a raw material addition during the media makeup stage.

Another important aspect of process simplification is minimizing the number of unit operations used. Typically, efficiency (related to yield) of manufacturing drops with the addition of unit operation. There are several reasons for this: loss during transfer, operational efficiency, energy efficiency, equipment failure, and the like. This efficiency versus the number of unit operations curve will vary for different operations (see Fig. 2.9).

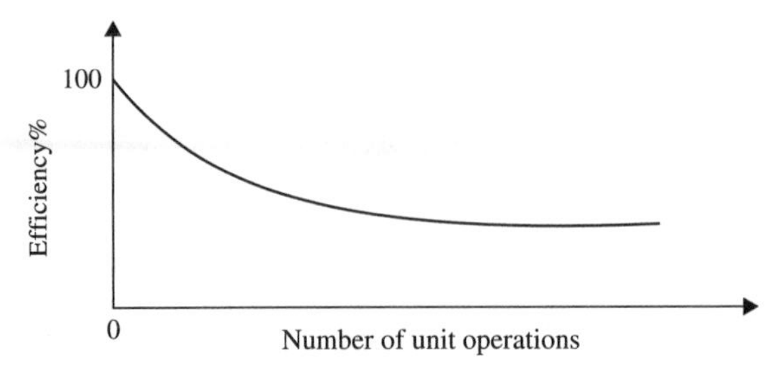

Figure 2.9 Number of unit operations versus efficiency.

Reducing or integrating the process equipment could help enhance the efficiency. For example, reducing or eliminating the number of storage tanks used in the process increases the efficiency. Another important aspect of unit operation minimization is integrating multiple unit operations in fewer pieces of equipment; for example, combining filtration and drying in one equipment set called a *filter dryer* could enhance the efficiency in case study B. Another example could be the use of static mixers for on-line mixing during the transfer stage eliminating the need for a mechanically agitated mixing tank. Or, for case study A, the bioreactor could be used for the media makeup stage, the sterilization stage, and the bioreaction stage. In process industries it is common to have distillation columns integrated with the reboiler, the overhead condenser, and the like to optimize energy usage.

Material handling also offers avenues for process simplification. Minimizing the transfer piping length and resident volume is a good practice. Another aspect is appropriate spatial location of equipment to maximize use of gravity feed, and to utilize differential pressure between equipment to minimize use of pumps. Another aspect of material handling is the understanding of the materials' physical properties in relation to the level-of-handling challenges. A few examples of level-of-handling challenges include:

Flowablility of powder is an important attribute for charging through a chute.

A highly viscous fluid would be difficult to pump.

A slurry with high-solid loading is susceptible to settling in transfer lines.

There are numerous design articles in the literature that list the rules of thumb for good design practices (Koolen, 2001; Peters et al., 2002).

It is also important to appreciate that oversimplification can inhibit flexibility and versatility. Therefore, a good balance between flexibility and simplification must be struck.

2.5 Process Flow Description

Once the integration and simplification exercises are completed the next step is to compile this critical information in a structured manner. *Process flow description* (PFD) is a systematic way of capturing manufacturing process information. PFD is the language of engineers and uses engineering symbols and conventions. PFD has three elements:

1. Process flow chart, sometimes referred to as equipment flow chart is a methodical illustration of equipment set and unit operations needed to execute the manufacturing process.
2. Control strategies required to run a process capable of producing quality product within specification reproducibly.
3. A narrative of the process.

A process and equipment flow chart for case study B is shown in Fig. 2.10, which summarizes the whole process. It captures process steps, material input and output, waste streams, unit operations, equipment used, process parameters, and control setpoints. The process parameters include operating parameters, proven acceptable range, regulatory limits and in-process sampling regime.

Control strategy, on the other hand, identifies the control scheme required to run a capable manufacturing process.

Control strategy could be depicted as shown in Fig. 2.11.

There are various aspects of the control strategy: raw materials are tested in the quality control, equipment is qualified, environment is controlled by implementing manufacturing area classification, waste is treated prior to release, and the final product is tested in the *quality control* (QC) laboratory. The focus for manufacturing should not only be to test quality, but also to build appropriate quality in the process. The in-process control is discussed in detail in Chap. 3, the fundamentals of which include identification of critical operating parameters and *proven acceptable range* (PAR). Critical operating parameter is identified during the process development stage (Chap. 1) along with the PAR.

Process integration, along with control strategy forms the cornerstone of manufacturability, which is captured into PFD. The process flow for case study B is shown in Fig. 2.10. The next step is to study the robustness in the process design to enhance the assurance that the manufacturing process would be capable of producing quality product, thus minimizing risk.

Robustness of process design implies that a rigorous risk management process is invoked in assessing and mitigating risk associated with product quality. Thus, a key aspect of process development is to ensure that quality is built into the design. Dr. Genechi Taguchi is one of the

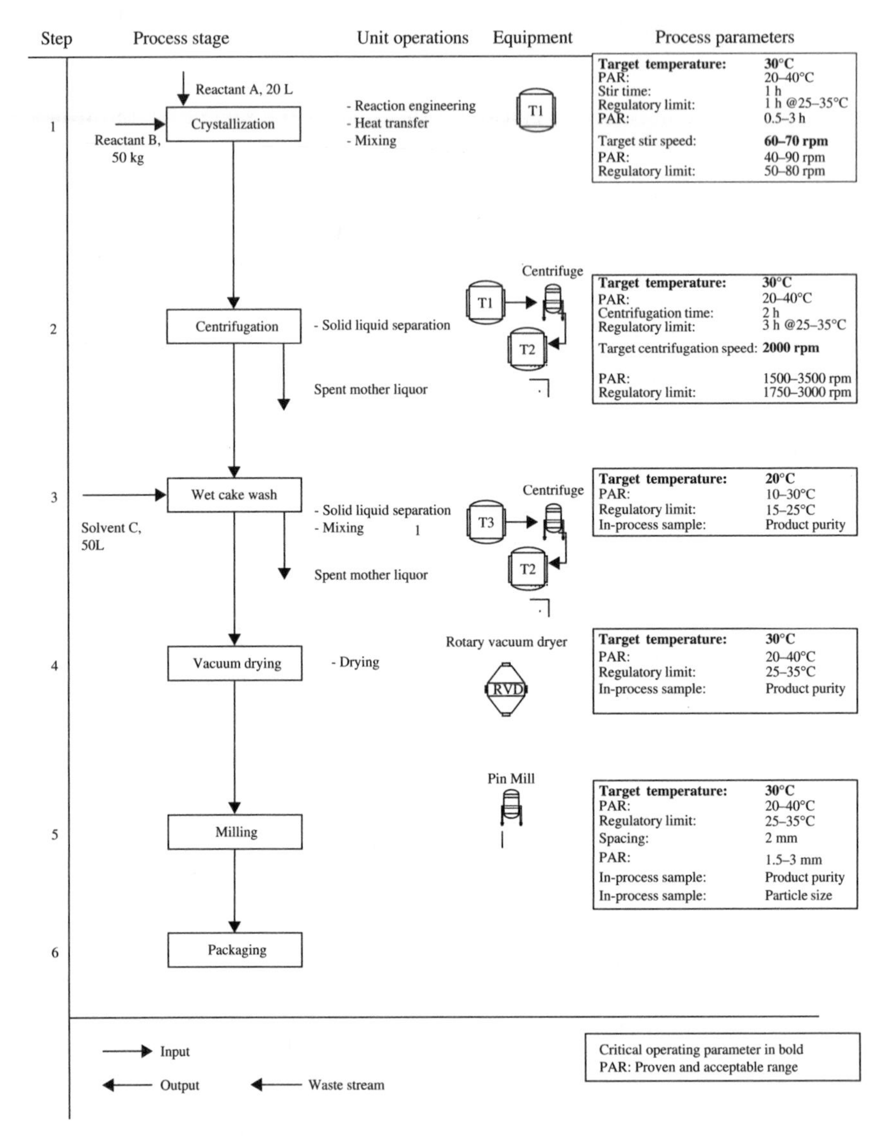

Figure 2.10 Process and equipment flow chart for case study B.

founding fathers of the modern quality revolution in manufacturing. He contended that the only way to improve quality permanently is to design it into the product. Typically, quality improvement efforts in most companies only involve inspection of a part after production. Most often, it is too late and too expensive to do anything about quality once the parts are produced. Instead, if quality concerns are addressed upstream in the

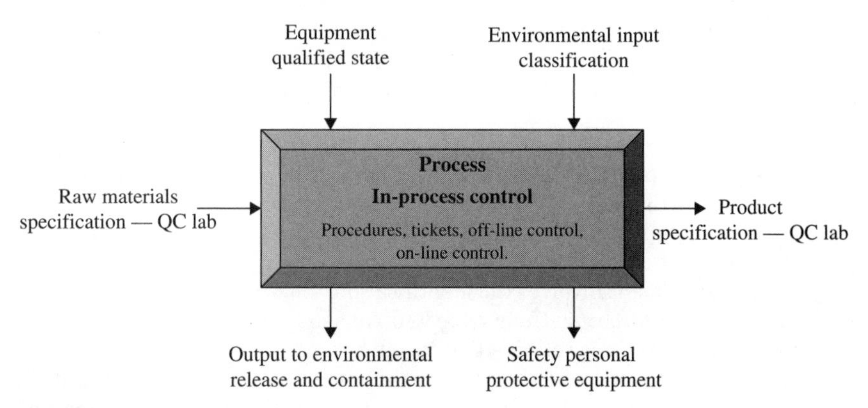

Figure 2.11 In-process control.

design of manufacturing (development stage) many improvements can be made less expensively. Thus, the mantra should be "design quality into the product." Return-on-investment in quality activities is much more in the design phase than in the production phase. The next section focuses on the risk-based approach to quality management of process design and operation.

2.6 Risk Management

Risk management could be defined as a risk-based decision-making process, based on risk analysis, risk mitigation, control strategy, and awareness. Much of the literature on risk, however, uses the expression "*risk management*" with two different meanings, depending on the context. Risk management is the making of decisions concerning risks and their subsequent implementation, and flows from risk estimation and risk evaluation. The risk management concept describes the overall subject area concerned with hazard identification, risk analysis, risk criteria, risk acceptability, and the risk management term, which describes the process whereby decisions are made. The British Government Center for Information Systems distinguishes between *risk management* and *management of risk*. The former "refers to planning, monitoring and controlling activities which are based on information produced by risk analysis activity," while the latter, "is used to describe the overall process by which risks are analyzed and managed" (Ayyub, 2003).

For the pharmaceutical industry the regulatory agencies have been setting expectations for risk management. In the United States, the

FDA guidance document (located in www.fda.com) indicates (for medical devices that could be generalized) that the role of risk management is to help identify, understand, control, and prevent failures that can result in hazards.

> Risk management is a systematic application of policies, procedures, and practices to the analysis, evaluation, and control of risks. It is a key component of quality management systems, and is a central requirement of the implementation of Design Controls in the Quality Systems Regulation. Risk management involves the identification and description of hazards and how they could occur, their expected consequences, and estimations or assessments of their relative likelihood. The estimation of risk for a given hazard is a function of the relative likelihood of its occurrence and the severity of harm resulting from its consequences. Following the estimations of risk, risk management focuses on controlling or mitigating the risks.

In summary, risk management has three elements (Fig. 2.12), namely, risk identification, risk assessment, and risk mitigation.

The techniques of risk identification are facilitative tools, intended to maximize the opportunity to identify all the risks or hazards inherent in a particular facility, system, or product. The tools may be categorized under broad headings of intuitive, inductive, and deductive techniques. Brainstorming is the main intuitive technique, involving a group generating ideas "off the top of their heads" with a philosophy of "nobody is wrong—let's get the ideas on the board." Although quick and simple, it lacks the comprehensive approaches of the more sophisticated techniques.

Inductive (what if?) techniques include preliminary hazard analysis, checklists, and human-error analysis. By far the two most common techniques, however, are *hazard and operability studies* (HAZOPS) and *failure modes and effects analysis* (FMEA).

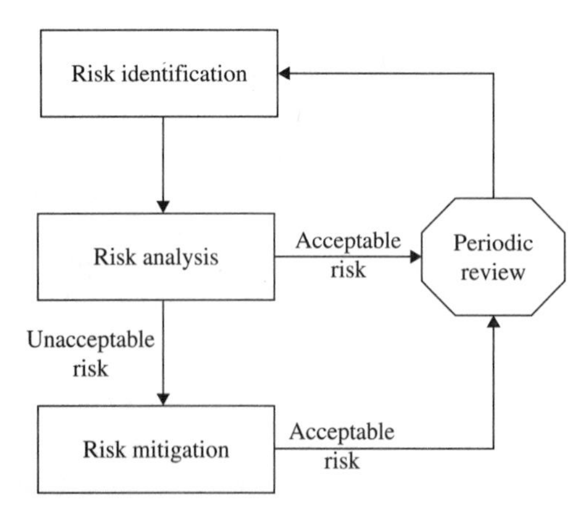

Figure 2.12 Risk management approach.

The benefits of hindsight inform the deductive (so how?) techniques of accident investigation and analysis. Event and fault trees, although primary tools in risk estimation, can also be employed in a purely qualitative way—event trees as inductive, and fault trees as deductive tools.

Once risk is identified the next step is risk assessment. Based on the set criterion acceptable and unacceptable risks are differentiated. A risk-mitigation plan is then defined to address the unacceptable risks. The risk is periodically reviewed to identify any new risks.

There are various structured methodologies for risk management that will be discussed in the following section with major emphasis on FMEA.

2.6.1 Failure mode and effects analysis

While various techniques are available to address quality, FMEA provides an excellent quantitative approach. A FMEA is an engineering technique used to define, identify, and eliminate known and/or potential failures, problems, errors, and so on, from the system, design, process, and/or service before they reach the customer.

The FMEA technique has been used in the safety arena for more than 30 years; recently it has gained widespread appeal mainly due to automotive industry application and its QS-9000 supplier requirements. The QS-9000 standard requires suppliers to the automotive industry to conduct product and process FMEAs in an effort to eliminate failures before they happen. One of the advantages of the FMEA tool, unlike other quality improvement tools, is that the technique does not require complicated statistics, yet it yields significant quality enhancements.

There are four types of FMEA, namely, system, design, process, and service. System or concept FMEA focuses on potential failure modes between the functions of the system caused by system deficiencies. Design FMEA focuses on failure modes caused by design deficiencies. Process FMEA focuses on failure modes caused by process design. Service FMEA focuses on failure modes caused by system or process deficiencies.

The purpose of process FMEA is to prevent process and product problems before they occur. The application of the technique during the design phase will increase the chances of a robust manufacturing operation and lead to the reduction or elimination of after-the-fact corrective action and late stage "firefighting." Typically, manufacturing resources are tied up in the so-called firefighting efforts—playing a catch-up game rather than continual enhancement.

Engineers have always analyzed processes and products for potential failures; the FMEA process standardizes the approach and establishes a common language that can be used at various levels within and between companies.

The FMEA differs from a somewhat similar tool, the cause-effect (Ishikawa) diagram. The cause-effect diagram aids in recognizing all inputs to a process that lead to the effect of that process. In contrast, the FMEA seeks to determine every possible way (mode) a process may fail, the effects of those modes, and the "criticality" of modes and effects. Both tools require the participation of a team made up of those involved in carrying out the work of the process being studied.

Hazard analysis and critical control points (HACCP)(see Sec. 2.7.6) and FMEA rely on teams to identify potential failures or hazards that are likely to cause harm, determine which are the most critical in causing harm, develop a failure/hazard proofing action plan, and monitor the results.

FMEA process overview is shown in Fig. 2.13, and its application could be best demonstrated by an example.

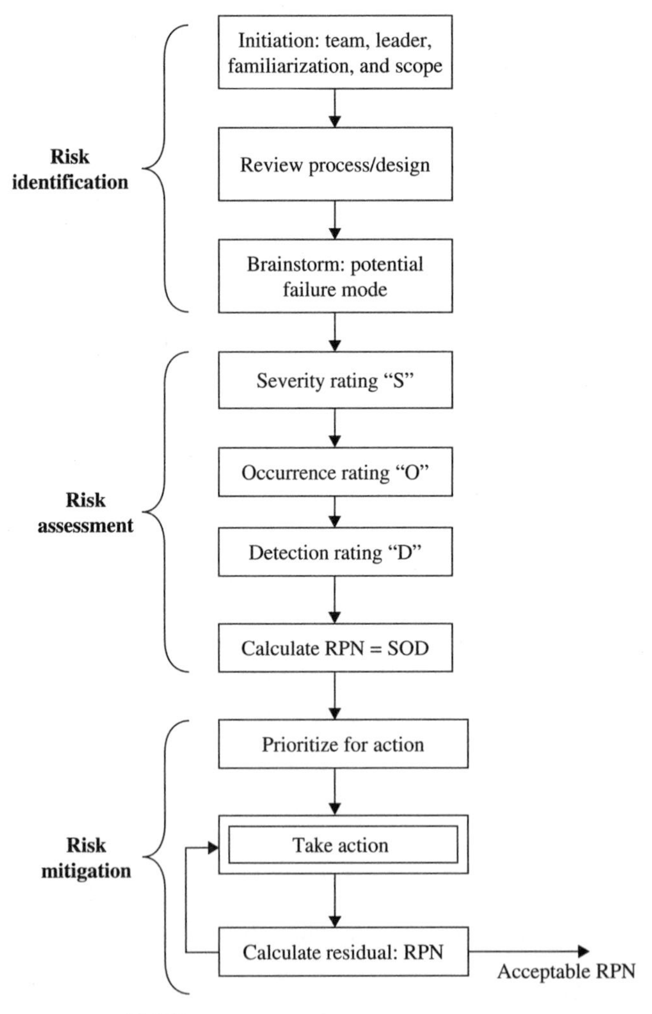

Figure 2.13 FMEA process overview.

Process FMEA case study—aseptic bioprocess design. In a fermentation process (bioprocess) producing valuable antibiotics, one of the key concerns is contamination of a fermenter (bioreactor). Contamination invariably leads to rejection and rework. Regulatory agencies mandate strict control over sterility assurance in a bioprocess. In this case study process FMEA applied to the fermentation process is discussed.

FMEA team members. The first step in the FMEA process is to identify the team members who will be participating in the process. Team membership should represent different aspects of the process to ensure the holistic view. In the case study on bioreactor contamination representation from different functions operating on the process was sought. These functions included operations, engineering, maintenance, technical services, development, and technical management. Also, to add external focus an external consultant was included.

FMEA team leader. The team leader should either be appointed by management or selected by the team. The team leader is responsible for coordinating the process by setting up and facilitating meetings, and is a point of contact with management ensuring appropriate resources and assisting the team toward completion. The leader should not dominate the team, instead should enable the team to succeed. For the case study the process engineer was appointed as the team leader.

Familiarization. The team should be aware of the FMEA process. A formal training session by a FMEA expert could help set the stage for the process. Knowledge of consensus-building techniques, team project documentation, and idea-generating techniques, such as brainstorming, are necessary for FMEA team members. In addition, team members should be comfortable using problem-solving tools such as flowcharting, data analysis, and graphical techniques.

FMEA scope. A specific and clear definition of the process or product to be studied should be written and understood by everyone on the team. Team members should have an opportunity to clarify their understanding of the scope, if necessary, and those clarifications should be documented. This will help prevent the team from focusing on the wrong aspect of the product or process during the FMEA.

The FMEA scope for the case study was:

> Design the equipment and operation for high-sterility assurance of the bioprocess, with an acceptance level of contamination = 1 in 20 batches.

Along with scope, the project budget, resource requirements (including constraints), timeline, and communication strategy must also be determined at the outset.

Review the process. The team should thoroughly review the design/process around the scope of the FMEA. For the case study, the team reviewed: process flow diagram, process measurement and control, workflow, raw material management (including media makeup operation), standard operating procedures, maintenance practices, and business processes.

Brainstorming Potential Failure Modes. A brainstorming approach is very useful in the FMEA process to ascertain potential failure modes. While some prefer unstructured brainstorming, an initial structure in brainstorming may help rapid convergence of diverse inputs. The initial structure could include five elements of a process: people, materials, equipment, methods, and environment. A fishbone diagram may assist in capturing the discussion. The first step is to identify the "potential sources of failure."

In the case study, the team brainstormed on the five elements of the process using the fishbone diagram. A sample of the discussion of potential sources of failure is captured in Fig. 2.14.

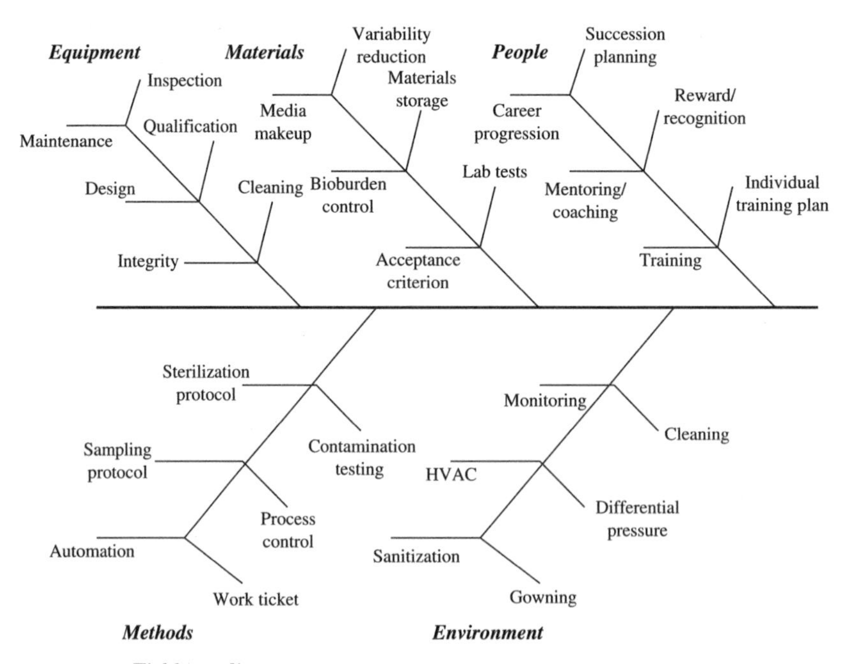

Figure 2.14 Fishbone diagram.

The second step then is to identify for each source the modes of failure. For example, cleaning of equipment in the case study is a potential source of failure; that is, inadequate cleaning may lead to contamination of the product. The potential modes of failure that would lead to inadequate cleaning could be: inadequate *cleaning in place* (CIP) system, selection of wrong cleaning agent, inadequate contact time, and lack of full equipment surface coverage.

Listing potential effects of each failure mode. This phase could be described as the worst-case scenario analysis phase, that is, "If the failure occurs then what are the consequences?" While there may only be one effect for certain failure modes others may have more than one. In the case study the potential effect of inadequate cleaning is "contamination" of the batch.

Severity, occurrence, and detection rating. These ratings are based on a 10-point scale (see Fig. 2.15). One is the lowest rating and ten is the

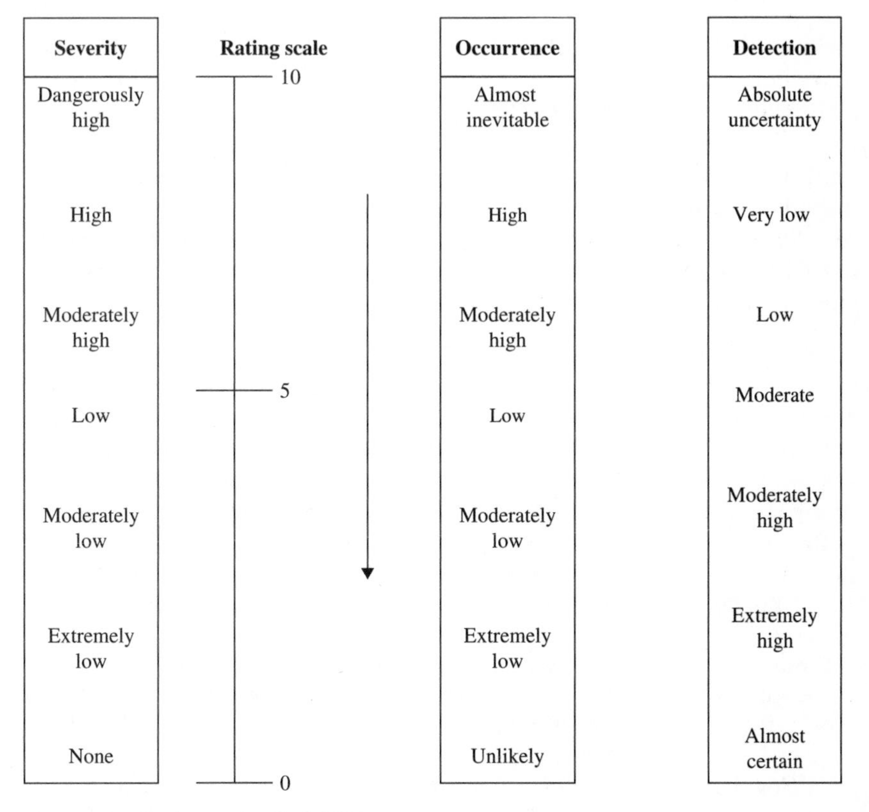

Figure 2.15 Rating scales for FMEA.

highest rating. It is important to have a common understanding of the rating system at the very outset of the process.

The severity rating is an estimate of the seriousness of the effect in the event of failure. Severity rating may be estimated based on either past data or on the knowledge and experience of the team members. Each failure may have a different level of severity; it is the effect not the failure that is rated. Thus, each effect should be given its own severity rating, even if there are several effects for a single failure mode.

Occurrence rating is usually determined using actual historical data. When actual data are not available, the team must estimate how often a failure mode may occur. Listing the potential cause of failure will assist the team in determining the occurrence rating.

The detection rating determines the likelihood of detecting the failure. The rating is determined by identifying the current controls that may detect a failure or the effect of a failure. For example, if there are no current controls, the likelihood of detection will be low, such as a nine or ten.

The *risk priority number* (RPN) is calculated by multiplying severity with occurrence and detection. The total RPN should be calculated by adding all of the RPNs. RPNs quantify the risk and help prioritize the course of action. Ideally the failure modes should be eliminated completely; however, in practice the risk is managed based on an acceptable RPN. One way to minimize risk is to enhance the probability of detecting failure and constructively intervene to avoid failure. Once action (based on priority) has been taken to improve RPN, new ratings for severity, occurrence, and detection should be determined and a resulting RPN calculated. Severity rating for contamination could be provided by HACCP.

FMEA worksheet. A rigorous FMEA analysis during the design phase is key to the success of the manufacturing design. The worksheet is designed to summarize the discussion in an objective manner. There are variations in the format of the worksheet; the worksheet discussed in this book is shown in Table 2.1. The table summarizes one of the several FMEA subprocesses for the design for aseptic processing.

To illustrate the rigor involved in the FMEA for the subprocess of cleaning, the following discussion occurs in the team setting leading to quantification of risk:

> Inadequate hardware/software design for the CIP system (nozzle design, valve sequence, flow rate, dead leg, and so forth) will almost certainly lead to contamination.

$$\text{Severity rating} = 10\text{; occurrence rating} = 8\text{, detection rating} = 8\text{: RPN} = 640$$

TABLE 2.1 Failure Mode and Effects Analysis Worksheet

Process: Design of aseptic processing in a bioprocess Timeline: 6 months Budget: $ 500K
Scope: "Design equipment and operation for high-sterility assurance of the bioprocess, with an acceptance level of contamination = 1 in 20 batches."

Subprocess: Cleaning FMEA No.: 248 FMEA date: 01-01-01 Revision No.: 1 Date of Revision: 02-02-01
Participating team members: Joe, Sally, John, Tim, Chris, and Sam Team leader: Sam

Potential failure mode	Potential effect(s) of failure	S	Potential causes of failure	O	Current controls	D	RPN	Action list	Owner, timeline, and budget	Action taken	S'	O'	D'	RPN'
CIP system	Poor cleaning—potential risk to aseptic processing	10	Valve sequence, nozzle design, flow rate, dead Leg	8	Process design review by experts	8	640	Simulation of cleaning operation	Joe, 3 months, $ 10K	3D simulation with experi-mental verification	5	3	3	45
Cleaning agent	Poor cleaning—potential risk to aseptic processing	8	Wrong cleaning agent	6	Process design review by experts	4	192	Small scale evaluation in the laboratory	Sally, 2, months, $ 30K	Experimental study conducted with various cleaning agent	2	2	1	4
Contact time	Poor cleaning—potential risk to aseptic processing	7	Inadequate contact time for cleaning	5	Process design review by experts	4	140	Small scale evaluation in the laboratory	John, 2 months, $ 40K	Small scale experimental study	3	2	2	12
Equipment coverage	Poor cleaning—potential risk	6	Pockets of unclean equipment surface	4	Process design review	3	72	3D simulation	Tim, 2 months, $10K	Computer simulation	3	2	3	12

Wrong cleaning agent may not remove all the contaminating sources.

Severity rating = 8; occurrence rating = 6, detection rating = 4: RPN =192

Inadequate contact time with the equipment surface during the cleaning process may lead to contamination.

Severity rating = 7; occurrence rating = 5, detection rating = 4: RPN = 140

Some part of the equipment may not be cleaned properly. In the presence of a contaminant in those unclean parts there is likelihood that contamination may occur.

Severity rating = 6; occurrence rating = 4, detection rating = 3: RPN = 72

This FMEA analysis helps identify and quantify risk using the RPN rating. The next step as illustrated in Fig. 2.13 is to prioritize the mitigation plan and take actions to mitigate higher RPN rated risks. The action taken lowers the risk to an acceptable RPN rating, and a robust design with a much lower risk is delivered to manufacturing.

Besides FMEA there are several other tools and techniques for risk management including HAZOPS, FMECA, fault-tree analysis, event-tree analysis, HACCP, and QFD. These tools and techniques will be discussed briefly in the subsequent sections.

2.6.2 Hazard and operability studies (HAZOPS)

The HAZOP technique was originally devised by ICI Ltd for use in risk identification in chemical plants. The HAZOP is a structured brainstorming exercise in which a multidisciplinary team of experts systematically considers each piece of equipment in the plant, defining its intention and using guidewords such as "not," "more," and "less" as cues to identify possible deviations from the intention. Such deviations can then be investigated to eliminate their causes as far as possible and minimize the impact of their consequences.

The HAZOP approach is flexible and can be used to identify potential hazards in buildings and facilities of all kinds at all stages of their design and development. For example, a HAZOP could be carried out on the detailed design drawings for a new plant, and any hazards identified could be designed out before the plant is built. Alternatively, a review of contingency plans at an existing facility could be more comprehensively informed by a HAZOP exercise, which could identify hazards not previously planned for. Further reading on HAZOP includes Hyatt (2003).

2.6.3 Failure modes and effects criticality analysis (FMECA)

Unlike the HAZOP, FMECA studies are usually carried out by an individual expert with a thorough understanding of the particular system under investigation. The FMECA is a step-by-step procedure for identifying failure modes or design weaknesses and the criticality of the consequences of failure. The technique can be either hardware oriented, concentrating on potential equipment failures; or event oriented, where the emphasis is on functional outputs and the effect of their failure on the system. Complex systems are often analyzed using a combination of these approaches.

The system under investigation is converted into graphical form by the preparation of logic block diagrams depicting its functional components and subsystems. Every component of the system is considered and each mode of failure identified. The effects of such failure are then considered at both subsystem and overall system levels. FMECA is considered as a variation of FMEA.

2.6.4 Fault-tree analysis as a qualitative tool

A fault tree depicts the way in which a particular system failure might occur. The failure mode, for example, a house fire, forms the top event of the tree. Working downward through the branches, using and/or logic gates, the analysis reveals the combination of events which themselves cause the top event to occur. Thus, the components of a fire would be fuel and oxygen, and a source of ignition. The source of ignition could be an electrical fault or a discarded cigarette. The tree stops when no further analysis is practicable or necessary.

Such an analysis leads to understanding how the failure could occur and what effect design modifications might have. In qualitative terms, the preparation of a fault-tree analysis for each of the failure modes or deviations identified through a HAZOP ensures that the possibilities of failure have been thought through and designed out as far as is reasonable and practicable.

2.6.5 Event-tree analysis as a qualitative tool

Whereas a fault tree depicts the causes of a particular failure, an event tree shows its consequences, asking sequential questions to which the answer is either yes or no. Taking our house fire as the event, one might ask questions such as: Are there smoke alarms fitted? Do they work? Are there people at home? Is the fire department contacted quickly? Following the alternative answers through the tree will lead to a variety of consequences, ranging from minor damage with no injuries to severe damage and possible fatalities. In the same way as a fault tree, the event tree has qualitative value as a comprehensive thought process, enabling attention to be paid to eliminating those paths that lead to more severe consequences.

The techniques described earlier are reliant on hindsight to identify risk. The FMECA analysts use their knowledge of what has gone wrong elsewhere as a predictor for the facility under examination. The HAZOPS team is doing the same thing, but has more chance of success because they share the collective experience of a multidisciplinary group. The inquiry team seeks to apply the lessons learned from a particular disaster to prevent similar recurrences. Hindsight cannot proactively predict the new risks.

Tools that can proactively and comprehensively predict new risks in a structured format include HACCP, QFD, and FMEA.

2.6.6 HACCP

HACCP (pronounced Hassip) is an acronym for *hazard analysis and critical control points.* This technique was developed in the food industry in the 1960s as a way to analyze a process, determine the high-risk steps, and control or monitor those steps, thereby ensuring that the process yields quality product. This tool is extensively used in the food industry to identify the risk of contamination. HACCP provides a rigorous analysis and documentation to ensure that a company understands its product and process well enough to control or monitor parameters that are important to produce quality products (Armbruster and Feldsien, 2000). HACCP consists of seven principles:

1. Analyze each step for hazards
2. Identify all *critical control points* (CCPs)
3. Verify limits for each CCP
4. Verify monitoring and testing of limits
5. Verify corrective actions
6. Verify operational procedures for CCPs
7. Verify that records of each CCP are documented in the batch record

CCPs are analyzed for (Rooney, 2001): risk level of the hazard; occurring risk of nonconformance; risk to a health care professional; risk to a patient; risk to a product; overall risk; and is this step a CCP. The levels of risk for CCPs are classified as: extreme (intolerable, possible catastrophic results, may be difficult to control or monitor, requires adverse event report); major (undesirable, monitor closely to determine exceeded limit, remove or reduce if possible); minor (not a severe hazard, not economical to reduce); nuisance (not considered a CCP, since efforts to correct outweigh the risk). The primary difference between FMEA or FMECA and HACCP is the way detection is treated. Once contamination is detected the quantification has little significance (i.e., higher or lower level of contaminant), whereas in FMEA/FMECA the level of detection is critically important (i.e., different values of pH, temperature for a process, and so forth). That is why FMEA is a widely used risk assessment tool across multiple industries (automotive, process, and pharmaceutical industries).

2.6.7 Quality function deployment (QFD)

QFD provides a systemic approach to determine, prioritize, and translate customers' needs to product design parameters. Such design parameters are frequently checked against customer needs throughout the product development cycle to ensure customer satisfaction with the end product. QFD is a system for translating consumer requirements into appropriate company requirements at each stage. This ranges from market studies, engineering, and manufacturing to marketing/sales and distribution. QFD systematically aims to ensure that important issues are not lost or misinterpreted as a project is handed off from functional group to functional group. Traditionally, it has been applied to product development, but QFD can be applied to any complex project.

Effective implementation of QFD hinges on forming the proper implementation team and employing the aforementioned QFD tools. Proper deployment of the implementation team has three phases: conceptualizing the subject issue, collecting the necessary data, and analyzing and reporting the results of the data gathered. To identify customer needs and wants, QFD employs the focus group approach as the data-collection method to ensure a comprehensive means to collect the customer requirements and expectations. Tools such as affinity diagrams, tree diagrams, and house of quality are used by the QFD team to understand the voice of the customer and assess the expected degree of success of new products (Bossert, 1991; Krueger and Casey, 2000).

The primary tool of QFD is the *house of quality* (HOQ). The HOQ is a useful tool for arranging facts so that important issues, relationships among these issues, significance of each, and their measures of success can be readily displayed (Hauser and Clausing, 1998). The HOQ uses a matrix to present what the customer requires (i.e., the "what" side) against how those requirements would be met (i.e., the "how" side). When fully developed, the matrix resembles a house.

As organizations strive to succeed in the highly competitive global economy, QFD is an important part of the product development process because it focuses the organizations' attention on the customer. QFD links customer requirements or "whats" with the appropriate engineering design characteristics or "hows" so the voice of the customer is translated into product designs and specifications. Building the house of quality is an important step in moving from customer requirements to production requirements.

QFD may also enable the firm to cut product costs and reduce time-to-market. The results show that QFD is able to simplify the manufacturing process, but overall product costs appear to be only slightly less when QFD is applied than when traditional practices are used. The reason for this small improvement in product costs and time-to-market may be a lack of experience with QFD. For many organizations in the survey, this was their first attempt at applying QFD. Also, in all but one case, this was the first application of QFD to that product. As companies gain experience with QFD and learn to apply it more effectively, product costs and time-to-market may decline. QFD is being replaced by *value function deployment* (VFD). Based on a similar concept as QFD, VFD focuses more on the value to customers. The primary question for any design modification in a process or equipment is: how would the customer perceive the value of the change?

QFD and FMEA are more than just technical tools; in practice, they are communication tools that act as the catalysts to spark teamwork. In doing so they enable a companywide, cross-functional, multidisciplinary network of teams that share like-minded goals to foster a broader, total quality management culture. The specific benefit of QFD and FMEA team cross-functionality is the ability to step outside the organizational structure and look at new product planning and problem-solving issues, perhaps more objectively, than organizationally structured teams. The earlier argument supports the use of both QFD and FMEA as drivers to bring the voice of the customer and voice of the engineer closer along the "performance quality line."

Operating QFD at least up to the point where FMEA starts, or beginning an FMEA at the point where QFD trails off, enhances a seamless handover of both teams and disciplines. This in turn will improve design reliability and robustness efforts, with genuine customer-driven satisfaction. It will also avoid the dilution or distortion of the true voice of the customer once component specifications are set. It would also reduce the lack of process foresight that may result in a compromised design solution. If both of these tools were brought into direct contact in the early to late stages of the commercialization cycle, improvements would include: process improvement, long-term cost reductions, standardization of the target setting process, fuller understanding of all the customer quality requirements, more balanced system trade-off processes, improved product quality, and higher customer satisfaction.

2.6.8 An integrated approach—integrating risk assessment tools

A multitude of tools and techniques for risk management exist (as is evident from the previous section), the challenge is to identify the right tool for the right application. There is opportunity to integrate and engage the most appropriate toolset to maximize the effectiveness and value of risk management. The toolsets described are complementary with different tools and methodologies having different focus. FMEA is recommended as the focal point for risk management and the overarching methodology for integrating other important toolsets. QFD and VFD could be used to set the target for equipment or process design by spelling out the expectation of the customers. This can help set the acceptable RPN number or the acceptable level of risk. Fault tree and event analysis could be used for structured methodology for risk identification feeding into the FMEA worksheet as potential failure modes, potential effects of failure, and potential causes of failure (see Table 2.1). HAZOP could be used to identify risks related to health, safety, and environment while HACCP could be used to determine the contamination-related severity rating.

The approach indicated in Fig. 2.16 though rigorous is resource intensive (people and time), and it is recommended that this approach be used for risk-critical aspects of the process or equipment design. Fault tree or event tree could be used as a quick guide to risk assessment as a less intensive approach. Such a decision to engage the level of rigor in risk management would entirely depend upon a company's perception of risk. In some process industries, such as the food and pharmaceutical industries, regulatory agencies expect rigorous risk management.

Risk management should be an integral part of process development; the risk should be mitigated by designing and building quality in the process. This will help set a successful foundation for manufacturing. Typically, in a bid to achieve speed during commercialization of new products the rigorous process design review is overlooked. In many cases process equipment cost is a small part of the overall capital spent as compared with other costs—building, utility, containment, and the like. Thus, the decision makers tend to ignore this aspect. Also, in the bid to outsource the new manufacturing facility creation to engineering firms, the process scale-up aspect is included as a part of the outsourcing.

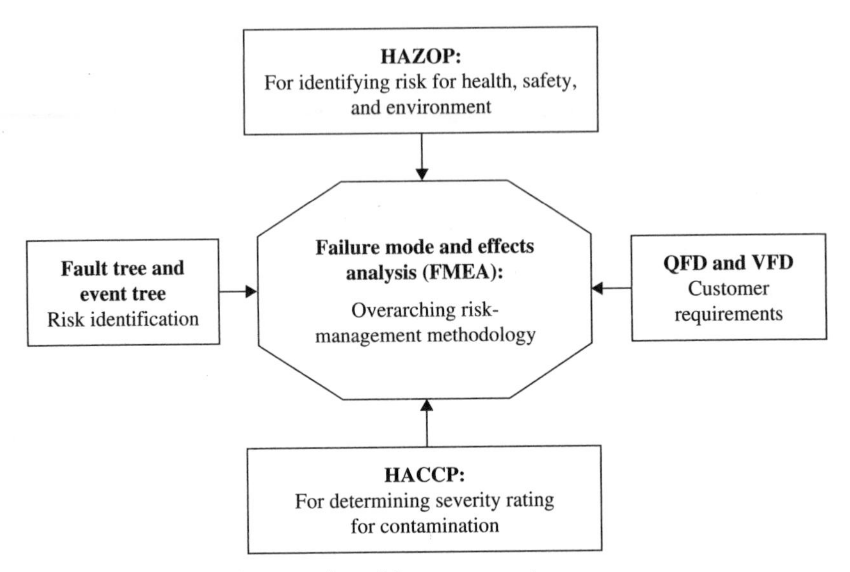

Figure 2.16 Integrated approach to risk management.

As a robust manufacturing process emerges it becomes important to understand the overall supply chain dynamics to reap the benefit from a robust manufacturing operation. Modeling provides the environment to capture various elements of manufacturing and will be discussed in the next section.

2.7 Modeling—Effectiveness in Handling Variable Capacity Opportunity

To achieve greater efficiency and market responsiveness, manufacturing companies must unite the work flow, from innovation through scale-up, manufacturing, and distribution. The fluctuation in market demand puts significant pressure on manufacturing. The opportunity is to rapidly adapt to the fluctuations to ensure that the supply is maintained and the asset utilization is optimized.

Modeling offers the opportunity to see the manufacturing system or big picture, which consists of numerous work flows of the process life cycle. The work flows could be interpreted in terms of hierarchy of the manufacturing system that converts raw materials into finished products for customer (see Fig. 2.17). The manufacturing system should have the capacity to meet customer demands. With capacity and process understanding as the overarching driver for modeling, the manufacturing system hierarchy could be categorized as equipment, process and infrastructure, and operations and organization.

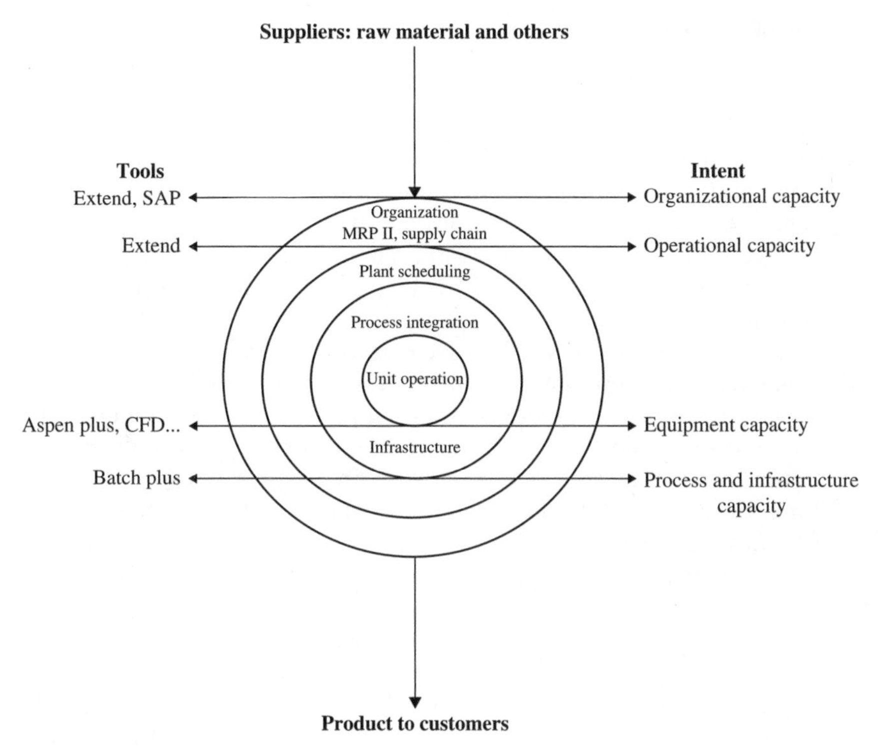

Figure 2.17 Modeling hierarchy.

Capacity-modeling tools fall under two broad categories—static and dynamic. *Static modeling* is the most common and the easiest to program. Static modeling can be very effective as a first-pass analysis tool where future manufacturing requirements are still wildly variable. The assumptions on a variety of planning metrics have a time-independent perspective. For example, with static modeling, one uses monthly demand to calculate the labor and equipment needed to support the required volume. *Dynamic* (simulation) modeling, although more complicated to build and use, provides a more realistic tool for planning a production area. Dynamic models, by definition, are time dependent and analyze how systems or areas will react to changes over time. Instead of examining the production line and resources on a monthly basis, the model simulates individual lots moving through the line based on a realistic production schedule. This model can then provide a picture of production during the month, including achievable cycle times, sources of delays, inadequate production-staffing levels, and shifting bottlenecks. A broad array of powerful tools, tailored to facilitate the different types of analysis and streamline the data handling, are needed for the chosen work processes. This includes chemical route selection, design of experiments, data collection and analysis, equipment design, production planning and scheduling, process modeling and simulation, materials management, writing operating instructions, to name a few. The "computing engines" in the tools should include expert systems technology to deploy "rules-of-thumb" for process scale-up and equipment design, check that equipment contents are compatible with materials of construction, examine reactive chemistry issues for mixtures, use heuristics to develop good schedules, and so forth. Some of the tools indicated in the Fig. 2.17 will be mentioned in the subsequent paragraphs.

At the core of the modeling hierarchy is unit operation modeling with the objective of capturing and optimizing the knowledge of the unit operation including the synergy between process and equipment. The unit operation model focuses on equipment design level and leads to valuable knowledge capture, resulting in better-informed decision making for scalability and process improvement. It also must include mass and energy balance.

Unit operation modeling has three aspects—mechanistic (first principle), empirical, and hybrid (mechanistic + empirical). Mechanistic models engage the first principle understanding of the process; for example, "computational fluid dynamics" modeling, involves defining finite volume grids and solving of Navier-Stokes equations with appropriate boundary conditions. For case study A (Chap. 1), it was important to understand and predict cell-suspension behavior at the manufacturing scale. A computational model was created and verified at the small scale with experimental data and later the model was scaled up to predict the cell-suspension behavior at the manufacturing scale. The model indicated that the cell-suspension behavior would be similar at the two scales. For unit operations some key fundamental equations (e.g., Navier-Stokes equations) do not differ among chemical processes. The physical and thermodynamic properties and chemical kinetics constants differ, and this is integrated into the computer simulator as an input. Data-based modeling is another important aspect of unit operation modeling, which will be covered in detail in Chap. 6.

Process models capture process knowledge from various sources to generate a holistic view of the process. One such modeling environment is Batch Plus from Aspen. The process model evolves from the development stage to the manufacturing stage and is the platform for capturing further process enhancements for the manufacturing process. The process industry is realizing the value of this approach in reducing costs of new facilities and also in sustaining process improvements at the manufacturing scale.

Process models include the process flowsheets, which are the language of chemical processes. They describe an existing process or a hypothetical process in sufficient detail to convey the essential features. *Simulation* is the tool chemical engineers use to interpret process flowsheets, to locate malfunctions, and to predict the performance of the processes. The heart of the analysis is the mathematical model, a collection of equations that relate the process variables, such as temperature, pressure, flow rate, and composition to surface area, valve setting, geometrical configuration, and so on. The steady-state simulations solve for the unknown variables, given the values of certain known quantities.

There are several levels of analysis. In order of increasing complexity, they involve: material balances, material and energy balances, equipment sizing, and profitability analysis. Hence, it is possible to prepare one or more equation-solving algorithms for each process unit to solve the material and energy balances and to compute equipment sizes and costs. A library of subroutines or models, usually written in FORTRAN or C, automates such equation-solving algorithms. There are different types of simulators in the process industry, such as Aspen Plus, HYSYS, CHEMCAD, Batch Plus, and Super Pro Designer. Process model should be integrated with the infrastructure, which includes utilities, waste treatment, and solvent recovery capabilities. The waste stream and utility load need to be integrated with the process stage creating a holistic view of the equipment and process flow. Tools like Batch Plus and other Aspen tools integrate the infrastructure in the modeling environment.

The next layer in modeling hierarchy is scheduling. Plant scheduling is focused on the overall operational logistics of the manufacturing unit. This operational logistics is described in detail in *Factory Physics*, by Hopp and Spearman (2001). Process and equipment work flow modeling along with scheduling are integral parts of plant scheduling. The intention is to maximize the value of the manufacturing platform as a whole. The focus is economics, in particular—cost, and capacity of the manufacturing unit. One of the most important aspects is cost of product (see Chap. 1), which must be included in any capacity discussion. The interaction between cost and capacity creates interesting dynamics in that there are various levers of cost and capacity. Ideally enhancement of capacity with reduction in COP is a preferred option. Capacity versus COP is covered in Chap. 8.

At the plant and organizational level, *manufacturing resource planning* (MRP II) and supply chain management are the central philosophies for capacity modeling. Such a type of capacity modeling includes forecasting, production planning, just-in-time, Kanban, managing *work-in-progress* (WIP), master production schedule, material requirements planning, job release, and job dispatching. Kanban is a pull production system that operates with zero to very low inventory.

As an example for a pharmaceutical plant, the supply chain focuses on procuring raw materials (e.g., reagents and solvents), converting the reagents to the drug substance, adding the additional ingredients to make the dosage form, allocating the resources to make the drug substance and dosage form (planning and scheduling of facilities and labor), and delivering product. The details of supply chain, plant, and organizational-level modeling are outside the scope of this book. Books that cover these aspects include Hopp and Spearman (2001) and Womack and Jones (1996). It is critically important to highlight that the essential building block of these high-level models are the process modeling activities at the process stage and infrastructure levels, which must be a major deliverable for the late-stage process development.

2.7.1 Model integration and optimization

As indicated in Sec. 2.7, the steps are understanding the work flow in the supply chain, defining input and outputs, adopting standards for each work function, utilizing technology to facilitate technical transfer, and finally focusing on work flow compression. Achieving work flow compression is essential for achieving market responsiveness and corporate efficiency. In the continuum of modeling hierarchy, Batch Plus can be used to integrate unit operation, infrastructure, and operational capacity with MRPII as the overall supply chain model (in SAP). Such a model could also integrate the commercialization cycle, that is, process design and implementation. Further, it can also provide opportunity for work flow compression assisting in speed to market. A recipe or manufacturing formulae could be used as an integrating element. It is important to note that some process simulators are limited in their ability to incorporate variability in the models. Modeling tools such as "extend" could be used to model work flows across multiple levels of hierarchy incorporating variability.

As technical integration of automation technologies becomes feasible, one searches for practical approaches to the life cycle implementation for recipe-based industries. Areas such as specialty chemicals, pharmaceuticals, biotechnology, consumer products, and food and beverage all share the same product development and manufacturing structure that is essentially the recipe model. S88.01 provides a standard structure to develop and evolve recipes through the stages of process life cycle—from a general recipe at the lab and pilot scale to master recipe and control recipe at the manufacturing scale. A recipe provides an excellent means to imbed the process knowledge, and both scientists and engineers understand the language. The batch process simulator, from AspenTech, Batch Plus, uses a single model of the process to perform material and energy balances, calculation of streams and vessel contents, and design and scale-up calculations, and the simulator applies S88 to represent the process.

2.7.2 Model verification and maintenance

Once a model is created, it should be put through a rigorous verification process to guarantee its performance and accuracy. One good verification technique is to use historical performance data to test the model. After plugging in previous production plans, the user can then compare model results with actual production performance. A simulation model considers variability that can randomly sway outcomes; a successful model should not be expected to reflect real-life performance case-by-case. Rather, it should show an accurate trend or average performance. For example, the cycle time for a specific lot may be different in simulation than on the floor due to built-in variability, but the average cycle time for several lots should be more accurate. Key inputs for such models are not only average labor and equipment processing times for a given product, but the variability of those times. As in real life, the average is rarely, if ever, achieved. A model that accurately accounts for this reality can provide a better picture of the production needs.

Over time, the modeled operation will change, and a formal process should be put in place to update the model. A valuable technique to make this manageable is to assign a model owner in charge of tracking all desired changes. These changes are reviewed semiannually or annually in the context of future operational strategy, and a short design process is initiated to clearly define the scope of the updated model. Additionally, process data should be collected on an ongoing basis in order to maintain model integrity. Systems should be put in place to capture actual processing characteristics (averages and standard deviations) of key model steps. These figures should be periodically reviewed, and the model should be updated accordingly. Modeling and simulation are powerful tools that can offer many benefits if done right and used appropriately. Like most powerful tools, they also need to be handled with care. Knowing what the model is expected to achieve and correctly identifying the relevant variables and their behavior are keys to ensuring meaningful results.

2.7.3 Benefits of modeling

It is important to understand the benefits of modeling that has become an integral part of the commercialization process and offers numerous benefits. Some of the potential benefits of modeling include:

- Communicating process information throughout the company
- Work flow management of process engineering, scale-up, and process analysis
- Faster conceptualization of new processes and facilitates generating and optimizing scenarios for optimal implementation

- Generating operating instructions, raw material orders, and hazard analysis
- Enabling optimal equipment selection and process optimization
- Analysis of waste and air emissions, which is now the expectation of environmental agencies
- Process integration—integrating a complex process and its various unit operations providing a portal for scenario analysis and understanding and optimizing the whole system off the ground
- Opportunity to integrate various stages of commercialization—process design, scale up, manufacturability, optimization, commissioning, and qualification

As companies in the batch industries globalize their research, development, operations, and strategies, a new look at the potential for overall optimization is required. True batch optimization is a realization of the true potential of the batch industry's operation. This means that hierarchies of modeling (as indicated in Fig. 2.17) are working independently and in concert with each other. Such integration will allow corporations to identify a hierarchy of capacity levers. For example, if a Batch Plus model indicates that the drying capacity in process flow is limiting the throughput, then a process stage/unit operations model could be used to understand the best way to debottleneck this stage. Once debottlenecking takes place, appropriate changes are made to models at plant and supply chain level to reflect this increased capacity. The next step then is to explore the next capacity lever, which may be supply of a raw material and so on.

Modeling can also feed into describing user requirements for construction and commissioning as process knowledge is embedded in the models, which are created throughout the evolution of the process.

2.8 User Requirement Specification (URS)

One of the key inputs in process and equipment design is *user requirement specification* (URS). URS is a documentation of design intentions (the what?) and sets the functional requirements. It should include all stakeholders' system performance requirements, such as: operations, maintenance, environment health and safety, engineering, validation, quality control, and automation. Requirements that are critical for product quality as well as critical business drivers should be included, along with critical process parameters and critical operation parameters (when applicable). If the actual value of a parameter is unknown, identify the parameter with a placeholder. When possible, refer to the critical process parameter or critical operating parameter in the process development documentation. System or equipment conformance to user requirements must be measurable and not a subjective evaluation.

User requirements will form the basis for system or equipment *operational qualification* (OQ) and *performance qualification* (PQ). System design will form the basis for commissioning. User requirements should be precise and to the point. User requirements should NOT include design information. Exception to this may be materials of construction where they are prescribed by company policy or procedure (e.g., 316 grade stainless steel for water for injection systems). User requirements may have to be established by system or by functional area. Once a detailed design is complete, a final system boundary can be established. Design qualification and impact assessment/classification occur at this point. The following section presents an integrated picture of commissioning, qualification, and validation.

2.9 Execution of the Manufacturing System

Following the process design and the consequent user requirements, the manufacturing system execution involves system classification, commissioning, qualification, and validation. It is important to understand the distinction between system classification, commissioning, qualification, and validation. While there would be a significant overlap in these activities, for simplicity, the execution of the manufacturing system could be represented as in Fig. 2.18.

User requirements indicate design intentions, which lead to detailed design specifications of the manufacturing system with functional specification. This conceptual manufacturing system is then classified into direct, indirect, and no impact system, which sets the focus for commissioning and qualification. Following qualification the system is validated to provide a high degree of assurance that a specific process will consistently produce a product meeting its predetermined specifications and quality attributes. Validation is a documented evidence of manufacturability of the production system.

2.9.1 System classification

Classification of system is an important activity during the initial stages of commissioning and qualification. This classification allows appropriate effort and focus to be concentrated on the quality-impacting systems and components.

Systems could be classified into direct, indirect, and no impact systems. *Direct impact system* has direct impact on product quality whereas the *indirect system* is not expected to have a direct impact on product

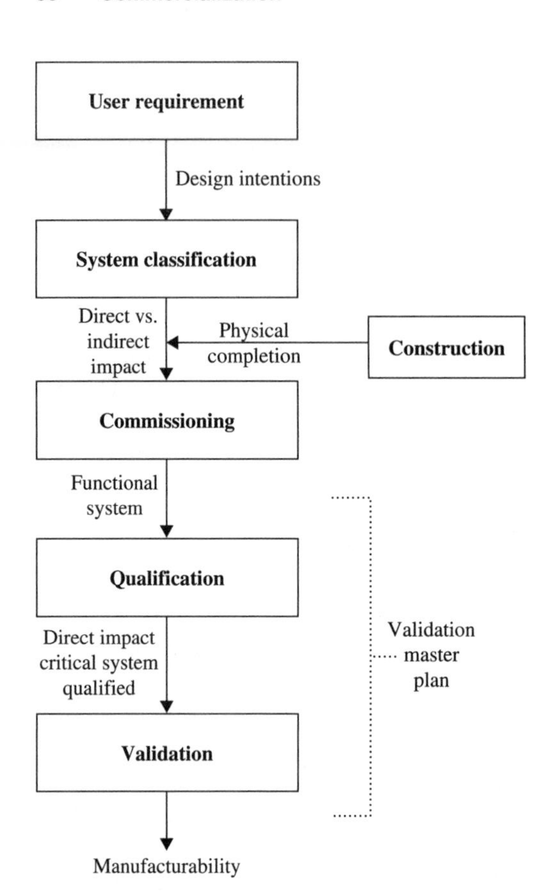

Figure 2.18 Execution of manufacturing system.

quality. *No impact system* has no impact on product quality. For example, an office complex in a manufacturing facility is a no impact system; Utility systems—cooling water, chilled water, and compressed air are indirect impact systems; and *water for injection* (WFI) for parenteral product is a direct impact system.

A system could have critical and noncritical components. Critical components have direct impact on the quality of product whereas noncritical components have no direct impact on product quality. For example, "a still (equipment that generates steam followed by condensation)" is a critical component in the WFI system. For system classification the first step is system identification followed by setting the system boundary; thereafter the system is classified as shown in Fig. 2.19. It should be noted that no impact or indirect impact systems are comprised of noncritical components only. An indirect/no impact

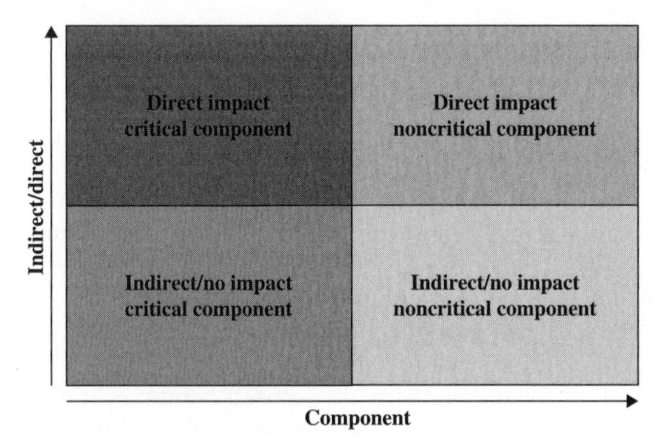

Figure 2.19 System classification.

system with one or more critical components implies that either the system has been misclassified or the component wrongly assessed. Therefore, this exercise allows critical assessment around system boundary and classification.

This classification is critically important for commissioning and qualification; it helps direct focus for efficient execution. According to the *International Society for Pharmaceutical Engineering* (ISPE) *Commissioning and Qualification Guide* (ISPE, 2001), only direct impact systems should be qualified with focus on critical components. All systems must undergo rigorous commissioning exercises.

2.9.2 Commissioning

ISPE Commissioning and Qualification Guide (ISPE, 2001) defines *commissioning* as "A well-planned, documented, and managed engineering approach to the start-up and turnover of facilities, systems and equipment to the end user that results in a safe and functional environment that meets established design requirements and stakeholder expectations."

Commissioning includes the start-up activities and typically occurs between construction (physical completion) and turnover to either the operations end users or the validation team. Typical start-up activities would include: predelivery inspection, *factory acceptance test* (FAT), functional tests, *site acceptance test* (SAT), shakedown runs, and the like. Where possible, advantage should be taken of the opportunity to inspect (predelivery inspection) and test (FAT) systems or major system components before delivery to site (ISPE, 2001). This allows a quicker and more efficient remedy of any failings, and avoids the delays to schedule that would result from discovering problems later, on site. Where such systems have a direct impact on product quality, every effort should be made to incorporate the predelivery inspection, factory acceptance test, and site acceptance test into the qualification effort, through the application of qualification practices (ISPE, 2001). Testing performed during commissioning for a direct impact system may be used to support validation.

The tests performed during the commissioning phase should be designed to provide the assurance that the system has been constructed to, and will perform to, the intended design criteria. The tests would include functional and performance tests. A successful commissioning forms a firm foundation for an efficient qualification and validation.

2.9.3 Master validation plan

Validation is a structured approach for translating user requirements, design specifications, and quality aspirations into a manufacturing facility. A validation plan is a comprehensive document describing the applicable validation requirements for the facility and providing a plan for meeting those requirements. For the pharmaceutical industry, the FDA defines process validation as "establishing documented evidence that provides a high degree of assurance that a specific process will consistently produce a product meeting its predetermined specifications and quality attributes" (available at www.fda.gov/cder/guidance/ pv.htm). A key phrase in this definition is *documented evidence*. Documentation represents validation studies to regulatory agencies.

The proof of validation is obtained through rational experimental design and the evaluation of data, preferably beginning from the process development phase (Chap. 1) and continuing through the commercial production. A well-thought-out process design with well-planned and well-executed small-scale process development is not just a good practice but a precursor to the success of validation efforts.

Regulatory agencies expect a company to demonstrate an understanding of the parameters that will affect the critical quality attributes of its product in the event of a manufacturing deviation. Once critical process parameters are defined (during development), companies must perform studies to test parameter ranges. Data to substantiate the ranges for critical process parameters should be obtained in most cases from laboratory- or pilot-scale batches, unless a specific parameter can only be determined from a production-scale batch.

Critical parameter studies should support at least two ranges of each parameter: operating and validated. The operating range—the range in which the product is manufactured—is the narrowest. The validated range falls outside of the operating range and is the range in which the process will still perform adequately. Although the validated range is not recommended for manufacturing, should a deviation occur during manufacturing that falls outside of the operating range, documented studies that demonstrate adequate performance will help determine whether the quality of the product has been compromised. If the parameter range fails during the study, document that failure. Although it may be unnecessary to test to failure, a failure provides the absolute range for that parameter. This demonstrates to a regulatory reviewer that the critical parameter ranges are fully understood. Should a manufacturing deviation occur, such documents will support the decision whether or not to fail the product lot.

Once the critical process parameter ranges are determined, the critical product/operating parameters (such as reaction times, reaction temperatures, and impurity levels as well as reactant ratios, concentrations, pressures, and pH) should be controlled and monitored during process validation studies. Repeating validation studies because incorrect parameter ranges were used is costly and time consuming. Process validation supports the operating parameters that will be used in the final to-be-marketed manufacturing process, and it forms the foundation of the regulatory submission.

A process *validation master plan* (VMP), see Fig. 2.20, is simply a list of all the studies that will validate a manufacturing process. The question of what and how much to validate arises when creating the master plan. The VMP should fully represent the to-be-licensed manufacturing process. To accomplish this, validation studies must cover all major aspects of that process. For example, a process VMP for a biological product would most likely include the chromatographic resin lifetime, cell culture performance studies, viral validation, filter extractables, column cleaning (both the equipment and the resin), in-process product and buffer hold times, and impurity and contaminant removal (DNA, buffer components, host-cell proteins).

Many of the studies listed in the master plan can be performed at small scale. Such studies are a valuable tool in process validation. They can be performed prospectively for determining resin lifetimes and in-process hold times. However, it is critical that the small-scale studies represent the to-be-marketed manufacturing process. A representative study uses the same process parameters (such as the linear flow rate and load pH) and conditions (such as the raw materials, resin bed heights, and feedstock and buffers). For example, if the feedstock is diluted or conditioned before loading onto the column or membrane in the manufacturing process, then the same should be done to the feedstock used in the small-scale study. A table in the protocol can address what process parameters and conditions will ensure that the small-scale

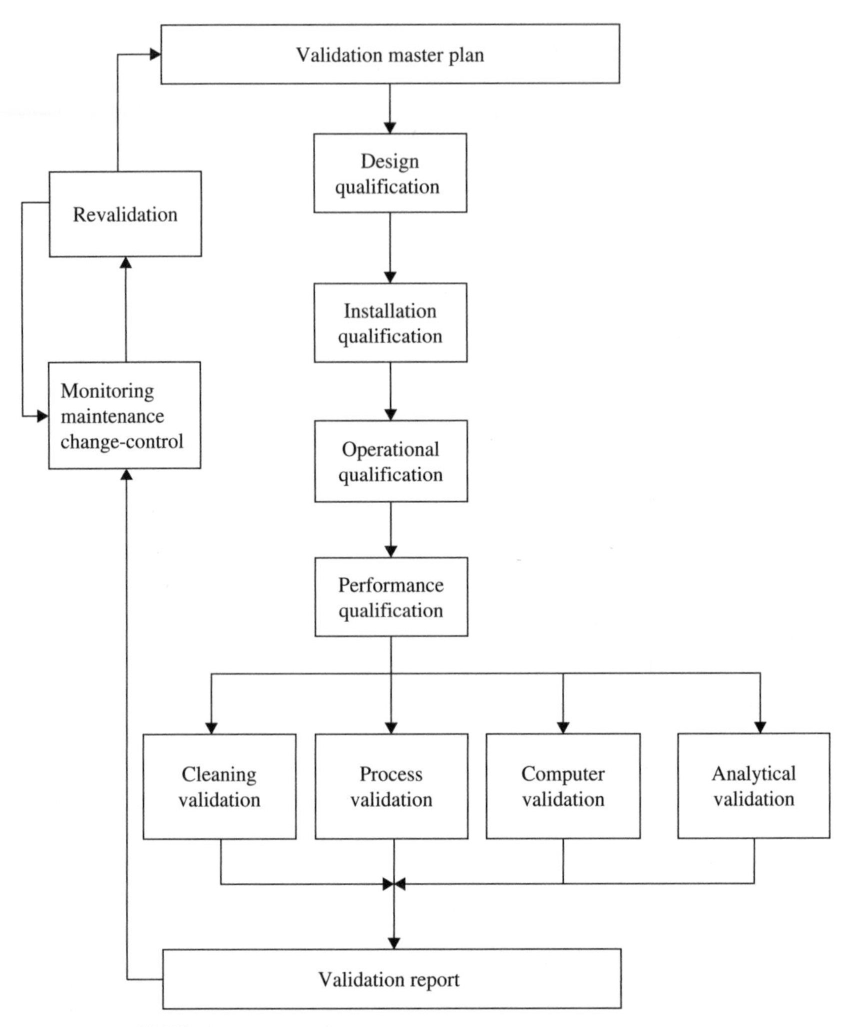

Figure 2.20 Validation master plan.

study represents the manufacturing process and how those parameters are to be appropriately scaled. The manufacturing representative on the process validation team ensures that the validation studies represent the manufacturing process.

After the requirement for each aspect is determined, the responsible engineers complete the design and the validation team again reviews it. After the approved designs are constructed and/or installed, the validation cycle continues with the preparation and execution of the validation documents. Validation is a systematic approach to gathering and analyzing sufficient data that will give reasonable assurance (documented evidence), based on scientific judgment, that a process, when operating within specified parameters, will consistently produce results within predetermined specifications.

Design qualification (DQ). Process design review is followed by rigorous evaluation of the equipment that is designed for the manufacturing operation. This is referred to as *equipment qualification* in industry, especially in the pharmaceutical sector. The intent of this exercise is to prove that the equipment is suitable and robust for the manufacturing process. "During design qualification (DQ), it is documented that the design aspects have been checked and approved. Design qualification contains a plant description and shows that the plant design agrees with the customer's design specifications. Design aspects that were not defined during specifications and that are not yet defined (i.e., when design qualification begins) must be listed and evaluated with respect to their influence on product quality" (ISPE, 2001).

The scale-up exercise leads to equipment specification for the manufacturing process. This equipment and the process flow are captured in a *process flow chart* (PFC), which is the summary of the manufacturing process and equipment requirements. These requirements are specified and documented in detail in the URS. URS forms the basis of the design qualification stage. There are two objectives of DQ: to ensure that the quality is built into the design, and that the design is consistent with the development and scale-up activity in the preceding stages of commercialization as specified in the URS.

The quality aspect of DQ involves *current good manufacturing practice* (cGMP) review of the design. This is documented evidence that the design complies with the published requirement and best practices of current good manufacturing practices document. Typically, this will involve audit of facility design, environmental considerations and waste management, equipment and material flow, automation, maintenance philosophies, and documentations. The documentation audit will typically include review of development history, URS, training plan, *standard operating philosophy* (SOP), and engineering studies relevant to the facility.

These reviews should be conducted by a joint team of design and process engineers, scientists, and quality (QA/QC) and operation personnel. The design qualification culminates at the FAT. The FAT is the first opportunity to test the manufactured equipment.

Installation qualification (IQ). The focus of the IQ is to ensure that the equipment is installed as intended in the design and in accordance with vendor's recommendations. In a cGMP environment this is documented under specific categories, i.e., system completion, security/utility connections, documentation inventory, equipment inventory, electrical power requirements, materials qualification, drawing validation, main equipment features, instrument calibration, and spares and maintenance. In the installation qualification process, written evidence is given that all the parts of the equipment are installed according to the equipment supplier and purchase specifications. The factory acceptance test will also be part of the IQ. It documents that the operating criteria for the equipment, as installed, are in compliance with the *process and instrumentation* (P&I) diagrams, plant functional specifications, and the process flow diagrams.

The IQ is the stage where the completeness and correctness of all required documents are checked. At this point, if necessary, documents must be completed and corrected.

Operational qualification (OQ). OQ provides the assurance that the equipment operates as intended throughout the anticipated operating ranges. This involves testing the equipment during operation using nonproduct materials such as water and air.

During the OQ process, documented evidence shows that all parts of the plant and equipment work within their specifications, and process parameters are within the acceptance criteria. Process controls that are part of the equipment (e.g., PLC—*programmable logic controller*) will be qualified during the OQ process. Computerized process controls (i.e., for complex processes) should be qualified in the *computer validation* (CV) process. To ensure that the systems tested during OQ are doing what they are supposed to do, a simulation of normal production conditions must be done.

Performance qualification (PQ). PQ is the evaluation of the overall equipment set to confirm that the system operates, throughout the anticipated operating ranges, as intended. The PQ evaluation is performed as close to the production conditions as possible and is performed without real product. The documentation of PQ may include: approved protocol, system description, purpose, sampling regime, testing regime, acceptance criteria, deviation, and corrective action. The PQ process provides documented evidence that all parts of the plant and the processes validated can operate as intended in the design. The PQ includes critical variable studies, for example, by simulating conditions of upper and lower processing, processing at the operating limits of the equipment, or circumstances such as worst-case conditions. It should show that such conditions do not necessarily induce process or product failure.

In contrast to the OQ procedures, where all parts of the plant and equipment are qualified separately, the PQ procedures qualify the entire plant with respect to the production process. The definition given for PQ is valid for retrospective validation as well as for prospective validation. While carrying out PQ processes, all necessary SOPs (e.g., for the use or cleaning of the plant) should be approved. Values of critical and noncritical process parameters recorded during PQ must be collected to evaluate the efficiency and performance of the plant.

Process validation (PV). PV is the ultimate assurance that quality is built into the process. It involves operation with real product (for the first time) evaluating the process and equipment's capability to deliver quality product. PV provides a high degree of assurance that the process will produce a quality product within the specification reproducible. PV ensures and provides documented evidence that processes within their specified design parameters are capable of repeatedly and reliably producing a finished product of predetermined quality.

A process validation program should include:

- Development history and a description of the product
- Control strategy and flowchart of the manufacturing process
- A list of all equipment required for production
- A schedule for PV test procedures
- A detailed description for all test procedures, including:
  Sampling procedure
  Labeling of the samples
  Test procedure
  Evaluation procedure
  Specifications for the intermediate and finished products
  Acceptance criteria
- Responsibilities

Types of process validation. Process validation may be conducted at different points during the life cycle of a product. The types of process validation are defined in terms of when they occur in relation to product design, transfer to production, and release of the product for distribution.

Prospective validation. Prospective validation is conducted before a new product is released for distribution or, where the revisions may affect the product's characteristics, before a product made under a revised manufacturing process is released for distribution.

Concurrent validation is a subset of prospective validation and is conducted with the intention of ultimately distributing product manufactured during the validation study. Concurrent validation is feasible when nondestructive testing is adequate to verify that products meet predetermined specifications and quality attributes. If concurrent validation is being conducted as the initial validation of a new process or a process which has been modified, product should be withheld from distribution until all data and results of the validation study have been reviewed, and it has been determined that the process has been adequately validated.

Concurrent validation may be conducted on a previously validated process to confirm that the process is validated. If there have been no changes to the process, and no indications that the process is not operating in a state of control, product could be released for distribution before revalidation of the process is completed. There is some risk to early release of product in that subsequent analysis of data may show that the process is not validated.

Retrospective validation. Retrospective validation is the validation of a process based on accumulated historical production, testing, control, and other information for a product already in production and distribution. This type of validation makes use of historical data and information, which may be found in batch records, production log books, lot records, control charts, test and inspection results, customer complaints or lack of complaints, field failure reports, service reports, and audit reports. Historical data must contain enough information to provide an in-depth picture of how the process has been operating and whether the product has consistently met its specifications. Retrospective validation may not be feasible if all the appropriate data were not collected, or appropriate data were not collected in a manner which allows adequate analysis.

Incomplete information makes it difficult to conduct a successful retrospective validation. Some examples of incomplete information are:

- Customer complaints that have not been fully investigated to determine the cause of the problem, including the identification of complaints that are due to process failures.
- Complaints were investigated but corrective action was not taken.
- Scrap and rework decisions that are not recorded, investigated, and/or explained.
- Excessive rework.
- Records that do not show the degree of process variability and/or whether process variability is within the range of variation that is normal for that process, for example, recording test results as "pass" or "fail" instead of recording actual readings or measurements results in the loss of important data on process variability.
- Gaps in batch records for which there are no explanations. (Retrospective validation cannot be initiated until the gaps in records can be filled or explained.)

If historical data is determined to be adequate and representative, an analysis can be conducted to determine whether the process has been operating in a state of control and has consistently produced a product that meets its predetermined specifications and quality attributes. The analysis must be documented.

After a validated process has been operating for some time, retrospective validation can be successfully used to confirm continued validation of that process if no significant changes have been made to the process, components, or raw materials.

Statistical process control is a valuable tool for generating the type of data needed for retrospective analysis to revalidate a process and show that it continues to operate in a state of control (Chap. 3).

Cleaning validation. In a pharmaceutical and biotechnology industry cleaning is essential to ensure that the appropriate level of general cleanliness is maintained to prevent the accumulation of dirt and microbial contamination, which could impact the quality of the product. Cleaning is also important to minimize risk of cross-contamination of a multiproduct train. During the cleaning validation process, written evidence shows that specified cleaning procedures will lead to reliable and repeatable results in the cleaning of surfaces with and without contact with the product. The following criteria will be fulfilled if cleaning procedures are used as specified in cleaning SOPs:

- The concentration of active substances on product contact surfaces will not exceed specified limits.
- The concentration of highly active substances (e.g., hormones or cytostatics) on surfaces without contact with the product will not exceed specified limits.
- The concentration of other pharmacologically active substances (e.g., process and cleaning materials or disinfectants) in the product will not exceed specified limits.
- The number of germs on product contact surfaces will not exceed specified limits.

Cleaning validation involves executing the cleaning cycle and analyzing the "cleanability" by swab testing and other modern techniques, such as riboflavin testing. This ensures that the equipment could be cleaned reproducibly by following the cleaning protocol. Further details could be found in published literature such as Wingate (2003).

Computer system validation (CSV). CSV provides documented evidence that a computer system reliably and consistently does what it purports to do and can be expected to continue doing so in the future. CSV provides documented evidence that processes controlled by computerized systems are checkable and produce the specified product quality repeatedly and reliably.

Software verification and validation. FDA treats "verification" and "validation" as separate and distinct terms (www.fda.gov). *Software verification* provides objective evidence that the design outputs of a particular phase of the software development life cycle meet all of the specified requirements for that phase. Software verification looks for consistency, completeness, and correctness of the software and its supporting documentation, as it is being developed, and provides support for a subsequent conclusion that software is validated. Software testing is one of many verification activities intended to confirm that software development output meets its input requirements. Other verification activities include various static and dynamic analyses, code and document inspections, walkthroughs, and other techniques.

Software validation. The FDA considers software validation to be "confirmation by examination and provision of objective evidence that software specifications conform to user needs and intended uses, and that the particular requirements implemented through software can be consistently fulfilled." In practice, software validation activities may occur both during, as well as at the end of the software development life cycle to ensure that all requirements have been fulfilled. Since software is usually part of a larger hardware system, the validation of software typically includes evidence that all software requirements have been implemented correctly and completely and is traceable to system requirements. A conclusion that software is validated is highly dependent upon comprehensive software testing, inspections, analyses, and other verification tasks performed at each stage of the software development life cycle. Testing of device software functionality in a simulated use environment, and user site testing are typically included as components of an overall design validation program for a software-automated device.

The life cycle for computer system validation contains functional requirements, design, testing, security, source code review, change control, planning, specification, programming, testing, start-up, documentation, operation, checking, and changing. Further details could be found in published literature such as Bismuth and Neumann (2000).

Analytical validation. Analytical methods and tests play a significant role in the GMP manufacturing operations. These analytical methods are used to generate data assuring that the critical product quality attributes have been met. The main objective of validation of an analytical procedure is to demonstrate that the procedure is suitable for its intended purpose. Typically validation approaches are based on the overall knowledge of the capabilities of the analytical procedures, for instance, specificity, linearity, range, accuracy, and precision. FDA has issued guidance documents on validation of analytical procedures (www.fda.gov). The essence of the validation approach is to ensure that a reliable dataset is generated so as to be able to make sound GMP decisions. The three main pillars for generating a reliable dataset are analytical methods, equipment calibration, and training. These aspects are discussed in greater detail in various books such as Chan et al. (2004) and Swartz and Krull (1997). Further, *process analytical technology* (PAT), as discussed in Chap. 3, initiative by the FDA has helped encourage the industry to adopt newer technology for better understanding and control of product quality.

Revalidation, monitoring maintenance, and change control. As long as the process operates in a state of control and no changes have been made to the process or output product, the process does not have to be revalidated. Whether the process is operating in a state of control is determined by analyzing day-to-day process control data and any finished product testing data for conformance with specifications and for variability.

When changes or process deviations occur, the process must be reviewed and evaluated, and revalidation must be performed where appropriate (www.fda.gov). Review, evaluation, and revalidation activities must be documented.

Processes may be routinely validated on a periodic basis; however, periodic validation may not be adequate. More important is appropriate monitoring, so that if problems develop or changes are made, the need for immediate revalidation is considered.

2.10 Quality System

It is important to conclude this chapter with a brief discussion on the quality system, which is designed to test the success of quality by design, manufacturability, and to encourage strategy for continuous quality improvement. This section is an adaptation of *Quality System Inspection Technique* (QSIT) developed by FDA for medical device manufacturers (www.fda.gov). This adaptation is aimed to extend this concept to the larger pharmaceutical and biotechnology manufacturing industries.

Figure 2.21 illustrates a quality system aimed at testing quality by design and ongoing continuous quality improvement with management control, ensuring robust and continuously improving product quality.

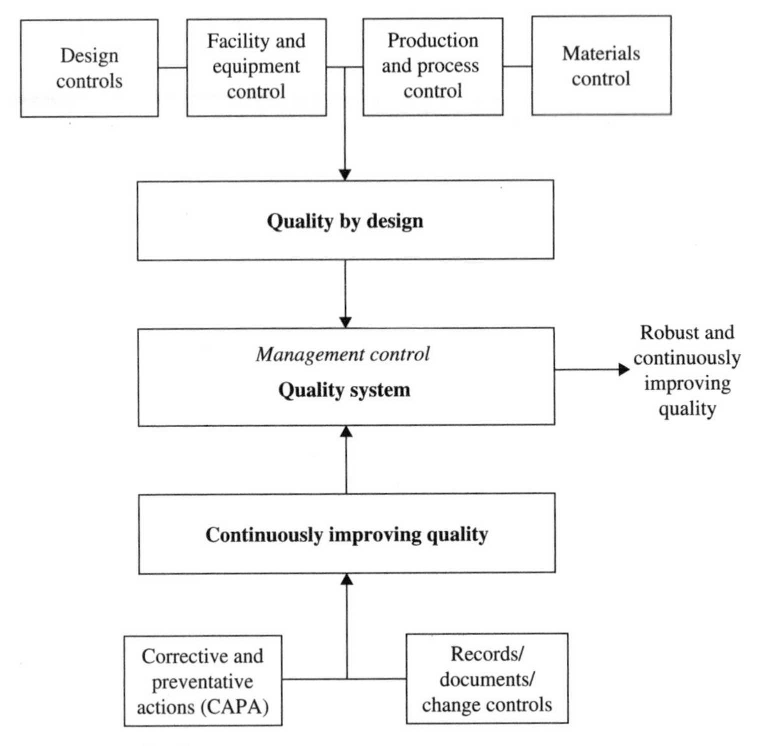

Figure 2.21 Quality system.

2.10.1 Testing for quality by design

The elements of testing quality by design are manifested in various controls namely: design control, facility and equipment control, production and process control, and material control.

The purpose of the design control subsystem is to control the design process to ensure that the product meets user needs, intended uses, and specific requirements. The design control could be subdivided into three key areas:

1. Ensures that rigorous PDF (Sec. 2.2) has occurred to ensure that engineering and scientific rigor and quality are built into the design philosophy.
2. Undertakes appropriate risk analysis/management to minimize or mitigate risk prior to implementation (Sec. 2.6).
3. Executes design qualification to document evidence that the design complies with the user requirements (Sec. 2.9.1.1).

Facility and equipment control focuses on construction, commissioning, qualification, and validation. The use of appropriate gate reviews is one such mechanism to ensure control over effective construction, commissioning, qualification, and validation in delivering a manufacturable system capable of producing quality product within specifications.

Material control focuses on testing and releasing raw material of appropriate specifications to the production process. The purpose of the production and process control subsystem is to manufacture products that meet specifications. Critical control strategy as outlined in Chap. 3 ensures that quality control is built into processes.

2.10.2 Continuously improving quality

Quality by design should be the fundamental objective of process design and implementation. Once the production process is in operation, then the focus should include structured approach to continuous quality improvement. Two important aspects of this continuous improvement are:

1. *Corrective and preventative actions* (CAPA)
2. Records, documents, and change controls

The purpose of the CAPA subsystem is to collect information, analyze information, identify and investigate product and quality problems, and take appropriate and effective corrective and/or preventive action to prevent their recurrence. The CAPA system provides a structured methodology (various software are available to assist with the structure) to address nonconformance as well as the root cause of the problem. Root cause analysis (Chap. 5) is a key tool for the CAPA methodology. Understanding and dealing with variability is also a key aspect of the CAPA system (Chaps. 5 and 6).

Records, documents, and change controls focus on procedural documentation with scientific rationale justifying the change. Such documentation forces the habit of critical scientific thinking.

2.10.3 Management control

The purpose of the management control subsystem is to provide adequate resources for process design, manufacturing, quality assurance, and distribution. The subsystem's purpose is also to ensure that the quality system is functioning effectively and make any necessary changes to enhance its effectiveness. Appropriate management and executive responsibility should be assigned to ensure an adequate and effective quality system. Management reviews and quality audits are a foundation of a good quality system.

Concluding Comments

Manufacturability has its roots in a rigorously executed design phase with the process development knowledge as the guiding principle. The key elements of manufacturability are people, equipment, and process. People need to be trained, the equipment needs to be qualified, and the process needs to be validated for successful manufacturability. Key aspects of process and equipment manufacturability include:

- Assembling the PDF team
- Process integration
- Process simplification
- Process flow description
- Risk management
- Modeling
- Process optimization
- User requirement specification
- Commissioning, qualification, and validation
- Instituting a quality system in place

References

Armbruster, D., and T. Feldsien (2000), "Applying HACCP to pharmaceutical process validation," *Pharm. Technol.*, **24**(10)(October), 170–178.

Ayyub, B. M. (2003), *Risk Analysis in Engineering and Economics*, Chapman and Hall, Boca Raton, FL, July 26.

Bismuth, G., and S. Neumann (2000), *Cleaning Validation: A Practical Approach*, CRC Press, Boca Raton, FL.

Bossert, J. L (1991), *Quality Function Deployment, A Practitioner's Approach,* ASQC Quality Press, Milwaukee, WI, pp. 47–48.

Chan, C. C., H. Lam, Y. C. Lee, and X. Zhang (2004), *Analytical Method Validation and Instrument Performance Verification*," *Wiley Interscience*, Hoboken, NJ.

Hauser, J. R., and D. Clausing (1998), "The house of quality," *Harvard Business Review*, **44** (May–June): 63–73.

Hopp, W. J., and M. L. Spearman (2001), *Factory Physics*, 2d ed., Irwin/McGraw Hill, New York.

Hyatt, N. (2003), *Guidelines for Process Hazards Analysis, Hazards Identification, and Risk Analysis*, Dyadem, New York.

ISPE (2001), *ISPE Commissioning and Qualification Guide*, **5**, International Society for Pharmaceutical Engineering, Tampa, FL.

Koolen J. L. A. (2001), *Design of Simple and Robust Process Plants*, Wiley-VCH, Germany.

Krueger, R. A., and M. A. Casey (2000), Focus *Groups: A Practical Guide for Applied Research*, 3d ed., Sage Publications, Thousand Oaks, CA.

Peters, M. E., K. D. Timmerhaus, M. Peters, and R. E. West (2002), *Plant Design and Economics for Chemical Engineers*, 5th ed., McGraw-Hill, New York, December 9.

Rooney, J. (2001), "7 steps to improved safety for medical devices," *Quality Progress*, **34**(9)(September), 33–41.

Swartz, M. E., and I. S. Krull (1997), *Analytical Method Development and Validation*, Marcel Dekker, New York.
Wingate, G. (2003), *Computer Systems Validation: Quality Assurance, Risk Management, and Regulatory Compliance for Pharmaceutical and Health*, CRC Press, Boca Raton, FL, December 18.
Womack, J. P., and D. T. Jones (1996), *Lean Thinking: Banish Waste and Create Wealth in Your Corporation*, Simon and Schuster, New York.

Part 2

Process Capability

A manufacturing process must be capable of producing a high-quality product within specification, reproducibly. A successful in-process control strategy is key to a capable manufacturing process. In-process control strategy implies that adequate quality control is built in the operation to deliver a capable process, and product within specification. Manufacturing should not only test quality but should also build quality into the process. This part focuses on in-process control to proactively minimize deviations in the manufacturing process. Another aspect of a capable manufacturing operation is its ability to manage knowledge and to put knowledge into action. Process capability has two elements: (i) critical control strategy and (ii) knowledge management.

Chapter

3

Critical Control Strategy

Build quality control in the process.

P. MOHAN

Objective

The aim of this chapter is to introduce a systematic approach to develop a critical control strategy for a robust manufacturing platform, with the objective of building quality control in the design (rather than testing for it).

Learning outcomes of this chapter will include:

- The control hierarchy
- *Process analytical technology* (PAT) and critical process parameter measurement
- Alarm strategy
- Data management
- Control of critical process parameters

The learning is then applied to a comprehensive case study to gain practical understanding of the application of the concepts.

3.1 Introduction

A manufacturing process must be capable of producing quality product within specification and reproducibly. A successful control strategy is key to a capable manufacturing process. A control strategy could be depicted as shown in Fig. 3.1.

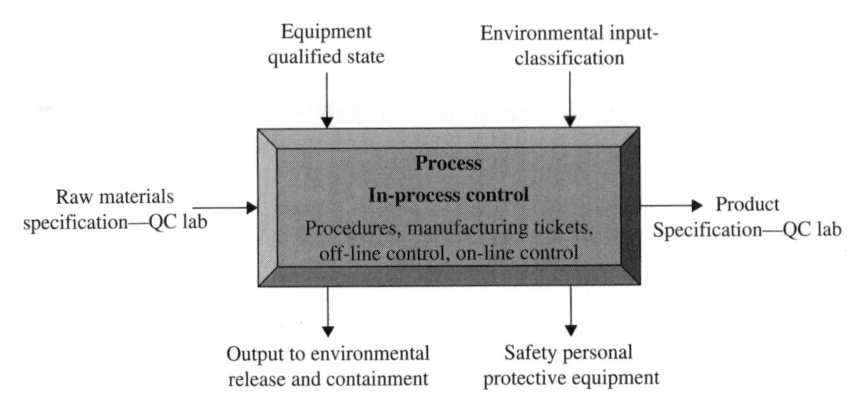

Figure 3.1 Control strategy.

There are various aspects of the control strategy. Raw materials are tested in the *quality control* (QC) laboratory, equipment is qualified (Chap. 2), the environment is controlled by implementing manufacturing area classifications, waste is treated prior to release, and the final product is tested in the QC laboratory. All these procedures when implemented effectively verify that "quality is built into the process."

Measurement of critical operating parameters is fundamental to understanding and controlling the state and health of the process. Data that are collected by various sensors are by-products of any operation. For example, a thermocouple measures temperature, a tachometer measures revolutions per minute, and so on. Some measurements are not so straightforward and require further interpretation before they can be of use. For example, *near-infrared spectroscopy* (NIR) produces a spectrum that requires processing to relate it to process variables. Most operation decisions are based on sensor output, thus for a robust operation an accurate reflection of process conditions by sensors is essential. Monitoring provides information on the health of the process and this information can be used for its control. Automation of the process can lead to the minimization of operator interference with the process and a reduction of human errors.

Process control and automation can be broadly divided into critical control strategy and advanced control strategy. *Critical control strategy* focuses on the control of critical parameters, which is essential for successful manufacturing and will be discussed in detail in this chapter. *Advanced control strategy* further enhances the capability of the process and will be discussed in Chap. 6. The hierarchy of methods is depicted in Fig. 3.2 where the critical control strategy underpins the advanced control strategy.

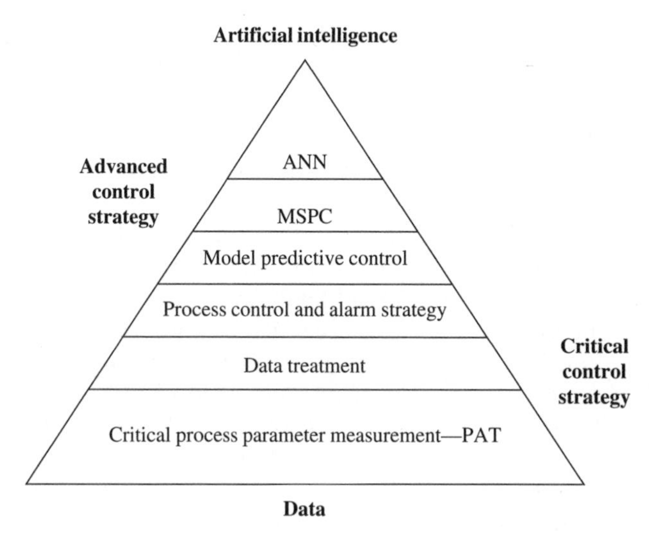

Figure 3.2 Control hierarchy.

Generally, a targeted automation is recommended with a validated and reliable automation system. Targeting should be based on sound financial decisions using cost-benefit analysis and taking into consideration the associated risks. Automation without process/quality consideration or financial justification is a wasteful exercise.

Process control strategies employed in the pharmaceutical industry have been traditionally very conservative. New technology has been viewed as raising regulatory uncertainty, and thus control of critical parameters is limited to those that have been proven over many years of operation. For example, in a bioreactor pH, temperature, and dissolved oxygen levels would be specified. In this case the measurements are reasonably reliable and easy to interpret, and adherence to the operating strategy is easy to verify. Furthermore, operational consistency has been predominantly considered on the basis of time (e.g., perform mixing for x minutes, transfer inoculums after y hours). This leads to consistency in processing operations but not necessarily to consistency in performance. To achieve consistency in performance it may be necessary to vary processing operations to compensate for natural variability. The problem here is the inability to measure performance variation with the necessary confidence. Process analytical techniques are available to provide an indication of performance, but regulatory implications have been unclear and thus they have not been widely used as an integral part of process control strategies.

3.1.1 Process analytical technology

While the pharmaceutical and biotechnology industries continue to spend generously to improve quality, they lag behind other automated industries, such as automotive assembly and semiconductor fabrication, in manufacturing efficiency and quality analysis. To remedy this situation, the U.S. Food and Drug Administration (FDA; Washington, D.C.; www.fda.gov) has launched PAT, an initiative that aims to foster improvements in manufacturing efficiency while harmonizing regulatory expectations. This scientific, risk-based framework is intended to help drugmakers develop and implement more efficient analytical tools for use during pharmaceutical development, manufacturing, and quality assurance. Although designed for the pharmaceutical industry, the objectives and methodologies of PAT can be applied to other process industries that undergo regulatory scrutiny, such as food and beverages, fine chemicals, and cosmetics.

The FDA defines PAT (www.fda.gov) as a system for the analysis and control of manufacturing processes based on timely measurements, during processing of critical quality parameters and performance attributes of raw and in-process materials and process, to ensure acceptable end-product quality at the completion of the processing. The FDA also states that PAT involves the optimal application of *process analytical chemistry* (PAC) tools, feedback process-control strategies, information management tools, and/or product-process optimization strategies for the manufacture of pharmaceuticals. In summary, PAT can be defined as the optimal application of PAC tools, feedback process-control strategies, information management tools, and/or product-process optimization strategies for the manufacture of pharmaceuticals and bioproducts.

Although PAC is not a new approach, many of the related techniques have been tested and used only on a limited basis by a very small percentage of the pharmaceutical industry. Although process analyzers are potentially the vital tools, the PAT initiative is essentially about process understanding, variability reduction, predictability, and efficiency. Associated benefits such as faster development of new products, shorter manufacturing cycle times, higher yields, reduced waste materials, and fewer product recalls should also be achieved by the greater understanding of the process gained by using this methodology.

PAT encourages the adoption of innovative technologies to increase quality without concern that a new approach will lead to validation risks and production delays. The key components of this approach are better understanding of the product manufacturing process, data analysis, process analytical tools, process monitoring, and continuous feedback during the manufacturing process.

Experience indicates that the highest-performing organizations are rigorous about their processes, especially as they relate to better understanding of manufacturing processes, process control, validation, and ongoing commitment to quality improvement. With an emphasis on process understanding, the PAT initiative is designed to allow companies to determine which variables are most critical to the final desired product, where controls should be inserted into the process, and what factors control sample degradation. The underlying premise is that quality cannot be tested into products; it should be designed or built into the process.

Pharmaceutical manufacturing is going to continue to evolve with increased emphasis on science and engineering principles. Effective use of the integrated systems and use of process knowledge—throughout the life cycle of a product—can improve the efficiencies of both the manufacturing and the regulatory processes.

One of the important outputs from the application of PAT is reduction of product variability, which can be summarized in Fig. 3.3. Process control strategies based on information from laboratory assays and supervisory computer systems [*supervisory control and data acquisition system* (SCADA)] are used to regulate process operation and correct for deviations resulting from raw material variations or production plant variations. If PAT can provide additional information on deviations,

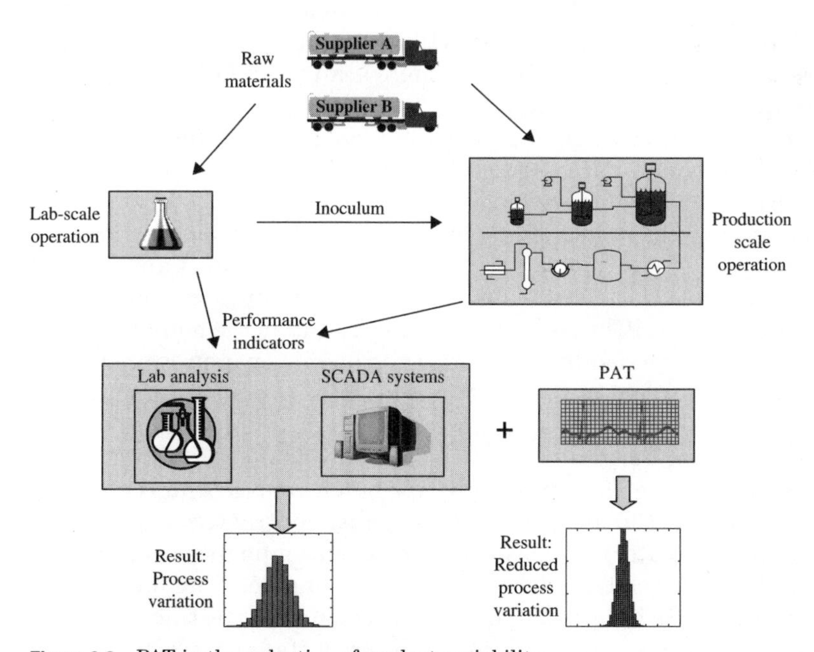

Figure 3.3 PAT in the reduction of product variability.

then the effects of disturbances can be reduced and quality control tightened. Thus, there is a desire to move from the current state where failure to exploit PAT is leading to a wider spread of product quality than would be achieved if PAT was used to compensate for natural variations.

It should be noted here that there is no shift in the belief that product quality *cannot* be determined by testing, merely that PAT provides information more representative of deviations and if reacted upon will reduce their occurrence. Considering Figs. 3.2 and 3.3, it can be seen that there are several aspects to achieving control of critical process parameters:

- Measurement of critical process parameters
- Verification of the information provided by measurements
- Regulation of the critical process parameters
- Parametric release

This chapter considers these issues with the motivation being to follow the FDA goal of *quality into products by design*: a well-designed system for critical parameter control leads to minimization of product deviations.

PAT regulatory agencies are also providing guidance on process control. For example, the FDA has issued guidance documented in 21 CFR Part 11, the Electronic Records and Electronic Signatures rule. The underlying philosophy of this rule is that a "state of control" is required to produce finished drug products so there is an adequate level of assurance of quality, strength, and purity.

A significant advantage of PAT is the huge potential for parametric release of a product. Parametric release is defined as a product release procedure based on effective monitoring, control, and documentation of a validated process in lieu of release based on end-product testing.

The purpose of parametric release is to improve product and process quality control by monitoring and utilizing on-line and in-line measurements at critical control points and phases in the manufacturing operation. This can result in economic and efficiency improvements for the industry. Currently, parametric release is a common practice for terminally sterilized and aseptically filled biologicals. Monitoring on-line process parameters provides a better assurance of consistency and quality, thus giving the opportunity for batch release and regulatory approval without the use of elaborate quality control tests.

European regulatory authorities have issued a guidance to encourage increased reliance on data derived from the manufacturing process and in-process controls as opposed to that furnished by lot testing. The FDA has used this approach primarily for sterile products, but seeks to increase reliance on release systems that provide assurance that the product is of the intended quality based on manufacturing process information.

The principle of parametric release may be applied during the stage of manufacture of different products resulting in the elimination of certain specific tests of the finished product. Parametric release has been traditionally applied to sterilization. For example, for a terminally sterilized product, approval for parametric release eliminates the requirement for a finished product sterility test as a condition for batch release. The release of each batch is dependent on the successful demonstration that predetermined, validated sterilization conditions have been achieved throughout the load. All relevant sterilization parameters, for example, temperature, pressure, and time must be accurately controlled and measured.

Conventional process control systems, especially batch production and laboratory information management systems have been useful to ensure control of production units. However, leveraging the most recent process control developments to improve product uniformity, cut costs, and ultimately boost plant profitability may require integration with other regulated systems. This integration requirement has compliance implications. The FDA developed Part 11 to allow pharmaceutical manufacturers to take advantage of advances in automated process control and computer systems in achieving good manufacturing processes. A shift to paperless or electronic systems can not only streamline the manufacturing process, but can also foster greatly improved document management, automated standard report generation, and rapid retrieval of archived process data. Moreover, compliance with 21 CFR Part 11 will require pharmaceutical engineers to rethink the way they approach process control.

Process control systems such as those provided by Emerson, Invensys, Honeywell, and ABB automate pharmaceutical plant batch production control activities and integrate recipe management, materials tracking, process management, and production information as well. Before Part 11, companies justified investments in process control and batch automation by document management, quality, consistency, and other performance improvements. More recently they have considered the ability of the system to generate data that can optimize processes, increase plant efficiencies, lower production costs, and reduce the time to market. Although the operations were automated, validation essentially continued to be a manual process. With 21 CFR Part 11, however, companies can begin automating the validation process as well as their operations and can avoid future compliance costs and woes by adopting a more structured approach.

In today's pharmaceutical enterprise, process control is increasingly becoming synchronized with other operations, manufacturing control, and enterprise management applications. At the operations level, one consideration is designing batch processes that are easily validated with 21 CFR Part 11 and are compliant with the S88 batch standard advocated by the *Instrumentation Systems and Automation Society* (ISA). S88 is a standard protocol that defines recipes, models, and procedures to enable consistent batch design.

3.2 Critical Process Parameter Measurement

To ensure final product quality it is necessary to be knowledgeable of critical quality and performance measures of raw materials and process states early in development. The term "states" here refers to all those process attributes that govern future performance. In reality practical limitations mean that it is only possible to measure a subset of these states, but if the critical states are correctly identified in the development stage, then the major sources of variation can be compensated for. The longer that variation persists, the larger the deviation in product quality.

A complete review of specific process instrumentation for critical parameter measurement is beyond the scope of this book. This section will not consider devices themselves but will concentrate on what characteristics measurements must have if they are to be used in a critical parameter control scheme. Such indicators lead to important questions that must be answered prior to sensor specification, and protocols that need to be followed during sensor use. In this sense conventional instrumentation and emerging technologies that fall under the scope of PAT are no different. If a measurement is to be used for critical parameter control then the measurement must possess certain characteristics. If it does not then critical parameter control cannot be guaranteed, as the process measurement device is an integral component in the regulatory scheme. The key considerations for a sensor are:

- *Accuracy and resolution.* A useful sensor provides measurement at an appropriate accuracy for the control task. If, for example, a temperature is to be controlled in the range of ±0.1°C then the measurement must be significantly more accurate than that; if not, although it may appear that the process is controlled within this range, the actual process may be suffering larger deviations. An awareness of accuracy is therefore essential, but unfortunately the information is seldom presented with this consideration in mind.
- *Long-term precision.* In the longer term, the information which has been gathered must be consistent. For instance, sensor drift from calibration can cause deterioration in system performance because the

desired values are not achieved. Drift is often inevitable, so it is important to know the rates of likely drift so that recalibration can be performed as necessary.

- *Reliability.* Sensors provide information that is acted upon either by process operators in a "human in the loop" control scheme or directly by closed-loop control schemes. When operators use the information, there is opportunity for human interpretation of the results. Failed sensors are more difficult to detect in a hardware-based closed-loop scheme. If the information is essential and a sensor fails, then implications on operation can be severe. Lack of reliability and consideration of the importance of the measured information for control leads to the use of multiple sensors and various decision schemes. More information on reliability analysis and control consequences can be found in Cluley (1993). It is essential to assess the reliability of sensors and adopt planned maintenance programs to assure reliability.
- *Practicality.* The environment within a process may be particularly demanding, for instance, the sensors may be exposed to high sterilization temperatures. While a sensor may in theory measure the variable of interest in ideal conditions, the range of the operational environment could render it incapable of functioning or may influence reliability.
- *Off-line/on-line sampling frequency.* Theory dictates that for a measurement to be of value it must be sampled above a certain minimum frequency. If it is sampled slower than this then the information provided can be misleading. This situation is known as *aliasing* and is of concern when a process is sufficiently fast compared to the measurement rate so that significant deviations can occur between samples. Often instruments that are used on-line (say temperature or pH) can provide continuous information so that these problems do not arise. Other on-line instruments (e.g., a mass spectrometer measuring off-gas concentrations) are multiplexed to save cost, but the frequency of information supply is limited because the instruments must serve several vessels. However, it is in off-line sample analysis where problems with low-frequency measurement are most likely to arise. The problems can be compounded by delay due to sampling. Indeed, the situation could arise where the sample rate is so low or the time taken for analysis is so long that there is little point in taking the measurement for corrective purposes. The case study presented in Sec. 3.5 considers such a situation.
- *Cost.* Sophisticated instrumentation is now available for process monitoring with PAT, but the price can be high. The price may be even higher for a particular market where there may be low sales volume.

However, the benefits gained can be significant if sensor information leads to feed savings or increases productivity. A cost-benefit analysis should be performed to assess whether the instrumentation is appropriate, although this can be far from straightforward if assumptions on sensor operation relate to their ability to make process modifications that determine the savings to be gained. With this in mind, assessing novel instrumentation is a valid and valuable role for the industrial R&D laboratory.

- *Speed of response.* Whether the control loop is closed by a process operator or by a hardware device, sensors form an integral component in the closed loop. Their dynamic characteristics are important because they must respond significantly faster than the process. This may not be a problem. For instance, bioprocesses are generally relatively slow, but if a sensor has a long response time it may indicate an *average* value rather than the actual process value.
- *Location of sensors.* Depending on the scale or design of a process, the location of sensors may be particularly important. For example, by the nature of operation of an airlift fermenter, the circulating microorganism experiences a cycle in the level of dissolved oxygen. It is therefore important to consider exactly where sensors are placed to measure the critical conditions. In this case the lowest levels of oxygen in the cycle may be key. Location is also important in terms of minimizing delay, in that the more current the information that is acted upon, the lower the deviations in process output.

Given the above considerations, what are the practical implications for critical parameter measurement and control?

- Process operation is dependent on sensors returning measures of performance. What sensors and where they are placed must be considered early in process development and must be central in the process design stage of a production plant. Sensors and control systems must not be seen as an "add-on" after process design.
- Procedures that ensure that specified sensor characteristics are maintained must be put in place. Procedures for maintenance are a fundamental component to process validation. Furthermore, it is important to remember that measurements from sensors provide data. What is required for critical parameter control is information that is reliable. The steps taken to move from sensor data to process information are discussed in Sec. 3.3.
- Specified operating protocols must be realistic, given the sensor characteristics. It is unrealistic to specify a target operational band that is narrow compared to sensor capability. What is a realistic band

depends on a function of many variables among which are control capability, delay, and sensor accuracy. To arrive at a sensible band requires an appreciation of control system functionality. This is discussed in Sec. 3.4.

- More demanding operating windows can be specified as target objectives to ensure that wider compliance bands are satisfied. If operators act to tight targets then it is more likely that performance within the wider band will be achieved, given that it is realistic as discussed earlier.

Measurement errors must be understood and minimized for a robust process monitoring and control. It is important to understand the measurement uncertainty and calibration in a bit more detail.

3.2.1 Measurement uncertainty and calibration

A measurement system consists of a sensor, transmitter, and a monitoring system. A sensor detects a change in physical property, for example, a thermocouple detects a change in temperature, and a strain gauge detects a change in pressure. A transmitter converts the physical property into an electronic signal for transmission to the monitoring system. The monitoring system converts the electronic signal into a displayed and/or recorded value.

Uncertainty is inherent in each component of the measurement system. Measurement uncertainty could be defined as the departure from the true measured value. The difference between the true value and the actual value of a parameter is referred to as the error in the measurement process. This error has two parts, random error and systematic error.

Random error could be interpreted as a common cause variation. It occurs in a random way and could be seen if a large number of data are collected (without systematic error) where the data fluctuate around the mean in a random fashion. For example, conductivity probe impacted by random change in ambient temperature. Random error cannot be eliminated.

The systematic error component is due to special cause variation. Typically, a probe starts drifting from its true value over time. This could be due to change in sensor element sensitivity and/or transmitter and monitoring system suboptimal function. For example, a sensor element in a process stream may have some scales deposited over the sensing element leading to a drift in measurement. Temperature, humidity, or electrical/magnetic interference may impact the performance of the transmitter and the measurement system leading to an error in reading.

Periodic calibration of the measurement system helps reduce the systematic error component. The calibration process aims at a reduction in systematic error of the probe, which would be higher than that of the standard. The calibrated probe has the inherent uncertainty of the calibration standard. The calibration process provides traceability of instruments to known reference standards. These reference standards are typically accredited by the national standards laboratory, for example, in the United States the national metrology laboratory NIST drives instrumentation standards throughout the country.

Measurement uncertainty could be quantified by statistical techniques. The data are generated by a set of experiments and is statistically treated to obtain standard deviation leading to the quantification of the measurement uncertainty. Further details of the measurement uncertainty could be found in standard references such as Dieck (1997) and Kessel (2002).

Measurement generates data and it becomes important to manage the data appropriately to generate useful information and knowledge. The next section focuses on data management.

3.3 Data Management

This section discusses general procedures for obtaining reliable information from process data. Here the word *information* is fundamental to ensuring robust process operation. Recent years have seen vastly improved logging systems and more sophisticated instruments provide higher volumes of data. But, data in their raw form are not necessarily information. The sheer volume of data can in itself be a problem, but the complex spread of information content within it can tax the most sophisticated of algorithms. Some of the practical challenges are:

- In any particular process, data can reside on a variety of computer systems, probably networked but not in a common format. To extract data for analysis can require considerable effort in coordination of data from *laboratory information systems* (LIMs) and on-line process data historians.
- The information provided by the instruments is not necessarily a direct measurement of the variables of interest, for example, spectral information from an NIR probe providing a general fingerprint of the sample.
- The operating environment places severe demands on instrumentation resulting in high levels of noise on signals, outlying data, and potential signal failure. A filtering algorithm can smooth noise but also hide/delay real information.

- Varying sampling frequencies or infrequent sampling can mask system response characteristics, limit the quality of filtering, and make data analysis problematic.
- Data-compression algorithms are useful for reducing storage requirements but information loss can result when reconstruction of the original data is required.
- Process problems can be reflected in changes to a number of measured variables or alternatively changes to several variables will have complex effects on the operation of the process. Whichever is the case, looking for patterns in multiple data sets distributed over time can be difficult.

It is also important to consider how the information is to be used. Take for example one of the simplest requirements for data treatment—spike removal. If the data are to be used for historical batch analysis, then any spikes in the signals are immediately apparent, but if the data are to be used directly in a closed-loop control scheme then removing signal spikes (generally referred to as "outliers") is a significantly more difficult and crucial task. In this case, the chance of operators effectively intercepting erroneous data is much more limited.

Data treatment can have three main objectives:

- The removal of errors in data that are not truly reflective of real process performance (e.g., outlier/noise removal).
- The ability to fuse data from a variety of sources, measured at different sampling frequencies, and so forth. For control of critical parameters, this is presently not so severe a problem, but as PAT becomes more commonplace data will become more remote. Presently, assembling a single database that contains all relevant data is more of a problem for high-level control functions such as statistical process control, where considerable off-line analysis tends to be undertaken prior to implementation.
- The extraction of information pertaining to a bioprocess condition, whether this is for performance assessment or for fault finding. This is discussed further in Chaps. 4 and 6.
- Fundamental to information extraction is the availability of data that are truly representative of process behavior. Several issues need to be addressed if such data are to be obtained, uppermost being the problem of spurious data points.

3.3.1 Data filtering and outlier removal

Figures 3.4 and 3.5 give some indication of the problems that can be encountered in data filtering. The data shown is the type of raw process

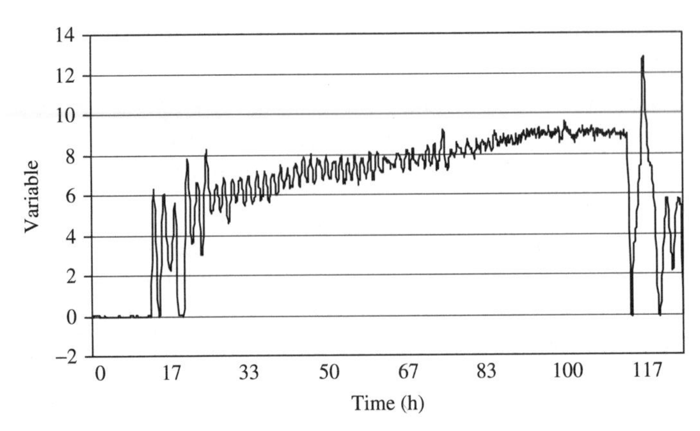

Figure 3.4 Example of noisy signal from on-line measurements.

data commonly logged from industrial processes. Figure 3.5 shows the signal outliers that have arisen from a number of sources. In addition to spikes, data dropouts occur when the signal is zero or no information is available. In this case, the instrument clearly is not calibrated correctly as there are some negative values in the flow rate signal.

In addition to the spurious points, the signal is obscured by noise. This can be seen more clearly in Fig. 3.6, where the section of data between 33 and 50 h has been magnified; the signal is corrupted with noise of a magnitude of ±0.5 units.

Before these data can be used to provide useful information or for control purposes, they must be "cleaned" by removing outliers and the noise level must be reduced by applying filtering algorithms.

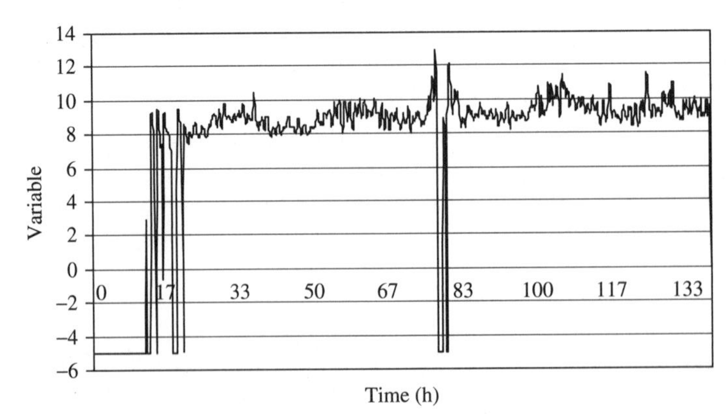

Figure 3.5 Example of noisy signal with outliers.

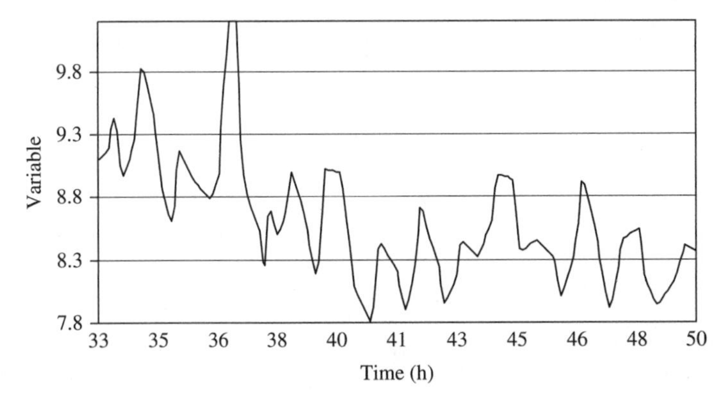

Figure 3.6 Noise present on raw signal.

3.3.2 Noise removal

Data filtering is a term that describes a broad range of techniques. There are many forms of algorithms available to remove noise from measured data. This section does not attempt to review all methods but presents an overview of commonly used approaches. Further information on data filtering can be found in Jazwinski (1970).

The standard technique for noise removal is to make use of "first-order" filters, which exponentially weigh past information. This produces a moving average, which places decreasing emphasis on older values. Such filters are typically found in commercial controllers. They provide the opportunity to reduce noise levels through setting the rate of exponential forgetting via the time constant. The larger this value the greater is the weighting given to old values. A physical analogy to a first-order filter is a stream flowing into and out of a fixed-volume, well-mixed tank. If the inlet concentration changes then the larger the volume of the tank relative to the flow rates, the longer it will take for the outlet concentration to approach the inlet concentration. In this case the volume of the tank would be analogous to the time constant of a first-order filter.

To demonstrate the effect of varying the time constant on noisy process data, a first-order filter is applied to the data shown in Fig. 3.4.

Figure 3.7 is a magnified section of the data in Fig. 3.4, and it shows that with an appropriate degree of filtering it is possible to remove noise effects from the data. An important lesson is that effective implementation relies on the user's process knowledge to decide what is short-term process noise and what is the underlying variation. In this case the oscillatory behavior could be true process operation or error values as a result of signal problems. The choice of filter time constant is not necessarily a straightforward task, but generally trial and error is successful.

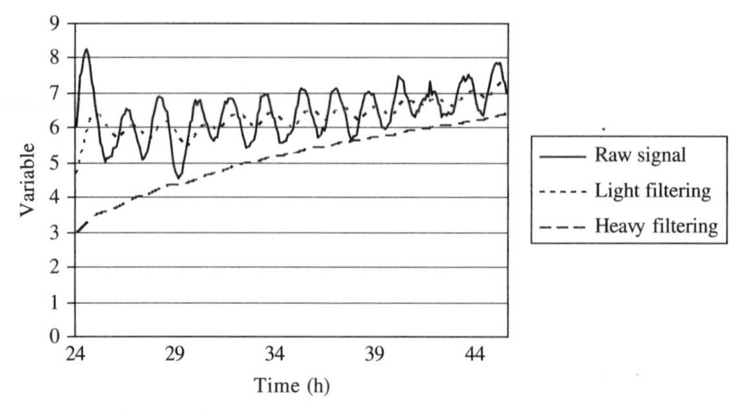

Figure 3.7 First-order filter application.

The facility to implement first-order filters is available in most commercial controllers and data-logging devices, but many users are unaware of this option.

Another important observation made in Fig. 3.7 is that heavy filtering causes information loss. The filtered signal lags excessively behind the true process response—this phenomenon is known as *phase lag*. This raises some important issues from a regulatory perspective. If heavy filtering is applied to a critical process parameter and if the data is then stored in the historian as the batch record, it is possible that deviations outside of the *proven acceptable range* (PAR) may have occurred and not been recorded. Clearly, heavy filtering is only necessary when noise is severe, but unfortunately process information can often have this characteristic. Phase lag is an error in the signal, so a filter that provides efficient noise removal while minimizing lag is therefore required.

The design of filters that minimize phase lag and enable effective smoothing has been widely researched. Gardner and McKenzie (1988) reviews the state of the art in exponential smoothing and describes a range of filters that are designed to minimize phase lag for predefined qualitative trends. Figure 3.8 shows the phase lag that results from implementing a first-order filter to noisy data, but if a linear-trend filter is applied to the same data, the phase lag is significantly reduced while the same level of filtering is achieved.

This filtering method can be applied to both on-line usage and subsequent off-line filtering of logged data. This is because the latest filtered value is a function of past filtered values and raw data; however, if filtering of off-line data is the only requirement then it is possible to use additional algorithms that make use of future values of variables. For example, the current filtered value can be made up of a

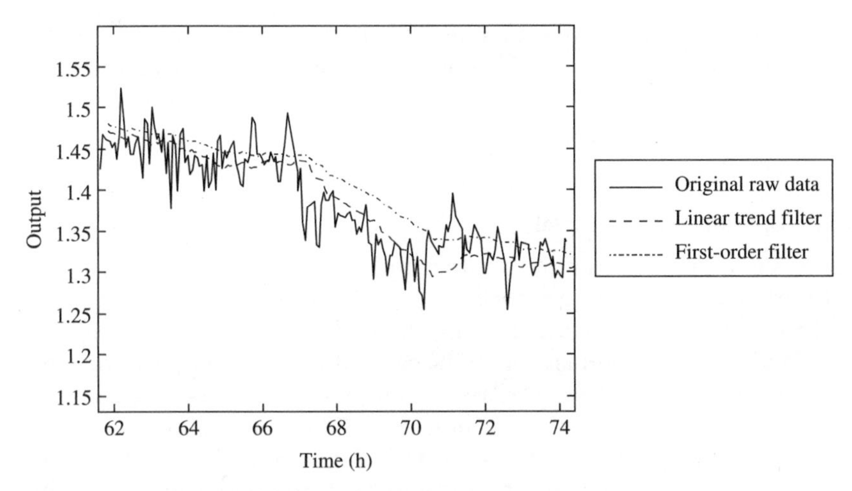

Figure 3.8 Linear-trend filtering of noisy data.

mean of raw values around it, thus a window of seven samples would give:

$$X_{tf} = \frac{X_{t-3} + X_{t-2} + X_{t-1} + X_t + X_{t+1} + X_{t+2} + X_{t+3}}{7}$$

The advantage of adopting this approach is that the phase lag is minimal. The length of the filter window must be chosen with care—too short results in noise persisting on filtered data, and too long results in an inability to follow trends sufficiently quickly. However, with off-line trial-and-error choice of window size, setting up such a filter is straightforward, and the performance is generally satisfactory. A minor variation of this method, using median rather than mean values in the filter window, reduces the influence of spurious data. Another method of off-line filtering is to filter a data set forward in time with a first-order filter (as described earlier), which will result in a phase lag. The resulting filtered value can then be filtered with the same time constant backward in time. This "double" filtering in reverse direction again acts to minimize the phase lag.

An important observation for both on-line and off-line use is the effect of the rate of data gathering on the achievable quality of filtering. As higher frequency information gathering decreases the ability to remove high-frequency noise, the ability to achieve an acceptable phase lag on the underlying signal becomes more limited. This implies that raw data should be gathered at the maximum frequency possible, with due consideration to the data storage requirements, to provide the greatest flexibility for filtering noisy signals. The preferred method of filtering, however, is to perform the action locally via analogue filters on the data-monitoring equipment, but it is essential that these filters should have been appropriately configured.

3.3.3 Spurious data removal

There should be an attempt to detect outlying points in a data set prior to filtering using techniques such as those described earlier. The reasons for this become clear by considering the results shown in Fig. 3.9.

An outlying data point is seen in the raw data for one sample at time 50. This may appear somewhat extreme, but experience suggests that this is a common occurrence in many industrial data sets. The raw signal has been filtered with a first-order filter with a time constant of 10, and the output of the filter is shown with and without the outlier removed. In this case the influence of the outlier remains for 30 samples after it has occurred. Decreasing the time constant can reduce the time frame of influence, but this would lead to a reduced filtering performance and a greater response to the outlying data point. Thus, the logical approach is to remove the outlier if possible.

It is obvious from looking at the data shown in Fig. 3.9 that the point at time 50 is an outlier; but if data are monitored on-line (rather than considering the batch off-line when the whole "picture" is available) this decision can be more difficult. The case shown in Fig. 3.10 relating to logged data from an industrial reaction process illustrates this.

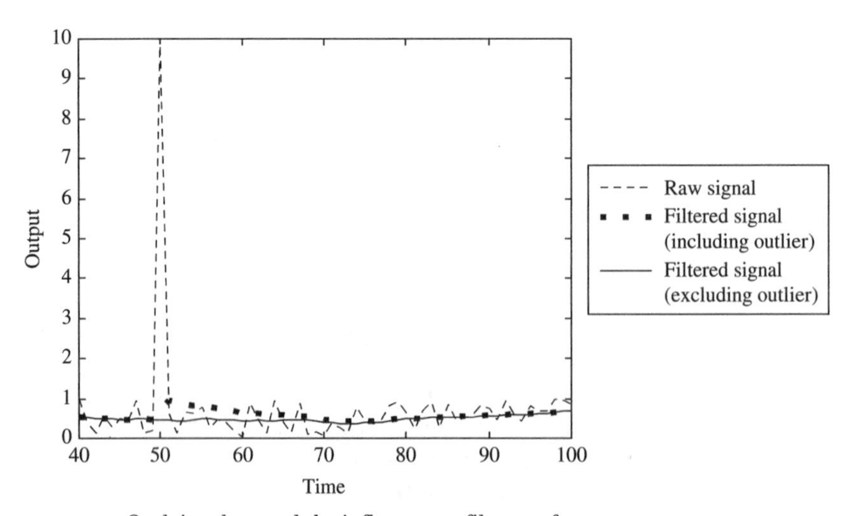

Figure 3.9 Outlying data and the influence on filter performance.

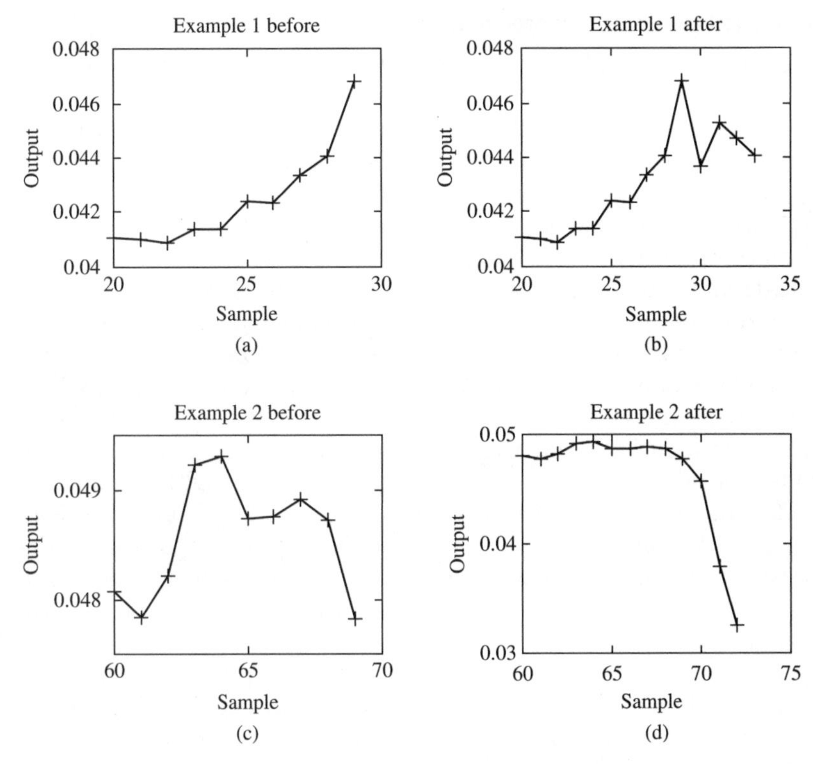

Figure 3.10 The problem of on-line data assessment for outlying data points.

In Example 1 (graphs A and B) there appears to be a significant change in trend from Sample 28 to 29, which might be considered the initial deviations of an outlier. It is only when the next four points are considered (graph B) that this clearly is not the case. The data shown in Example 2 (graphs C and D) show a similar change in trend between Samples 68 and 69 (graph C), but in this case the following two points indicate that an outlier may be present. This can only be confirmed with data reconciliation (i.e., has another process variable changed causing this deviation in trend?) and knowledge of process behavior (i.e., is it possible for the process to change at this rate?).

This example demonstrates that outlier removal is not necessarily an easy task, but methods are available (Tham and Parr, 1994) to detect them in on-line applications and remove them from the data set. An appreciation of the data quality and the likely speed of process response are necessary to implement this approach. Outlier detection procedures typically have to be configured using off-line data to simulate on-line performance, but with care they can ensure that minimal effects are observed in the treated data.

3.3.4 Issues with data compression

Within the process industries the prime motivation for data compression is the reduction of costs in computer storage. As the cost of computer storage decreases, it might be argued that the need to implement data compression is reduced. However, new instruments such as those based on spectral measurement are generating vast amounts of data, thus reductions in the cost of computer storage have not removed the need for data compression.

Industrial computer archiving systems make use of a variety of data-compression methods. For instance, compression using piecewise linear trending is in widespread use in industrial data historians. AspenTech uses an adaptive method based on the *box-car/backward slope* (BCBS) method, while the OSI PI data historian uses a type of swinging door compression algorithm. Further details on data compression are provided by Thornhill et al. (2004). In addition to giving insight into the fundamental procedures, important issues are raised by Thornhill et al. concerning the compression of data and the suitability of the reconstructed data for use. In the worst case, the reconstructed data may have lost important aspects, particularly when considering the higher frequency components.

If storage of data is for the purpose of long-term process productivity assessment, then standard data-compression methods are likely to be acceptable. Alternatively, if dynamic analysis of control system behavior is to be studied, reconstructed data may be unrepresentative of the original raw values. In this case it depends on how the data-compression method has been configured. Past experiences suggest that what appears suitable on configuration may not be suitable when historical data are analyzed at some point in the future.

Ultimately, data compression results in information loss, but if not employed storage costs are higher. As information is of a higher value compared to storage costs, Thornhill et al. offer the advice that data-compression algorithms are best avoided if feasible.

3.4 Control of Critical Process Parameters

The ability to measure or estimate process conditions is of little benefit unless this information is used to improve process understanding and the overall behavior of the process. An example is temperature control of a chemical reactor, which is regulated at a value specified in the manufacturing ticket. The controller maintains this temperature against internal and external reactor disturbances and so delivers the desired performance. In addition to standard loops such as temperature and pH, there are other more subtle examples of regulation. For instance, even when information is gathered for noncontrol purposes such as for quality assessment at the end of a batch, control actions can be instituted based on deviations from the desired value. Although the time frame over which changes are made differs considerably, a common theme is the assessment of process conditions using deviations from desired value. Changing one or more variables to force a return toward ideal operation is the job of the controller. The basic concept is termed *feedback* or closed-loop control. For example, when temperature deviates it is possible to adjust cooling water flow using feedback control to return the temperature to its desired value.

It might appear that feedback of information to correct for deviation would always result in process improvement. Sadly, this is not the case. The magnitude of the corrective action can be inappropriate given the process characteristics. In the worst case the actions of the controller can make the process condition worse than it would have been if no action had been taken, causing oscillation and instability. Surveys of industrial applications of process control suggest that such instances are commonplace. To reap the benefits of feedback control it is necessary to understand the function of the controller and its interactions with other process elements. Although the technology to achieve closed-loop control has been available for many years, the use of feedback systems for effective regulation can be problematic due to poor understanding of the function of the control loop and/or due to poor implementation and maintenance procedures. For this reason this chapter considers the basic concepts and traditional methods of control loop tuning and goes on to consider state-of-the-art techniques for feedback controller tuning, with advanced methods capable of tackling demanding problems and delivering improved performance.

3.4.1 Feedback concept

The concept of using an error signal to calculate the change to a process variable is termed feedback or closed-loop control. A conventional control loop is shown in Fig. 3.11.

The control loop consists of a *feedforward* and a feedback path. In the feedforward path at the point in the diagram labeled Ⓐ the desired value (also known as set point) of the output signal is compared with the measured process output, and the error is calculated using a summation (Σ) block. This error is used by the controller to calculate a change to the manipulated variable Ⓑ with the intention of driving the error to zero (i.e., forcing the output of the system to equal the set point). The algorithms used to achieve such an objective are discussed in Sec. 3.4.2. A typical output of Ⓑ is an electrical signal to a pump or a signal to move a valve position. Changes in manipulated variable are sent to the

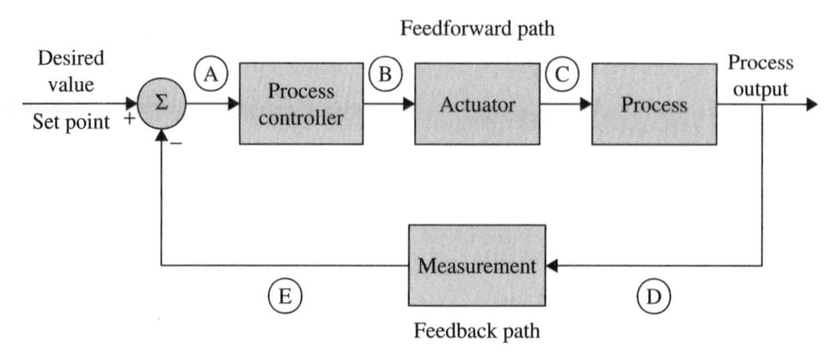

Figure 3.11 Feedback control loop and its constituents.

actuator. The signal at point Ⓒ is typically a flow rate. This influences the process output, and if the controller has been appropriately tuned (i.e., parameters selected correctly), the signal drives the output toward the desired value. The function of the feedback loop is to provide the information to calculate the deviation from the desired output. Thus, the measurement device (i.e., pH probe, thermocouple) is in the feedback loop. It is important to realize that the function of the control loop is to drive the measured value of output to set point and, if there is a problem with a measurement, this may not be effective in terms of achieving the desired value of process output. It is clear in Fig. 3.11 that the signal at point Ⓔ, the measured signal, rather than the actual value at point Ⓓ, is forced to the desired value. Thus, the measurement device is a key constituent of the control loop.

Experience suggests that the step of going from a physical process to a control scheme diagram can be difficult for those new to the control systems theory. By means of a simple example, consider the regulation of temperature in a jacketed tank by adjusting cooling water flow. Figure 3.12 shows the closed-loop control scheme.

Here the block representing "process cooling dynamics" describes how changing the cooling water flow influences the process temperature. It therefore combines descriptions of cooling jacket and vessel heat-transfer dynamics. An additional component has been added to Fig. 3.12, and it now contains a block representing the effect of process disturbances on temperature.

Considering the basic closed-loop diagram in Fig. 3.11, the purpose of the controller is to ensure that set point is achieved. But what causes the process to deviate from set point? In the simplistic case, only manual changes of set point would cause such deviations. In this case, a simple "look-up" table for manipulated variable settings to give

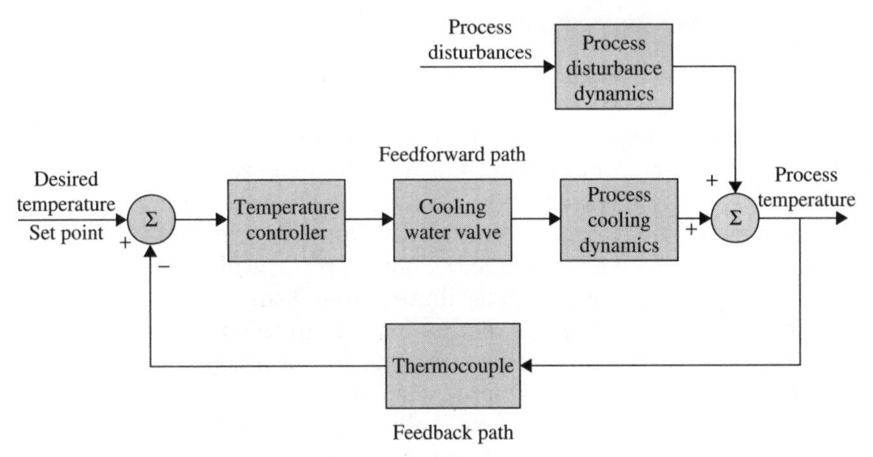

Figure 3.12 Reactor temperature control schematic.

a desired output value would be sufficient. This is unrealistic. Figure 3.12 presents a more comprehensive description of the closed-loop system. Process disturbances cause deviations in process temperature that the controller responds to. Process disturbances represent influences such as cooling water temperature changes, heat-production changes in the vessel, and ambient-temperature variations.

In establishing a diagram similar to that shown in Fig. 3.12 for a particular control loop, it is essential to consider the relative dynamics of the components within the loop. For instance, if the temperature-control system of a reactor is considered, then the response time of the valve may be irrelevant when compared to the time taken for a change in flow rate of the coolant to modify the temperature. Thus, the manipulated variable block is not influential in the closed loop and as far as analysis is concerned can effectively be ignored.

An important characteristic of the feedback loop is that it has the potential to cause instability in a system. Instability is when the magnitude of the system output increases over time. Simplistically, this occurs when the controller acts on the signal fed back around the loop and further increases the magnitude of the input to the system. The consequence is either increasing magnitude oscillations or exponential deviation, both being undesirable. If all the loop characteristics are known, then it is possible to predict controller conditions that will cause instability. This is not generally the case, so tests for controller parameterization (more commonly known as tuning) involve quantifying system characteristics in some way. Before going on to consider controller tuning, the structure of the control algorithm must be defined.

3.4.2 Algorithms for feedback control

On/off control procedures. The number of different procedures used to determine changes in the manipulated signal is limited to variations in a few control strategies in most commercially available systems, the simplest being an on/off algorithm. For a process where increased control effort causes a process output to rise, an on/off controller switches control action full on when the output deviates significantly below the desired value and off again when the output returns above the set point. As a result, the system output oscillates around the desired value. This may be adequate if all that is required is to maintain the level in a tank within a range, and then small oscillations may be acceptable. Foam control in a process tank is an example where on/off control is perfectly adequate with the average foam level being maintained within acceptable bands. A variation on the on/off control strategy is low/off/high action that is found in some process control systems for pH control. If the pH deviates below a particular value then the alkali pump is started, if it is above set point then the acid pump is started, and if it is in a band around set point both pumps are off. A band around the set point is required to prevent excessively frequent pump switching. If this band is too wide then this is clearly undesirable, and what is required is a controller that is able to maintain a steady value. The use of a three-term controller is the standard method of achieving this.

Three-term controller. Three-term controllers, commonly known as *proportional, integral, and derivative* (PID), are found in use in the majority of industrial systems. The basic PID control algorithm is:

$$u = K\left(\text{error} + \frac{1}{T_i}\int \text{error} + T_d \frac{d(\text{error})}{dt}\right) \tag{3.1}$$

where u = manipulated variable
K = controller gain
T_i = integral time
T_d = derivative time
error = difference between desired value and measured output

A common variant is to present the controller gain as the *proportional band* (PB), where PB is 100 divided by controller gain K. Thus, increasing gain translates to decreasing PB. The algorithm is not usually implemented in this fashion as simple transformations can improve its robustness but in its basic form it serves to demonstrate the fundamental concept. The proportional action (due to K times the error) is the most simplistic use of the PID algorithm but on most systems the use of the so-called P controller results in offset (i.e., a steady state error from desired value). This is shown in Fig. 3.13 when a system is controlled with a P controller. Several responses with different controller gains are displayed. The system has been chosen to be first order with a gain of 1 and time constant of 10 and can be seen in Fig. 3.13 in Laplace representation. The traditional control system block diagram is shown, as well as the closed-loop response. The diagram was developed in Simulink from Mathworks (www.mathworks.com) to simulate the closed-loop system. A step change in desired value at time zero leads to a dynamic response. It can be seen that an offset from set point results and the size of the offset decreases as the gain of the controller is increased.

Although in this simplistic case increasing gain causes the offset to decrease, using an excessively large gain to minimize the offset is not a general solution in real process applications. Nevertheless, the offset from set point is often undesirable, and it is for this reason that the integral term is added to the P controller, resulting in a PI controller. The effect of the integral action is to continue modifying the manipulated variable while an error exists, effectively forcing the steady state error to zero. The resulting dynamic response and closed-loop block diagram are shown in Fig. 3.14 for several choices of controller parameters.

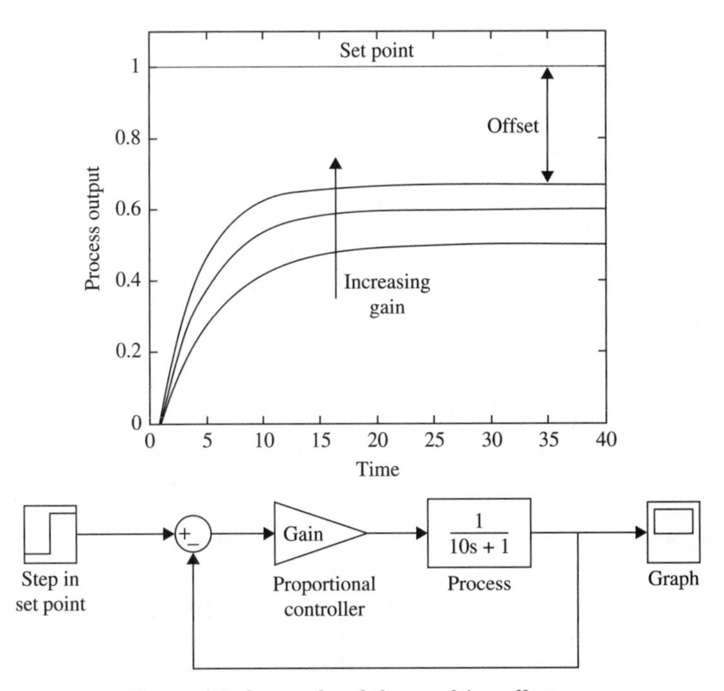

Figure 3.13 Proportional control and the resulting offset.

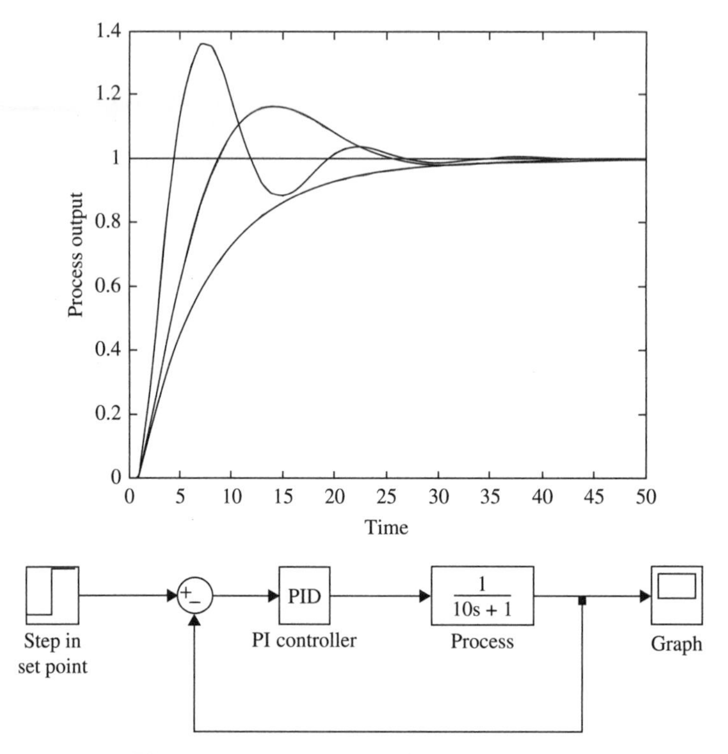

Figure 3.14 PI control eliminating the offset.

It can be seen that the system settles down to the set point but the track it takes to steady state can be quite varied. For any system, whether or not the response oscillates and the severity of the oscillation depends on the tuning (i.e., the value of K and T_i). Furthermore, the desirability of oscillation is something that must be considered. Notice that the oscillatory responses reach set point faster, but on some systems overshoot of set point could be a problem.

The choice of controller parameters is discussed after consideration of the third of the terms in the three-term controller, the derivative action. The reason for its inclusion is to react to a sudden change in error, in which case rapid changes in control signal are required to return the process to the set point. This may appear to be a desirable controller facet and indeed for simplistic systems, such as that considered in Figs. 3.13 and 3.14, it would improve the response characteristics. However, for real systems problems can arise. Noise added to the output of the system now serves to corrupt the signal, and this is particularly significant when the controller characteristics are considered. This is demonstrated in Fig. 3.15, where the responses for both a PI and a PID are compared.

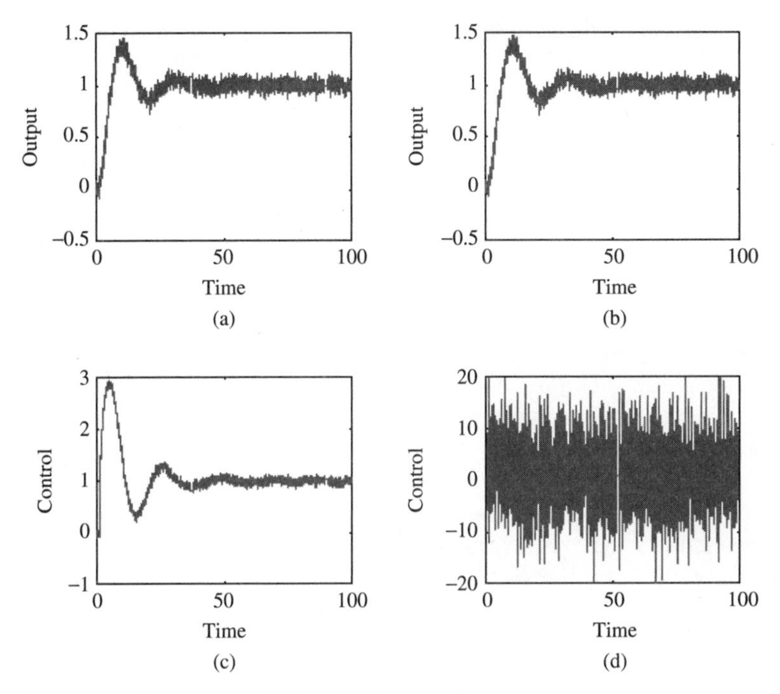

Figure 3.15 Comparison of PI and PID controller responses.

The important observation is that while the output responses are not significantly different, the control effort required to achieve the responses are very different. Figures 3.13*a* and *b* show the output and control signal required to achieve set point for the PI controller. The noise on the output is apparent, and this is carried through to small amounts of noise on the control signal. The control response that results from the PID controller is shown in Figs. 3.13*c* and *d*. While the output response looks similar to the PI controller, the control signal is extremely noisy. This results from the derivative action magnifying the noise on the output signal (i.e., the derivative of a noisy signal is an extremely noisy signal). In this case either the derivative time should be reduced or filtering should be employed to reduce the noise on the measured signal. The example serves to demonstrate how problems can arise. In theory, the use of derivative action improves the system response, but for practical reasons, PI rather than PID controllers are generally preferred.

3.4.3 Controller installation

When applying a PID controller for feedback control it is necessary to specify the parameters (K, T_i, and T_d) to achieve acceptable performance known as *controller tuning*. As a result of the widespread use of

PID controllers, much literature on tuning is available and research still progresses with, for example, conferences dedicated to PID control. The basic principles are well established, the first being essential to define what constitutes acceptable process performance. From the limited earlier examples, it can be seen that a variety of responses are possible, given the flexibility of choosing up to three controller parameters. Two basic approaches for controller tuning are possible: manual tuning using a variety of procedural methods or automatic tuning by in-built strategies for parameter determination. Whichever is chosen, the first decision must be to decide on the ideal response characteristic. Getting to set point as rapidly as possible is a potential objective given that achieving set point is the purpose of the controller. This is not the sole consideration since rapid process changes require large variations in manipulated variables, which may be undesirable or more likely not possible due to constraints (i.e., valves can only open to 100 percent or close to 0 percent). Also, rapid rises lead to overshoot, and overshoot may not be acceptable (i.e., will overshoot in temperature cause a process failure?). However, in other cases it may be more important to get close to set point, and oscillations around the value are of no consequence. The choice of ideal response is obviously one that must be made based on appreciation of the process operating objectives and should be considered before controller implementation.

In order to move from a reasonable response produced by the tuning methods to an ideal response dictated by process knowledge, further refinement of the controller parameters is required irrespective of whether manual tuning uses procedural or automated methods. Moving to this ideal response is achieved by manually "tweaking" the controller parameters. This requires some appreciation of how changing a parameter will change the response. This is something that comes with experience.

During manual controller parameter determination, visual inspection of responses is often the basis for controller tuning. This is so even though methods exist that are based on numerical values, which capture characteristic response, for example, the Ziegler and Nichols technique. Automated methods for tuning controllers are becoming increasingly common for commercial controllers, and the fundamental concepts have been the subject of considerable effort during the past three decades. There are two primary reasons why automatic tuning is desirable:

- The lack of expertise in controller tuning results in a less than ideal implementation of the algorithm. This commonly has financial implications in terms of deviations from desired value.
- Process characteristics change over time, so a well-tuned controller becomes less effective. For example, as a heat exchanger becomes

"fouled," the dynamic characteristics change, and to achieve the same controller behavior the PID settings must be modified.

Automatic controller parameter-determination techniques rely on disturbing the process in some way or making use of natural upsets and using the resulting output pattern to determine the PID parameters. To attempt to address both the question of tuning difficulties and process dynamic variations, self-tuning/adaptive controllers have been developed.

There are two basic strategies: "tune on demand" and automated, possibly continuous controller parameter updating. From a robustness perspective, "tune on demand" is attractive, bringing the best opportunities for improved loop performance. The problem with automated techniques is that guarding against controller modification in response to outliers is extremely difficult without human intervention. Strategies have been the subject of considerable academic interest, but there has been little uptake by control system vendors. The main problem is the lack of robustness of the model identification module. The algorithms used and the fact that they can react to outlying data and thereby significantly degrade controller performance have outweighed the benefits to be gained from controller algorithm improvements. Furthermore, the ability of the controller to adapt to process changes has raised regulatory concerns. Thus, a more conservative approach has generally been adopted.

Rather than allowing continuous adaptation of the model/controller parameters, it is possible to update parameters only when required by the user. Since this is an infrequent and supervised event, somewhat more active disturbances can be utilized, which ensures effective identification. Furthermore, since the user initiates tuning, it is possible to observe the effectiveness of the resulting parameters and hence verify the setup procedure. These facets act to considerably upgrade the robustness of the control system. This approach is normally termed *autotuning*.

One method of controller autotuning is based on perturbing the system with a relay perturbation of the manipulated variable. For the majority of systems this induces oscillations of constant magnitude and frequency. The resulting pattern can be used to determine the parameters of the PID controller. Rather like the Ziegler and Nichols tuning approach, the input/output pattern is used to determine appropriate PID controller parameters. The original method was proposed by Astrom and Hagglund (1984), and has subsequently been implemented in commercial products.

Other forms of autotuners are based on the use of rule-based expert systems to encode pattern-recognition procedures. One of the earliest and most notable is the Exact Auto-tuner (Kraus, 1984) from Foxboro, and others have followed. In general, products based on autotuning ideas are making PID controllers more straightforward to install for the nonexpert. This, therefore, goes some considerable way to overcome the need for familiarity with the details of the controller algorithms. However, it does not necessarily overcome the problem of changing process dynamics during a process batch because the severity of such problems is process specific and does not arise in all instances. In many cases, well-tuned PID controllers will be satisfactory but not necessarily optimal for the whole batch. In a minority of cases, the change in dynamic characteristics is significant enough to prevent a fixed parameter controller being effective. Such a system functions acceptably at the start of the batch, but becomes unstable or very sluggish toward the end of the batch. In such instances, gain scheduling may be an appropriate option.

Gain scheduling involves responding to changes in process dynamics by updating the controller parameters through a rule-based approach. For instance, in batch control as a batch proceeds through its control trajectory, it is also possible to implement preplanned variations in the controller parameters, with switches being initiated on operational events or batch times. This procedure is known as *gain scheduling* since it is usually sufficient to vary the controller gain to maintain control system stability and performance. The problem of determining the appropriate gains at particular instances can be addressed using the procedures outlined earlier. With a system that possesses only a few inputs this may be possible but with complex processes, the matrix of controller gains that must be established and maintained becomes unwieldy. If this is the case then it may be more appropriate to implement a function which supplies gain values, but for which function specification is difficult.

For a thorough review of the autotuning controller methods, the IFAC professional brief written by Leva et al. (2003) is comprehensive with a detailed literature review.

3.4.4 Advanced conventional control

In addition to the standard feedback control loops found in bioprocess plants, there are a number of modifications and extensions to the basic feedback scheme that can result in significantly improved process behavior. These are advanced conventional control methods. One of the most useful additions is feedforward compensation.

Feedforward compensation. In many cases feedback control results in less than ideal closed-loop behavior. Due to the nature of the algorithm, control action is not taken until an error in the process output is observed. If the cause of the error is unknown or at least not quantifiable, then it is not possible to improve on feedback performance. However, if it is possible to quantify the disturbance, then control action can be taken to correct its influence on the process output before it causes an error in the output. For example, when cooling water temperature increases, if no action is taken the vessel temperature would rise. Why wait for this to occur? Increasing cooling water flow rate would compensate for the increase of cooling water temperature, and if adjusted appropriately there would be no change in the vessel temperature. This concept is known as feedforward compensation that is normally introduced into a feedback controller. The principle of feedforward compensation is that the action taken by the feedforward element of the controller and the effect of the disturbance are opposite and cancel each other out, leading to no deviation in the process output from the set point. Figure 3.16 shows a schematic for a combined feedforward/feedback controller.

The important information required to establish a feedforward scheme is knowledge of the effect of the disturbance and control action variations on the process output. This translates in a practical sense into a process model. In reality it is not possible to define a completely accurate model, and a feedforward element alone will result in offsets from desired values. It is for this reason that feedforward compensation is generally implemented with a feedback loop. The feedforward compensator attempts to eliminate measured disturbances, while the feedback controller compensates for feedforward compensator model errors and unmeasured disturbances to the system. One important practical issue arising from theoretical analysis of the system is that the feedforward compensator element of the loop plays no part in the stability of the system, only affecting the precision of its performance. Thus, as long as

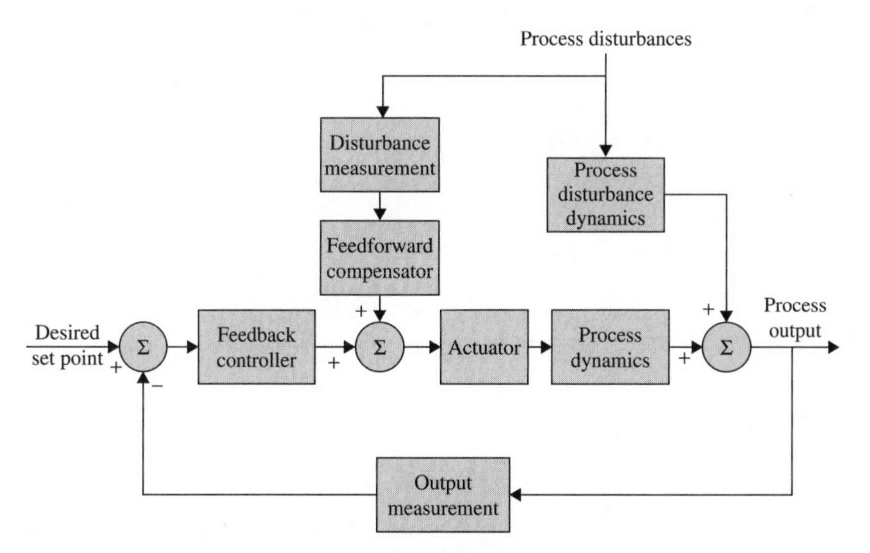

Figure 3.16 Schematic of a feedforward control system.

the model in the feedforward compensator is approximately correct, improvements in control behavior can be expected.

Feedforward compensation is a powerful addition to feedback control, which, if there are significant measured disturbances to a system, can result in considerable improvements to performance for little extra effort in implementation and tuning. Consider the vessel temperature control problem introduced earlier. The vessel temperature is controlled via changes made to the cooling water inlet flow. Flow modifications are made via movements to the cooling water control valve, and their influence on temperature is observed via the feedback scheme. If frequent temperature variations occurred in cooling water temperature, their influence would not be compensated for until the fermenter temperature deviated. The addition of feedforward improves control by measuring cooling water temperature and making modifications to its flow rate before fermenter temperature deviations occur. This is shown schematically in Fig. 3.17.

Cascade control. A situation similar to feedforward control arises in cascade control. In this case, since it is possible to control the variable subject to the disturbance, the feedback concept can be utilized rather than relying on a feedforward model. The basic methodology behind cascade control is best explained by considering an example. The basic feedback scheme might be perfectly effective in many cases, but consider a vessel temperature control scheme where cooling water maintains temperature.

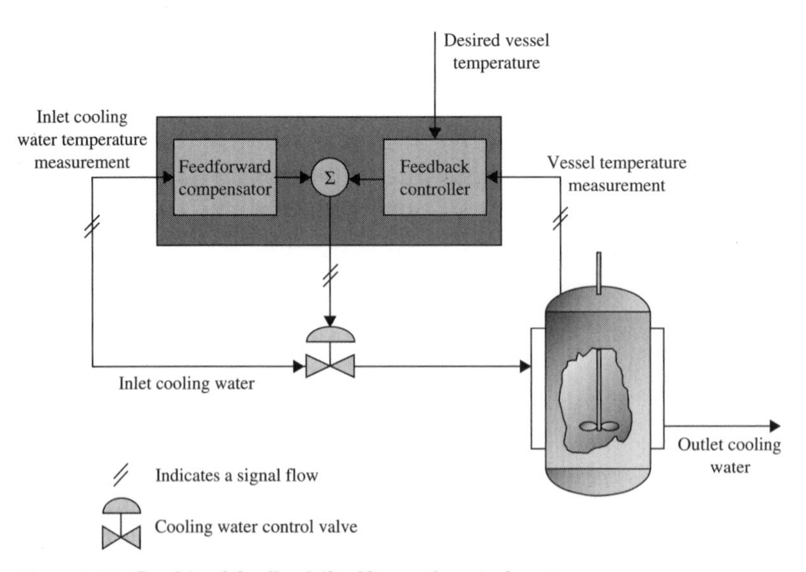

Figure 3.17 Combined feedback/feedforward control system.

In the event of a loss of cooling water supply pressure, cooling water flow will drop and thus the temperature in the vessel will begin to rise. A feedback control scheme will compensate for the temperature rise, but obviously not until an excursion in the temperature has occurred. A far better approach would be to regulate the cooling water flow rate directly in order to compensate for pressure variations. A cascade control system offering this potential is shown in Fig. 3.18.

The inner loop is the flow control loop where a measure of cooling flow rate is fed back to compensate for delivery pressure changes. Thus, to implement the cascade control system it may be necessary to install additional instrumentation. Rather than setting the cooling water valve position, the outer loop sets the set point for the inner flow control loop. Cascade control strategies are only effective in the situation where the inner loop is significantly faster than the outer loop. The tuning of the controllers is relatively straightforward with the inner loop tuned first and parameters fixed. The outer loop is then closed and tuned following the standard procedures.

Ratio control. In some situations it is desirable to maintain the ratio between two variables at a fixed value. For instance, this may arise when it is necessary to fix the ratio between two flows of reactants into

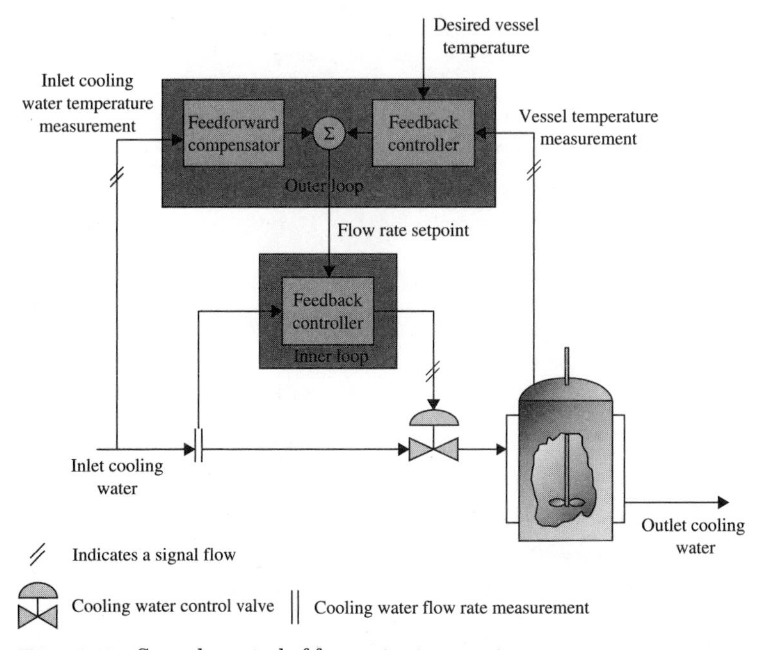

Figure 3.18 Cascade control of fermenter temperature.

a chemical reactor. In this case, one stream is termed the *wild stream* and the other the *controlled stream*. The wild stream varies and the ratio controller is responsible for maintaining the ratio between wild and controlled flows.

Figure 3.19 shows the concept where the flow rates of both the wild and the controlled stream are measured. The flow rate ratio is then calculated and compared with the desired ratio with the ratio controller adjusting the controlled flow rate until the desired ratio is achieved.

3.4.5 Rules of thumb and strategies for unit operation control

From the earlier examples it is apparent that control systems can be configured to give a variety of performance characteristics. Controller tuning techniques provide a structured approach for parameter determination but require refinement for ideal performance to be achieved. Thus, a degree of human involvement in the controller tuning is inevitable. In such circumstances assistance in predetermining the best controller structure and settings would be valuable. Such rules of thumb exist for the more common control loops such as level, flow, temperature, and pressure. For example, the following are typical:

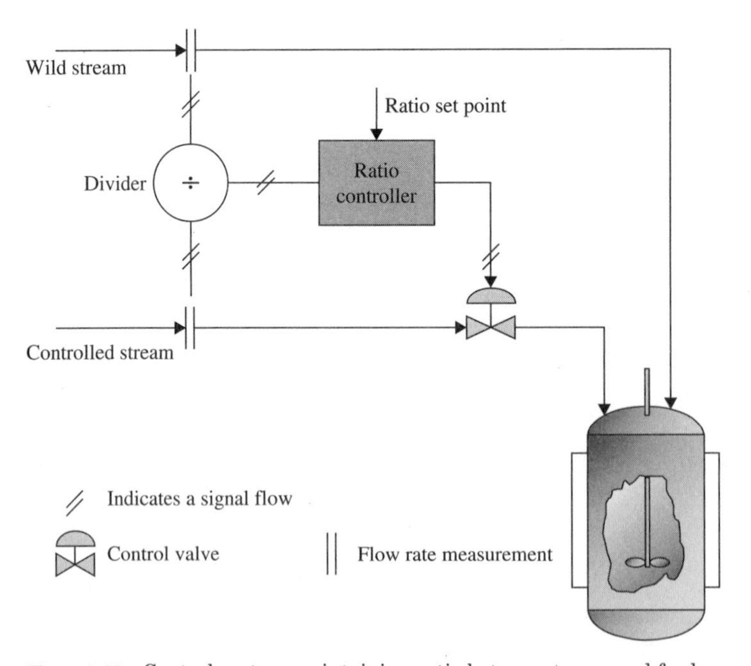

Figure 3.19 Control system maintaining ratio between two vessel feeds.

Flow control. PI controllers tend to be used with high proportional band (low gain) to avoid reacting to noise on measurements but with a low reset time (large integral action) to get fast set point tracking. This can be used as the loop dynamics tend to be very fast.

Level loops. The strategy adopted depends on the purpose of the equipment. If accurate control is not essential, such as in surge tanks, then proportional only control is the preferred strategy; and determining the settings of the controller becomes a straightforward task. There are circumstances where accurate level control is essential, for example, when regulating levels in a reactor, and in these circumstances a more complex scheme is justified.

Temperature loops. These loops tend to be quite slow because of the slow dynamics involved with process heat transfer. PID controllers are commonly used with a low value of proportional band, a reset time around the value of the process time constant, and a derivative time around a quarter of the process time constant (but could be less in noisy conditions).

Further details on rules of thumb for controller tuning can be found in many control texts (Luyben and Luyben, 1997).

It is important to understand the basic control design in relation to the measurement range of sensor, operating range, excursion tracking, and alarm strategy. The following section discuses this important aspect in detail.

3.4.6 Alarm strategy

Alarm strategy forms a critical part of the operator interface with the manufacturing system. In this section, alarms related to the manufacturing process will be discussed. Alarm systems provide vital support to the operators in managing a complex manufacturing process. *Alarms* can be defined as (EEMUA, 1999):

> Signals which are annunciated to the operator by an audible sound, some form of visual indication, usually flashing, and by the presentation of a message or some other identifier. An alarm will indicate a problem requiring operator attention, and is generally initiated by a process measurement passing a defined alarm setting as it approaches an undesirable or potentially unsafe value.

If these alarm systems do not work well the effects can be very serious; there are several instances of disaster (e.g., Milford Haven refinery) and near disaster (e.g., Three Mile Island incident—a nuclear plant that was close to disaster). The explosion and fires at the Milford Haven refinery (Health and Safety executive, 1997), which caused over $70M of plant damage and a major production loss, could have been prevented by the operating staff. They failed to do this partly because they faced a continuous barrage of alarms for the whole 5-h period leading up to the accident. Thus, it is critically important that a well-engineered alarm system be implemented in manufacturing.

EEMUA alarm guideline (1999) also sets the expectation for a well-engineered alarm system, according to which a good alarm should have the following characteristics:

- Relevant: not a nuisance, spurious, or of low-operational value.
- Unique: not duplicating another alarm.
- Timely: not long before any response is needed or too late to do anything.
- Prioritized: indicating the importance that the operator deals with the problem.
- Understandable: having a message that is clear and easy to understand.
- Diagnostic: identifying the problem that has occurred.
- Advisory: indicative of the action to be taken.
- Focusing: drawing attention to the most important issues.

The main purpose of a well-designed alarm system is to avoid deviation from a designed operating philosophy with the focus on safety, environment, product quality, and process capability.

Safety is of primary concern in any manufacturing operation. According to the international standard IEC 61508 (1999), an alarm system should be considered safety related if it is claimed part of the facilities for reducing risk from hazards to people to a tolerable level, and the claimed reduction in risk provided by the alarm system is significant. If any alarm system is safety related then it should be designed, operated, and maintained in accordance with requirements set out in the standard (IEC, 1999). The safety alarm system should be independent and separate from the process control system (unless the process control system has itself been identified as safety related and implemented in an appropriate manner).

Development of an environment-friendly process is a key objective during process development. Discharge and emissions are closely monitored in manufacturing, and deviations to set acceptable limit (based on regulatory guidance) are appropriately alarmed.

The deviations that result in impacting product quality and process capability have a high economic consequence. A product not meeting specification is discarded leading to factory loss. On the other hand, a process operating in a suboptimal fashion operates below its process capability leading to low productivity. The critical product parameter (see Chap. 2) is a measure of product quality. Critical product parameters are dependent on critical process parameters/critical operating parameters (see Chap. 2). The measurement of critical process parameters and setting alarm limits around the PAR are integral to a successful alarm strategy.

Figure 3.20 illustrates a basic control design. A high and low alarm is set around the control set point, the range between the high and low alarm is referred to as the operating range. The operating range should be well within the measurement range of the sensor and the PAR. A *high-high* and *low-low* alarm limit is set around PAR giving allowance for measurement uncertainty (see preceding section). If the parameter goes over the high-high/low-low alarm limit (over PAR) the product quality is impacted adversely and may lead to rejection of the material manufactured.

The range between high and high-high alarm limit and the range between low and low-low alarm limit is the window of opportunity for real time intervention by the operator or the control system to save the product. To further focus on this window of opportunity some advanced control systems have excursion-tracking capability that allows an analysis of the rate at which the excursion (or deviation) from the set point is taking place. Excursion-tracking capability helps calculate the time window within which appropriate action should be undertaken before the process parameter exceeds the critical PAR limits. Excursion tracking also provides the real time feedback on the impact of corrective action on the process parameter value with regard to the PAR limits.

Thus, engineering individual alarm limits are a critical piece of the design of an alarm system. Other aspects of a good alarm design include risk assessment (HAZOP or FMEA, see Chap. 2) to effectively prioritize the

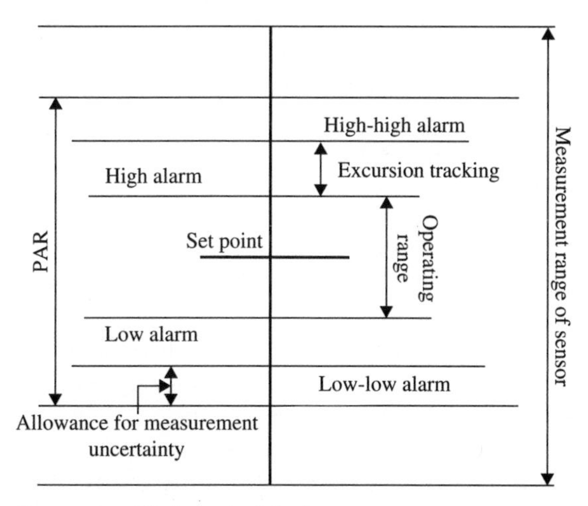

Figure 3.20 Basic control design.

alarms into high, medium, and low categories; testing of alarm sensors, hardware, and software periodically; documentation of the alarm strategy rationale and linking it to the process design documents; and allocation of roles and responsibilities for maintaining and managing the alarm system. This section briefly describes the basic alarm design principles; in Chap. 6 an advanced strategy for alarm design is discussed.

3.5 Process Control Case Study C

In selecting a case study to demonstrate issues related to feedback and more advanced forms of control, the most straightforward demonstration would be to describe a classical problem—an automatic controller in a feedback loop—highlighting deficiencies, and suggesting improvements. Appreciating problems arising at this level follows from the earlier discussions, but control loops arise not just in automated systems. The case study presented involves a control system where process operators are a critical component and as a result appreciation of the control strategy is more involved and provides a broader experience.

It is important to remember at the outset that feedback control is sufficient to handle most control problems, but this is not always the situation. The example demonstrates an all too common problem, when infrequent laboratory sampling causes feedback control to be ineffective. Nevertheless, with use of other control approaches it is possible to rectify ineffective control and deliver satisfactory performance. The stages outlined below show how the case study was progressed.

Stage 1: Process familiarization—understanding the process is paramount. The production plant manager had reported problems in adhering to operating targets and suspected that it was a result of poor control practice. The first task for the control engineer tackling the problem was to understand plant operation. Figure 3.21 shows the main units in the production line. Note that in this figure and in further discussions the timings given are used to clarify the discussion rather than being precise for the process under consideration.

Raw material is delivered to the process plant by different suppliers and stored in separate storage vessels. When demand requires, the raw materials are fed to the process plant until the whole batch is used before switching to the next batch load.

A key early finding was that raw material loads could have different quality. The extent of variation was determined on delivery by quality control laboratory tests prior to the raw material being sent to the storage vessels. It takes approximately 1 h to process a single load of raw materials. The processing involves treatment in a number of sequential unit operations and the final product is subsequently

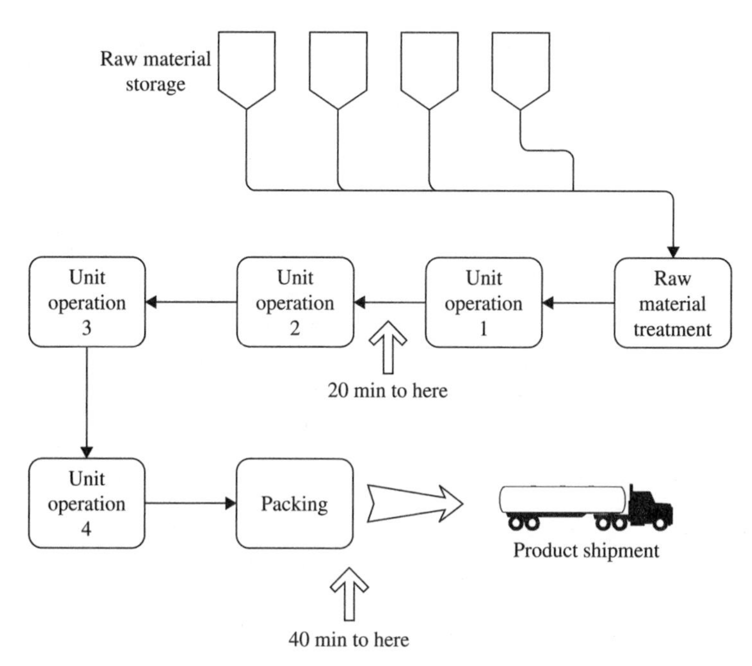

Figure 3.21 Overview of the production line.

packed for distribution to the customer. Quality control tests are carried out hourly on the final packed product to confirm that it meets customer specifications. There are two sources of information on the production: the quality control laboratory provides information on raw material and final product and the computer supervisory control system provides information on unit operation status. The paramount production objective is to manufacture product within specified targets but the target range changes with the customer.

Following the overview of the key plant features, the current plant operation was assessed. This involved looking at plant operating data and determining the current control strategy. Immediately the concerns of the plant manager became apparent. Figure 3.22 shows quality variations over a period of typical plant operation with the continuous lines signifying the target range.

Figure 3.22 gives the impression that plant performance is far from acceptable with many violations of the target bands. In this case, the plant operating target bands were set to be narrower than customer requirements to give some scope for deviation but provide a reasonable objective for the operators. Such a strategy is fine from a psychological perspective as long as target bands are achievable: in this case they were clearly not with the current strategy. Deviations from a customer perspective were not so severe as may appear at first sight, but from the

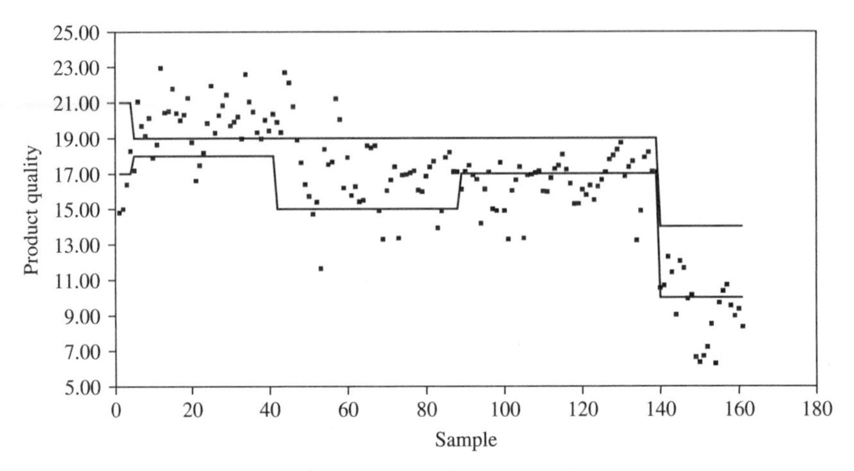

Figure 3.22 Historical trend of quality variation compared to target.

manufacturing standpoint these deviations were costing money and savings were possible. The cost issues will be discussed in Chap. 5 when the motivation for control improvement is presented.

Stage 2: Opportunity assessment—"how is it done now and can we make it better?" Although product quality is influenced by several unit operations, the main influence and therefore the prime control variables are within unit operation 2. This finding was arrived at through the knowledge of the engineering aspect of the plant and the experience of the operating staff. The operation of unit 2 is not an insignificant task and was under manual operation. Analysis of the existing control policy control revealed two important issues:

- The severity of control changes to the same deviation varied from operator to operator.
- The operators corrected process deviations using a feedback strategy based on information from the quality control laboratory.

While the first issue could be easily rectified, the second highlighted a fundamental control problem. To appreciate the control problem, it is necessary to understand the timing involved in process changes. Figure 3.23 shows the approximate process delays that exist in the processing line.

Two important conclusions can be drawn from the information presented in Fig. 3.23. First, feedback control is not a particularly effective means of controlling the process. Delays in the overall loop of 35 min at best are significant. This would occur if a sample were taken from the line and immediately a change reached the sampling point. In the worst case since product quality samples are taken only every hour, the delay could

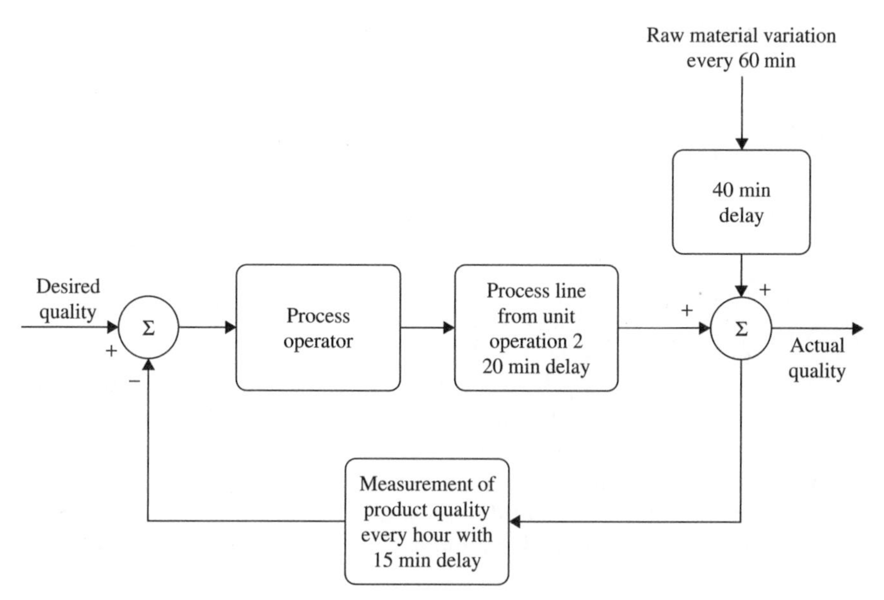

Figure 3.23 Approximate process delays in the control scheme.

amount to 95 min. When the line is producing many tons of products, this could amount to significant off-specification product. Of equal concern is that with significant disturbances coming from raw material variation, a change in product moisture takes at least 55 min to be observed. Corrective action could then be taken, but by this time a new load of raw material is fed to the line since it takes about 60 min to process a load. Such corrective action would therefore be completely inappropriate. Thus, it is clear that this scheme is fundamentally flawed.

In analyzing the existing control scheme it is apparent that the problems are a result of process and measurement delays and the sampling rate of the quality variables. Even if the sampling rate could be increased significantly, which given human resource requirements would be difficult, the fundamental problem of process delay remains. Overcoming the problem of delay requires a predictive control philosophy. With the answer to two fundamental questions, control performance could be considerably improved. The two questions are:

1. If a change is made to unit operation 2, how will the product quality respond?
2. If the raw material quality is known, can its effect on the product quality be predicted? If so, how much and when should unit operation 2 be changed to compensate for it?

If answers to both these questions could be obtained, an improved control scheme could be developed for the following reasons:

Answer to question 1. If the product is off target or a change to the operating target is required, information on how to change unit operation 2 to get the product approximately within range will avoid reliance on delayed feedback. Although predictive information will never be perfect, the predictive action will move the product quality close to the desired value, and feedback could provide fine modifications to the operation. Typically this will avoid well over an hour's worth of production potentially out of specification.

Answer to question 2. If it can be anticipated how a raw material change will influence the product quality, corrective action can be taken in a feedforward control sense to nullify the changes. It should be realized that perfect process information will not be available, but even approximate process information can serve to provide effective feedforward control, with feedback control again providing fine modifications.

The modified control strategy is shown in Fig. 3.24.

Two key control strategy parameters had to be specified for the scheme to function acceptably. First, the feedforward controller gain was determined from analysis of data produced from some simple plant tests. Observations of independent variations of unit 2 settings and raw material quality on product quality provided the necessary information to determine the feedforward controller gain. Second, inversion of the information on unit 2 settings/product quality provided the predictive

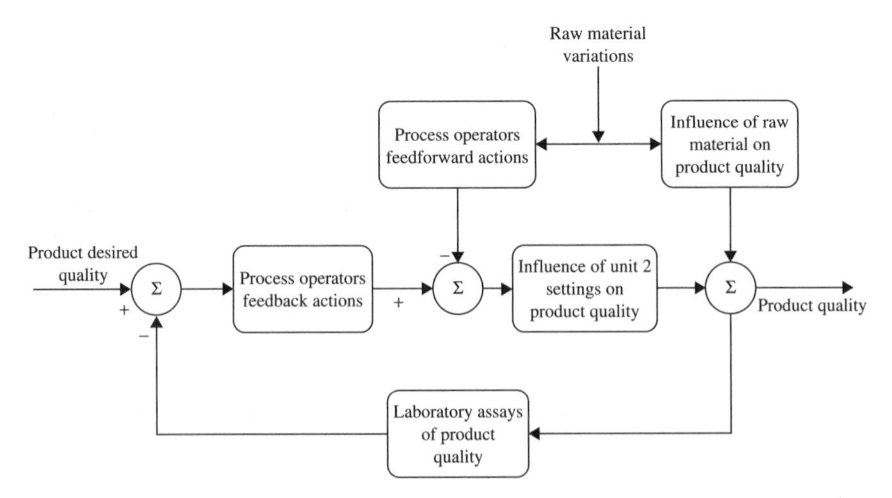

Figure 3.24 The modified control strategy for product moisture.

information to determine by how much to change unit 2 settings to correct product quality deviations.

Stage 3: Commissioning of control scheme. Trials of the new control scheme took place over a number of days of operation. From a practical perspective it is important to note that no new instrumentation was required and few, if any, extra laboratory analyses undertaken. The essential aspect of the new control philosophy was to use the available information but respond at appropriate times using knowledge of the likely outcomes of process changes. This follows from the fundamental controller design philosophy to determine where the disturbances to process operation are occurring and nullify their influence. A demonstration of the procedures can be seen in Figs. 3.25 and 3.26 that were the first trials of the new scheme. Figure 3.25 shows the effect of changes in raw material quality on the product quality.

The top left graph in Fig. 3.25 shows the product quality and the target limits as two parallel lines. Initially the quality is in the target band. Here the error bars on moisture refer to measurement accuracy alone. A change in raw material quality occurs as observed in the top right graph. No significant action was taken with unit 2 operation (as shown by the continuous line in the bottom graph). Here the previous

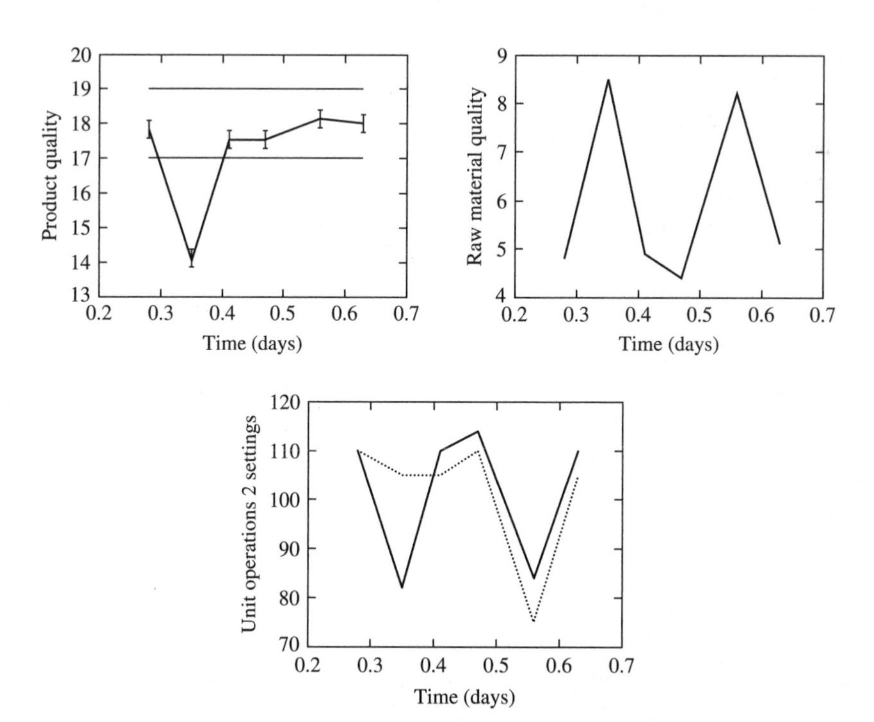

Figure 3.25 Example of raw material quality disturbance rejection.

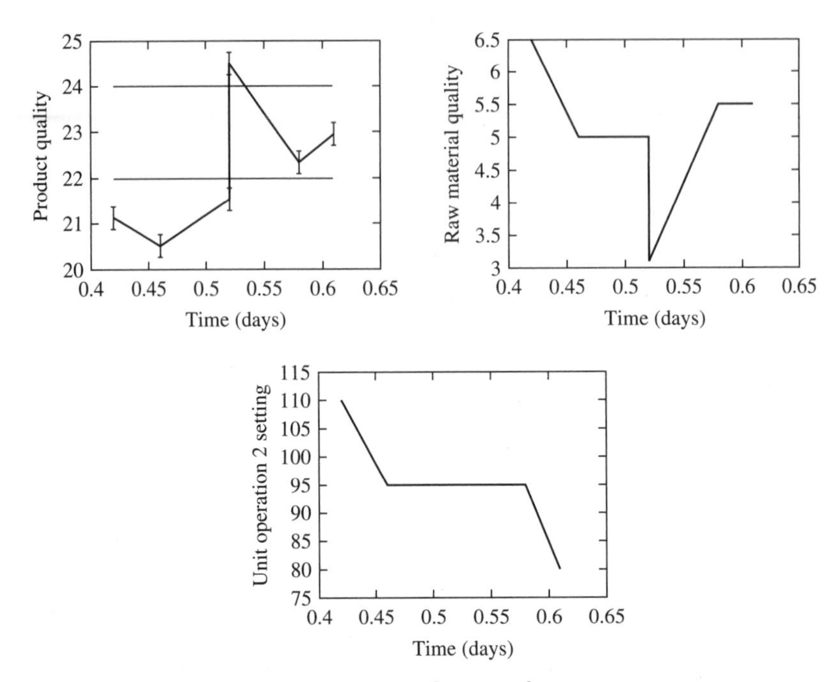

Figure 3.26 Control of product moisture when out of target range.

feedback control philosophy is adopted and as a result a significant deviation in product quality occurs, taking it outside the target range. The dotted line is the unit 2 settings as suggested by the new control scheme and in this case is not used. A further change in the raw material quality at around 0.4 days brings the product quality back within specification. At 0.55 days the raw material quality changes, but now preventative action is taken and unit 2 settings are decreased. The graph indicates that the operator took slightly more severe action than the new control scheme suggested. Only a minor change in the product quality can be observed. Again, when the raw material quality falls and unit 2 settings increase (following the new feedforward control scheme) product quality remains predominantly unaffected. These results clearly demonstrate that feedforward control can compensate for raw material disturbances effectively.

This test demonstrates disturbance rejection, but the product quality is not always within the target range. Take for example the situation shown in Fig. 3.26. This was one of the early tests carried out and some of the early control scheme modifications can be seen.

It can be seen that the product quality is initially outside the target range. Control with the new scheme commenced at around 0.47 days (11 a.m., the second sample on the graph). A fall in the raw material quality and a reduction in unit 2 settings brought the product quality up toward the target range. The predictive aspects of the control scheme are demonstrated at around 0.525 days (12:30 p.m.). The product quality was under the target range and therefore needed to be increased. A change in the raw material quality suggested that the product quality would increase by around 2, so unit 2 settings were left unchanged. In fact, the product quality increased by 3 and overshot the target range. Thus, the assumption that a decrease of 2 in solids leads to an increase of 2 in the product quality is not correct and the gain should be increased. Using this new gain, when the raw material quality increases by 2.5 at around 0.57 days, if no change to the unit 2 settings is made, a drop of almost 4 could be expected in the product quality. This would take it well below the target range. Using the feedforward control rule, unit 2 settings are decreased to result in the product quality falling within the target range.

Stage 4: Long-term performance assessment. The preceding results were obtained in a series of process tests undertaken by the engineering team in collaboration with the process operational staff. During such tests, closer attention than normal is obviously paid to the process plant operation. The worry is therefore that although plant improvements are indicated, in the long term when normal day-to-day operation resumes, without a specific focus on the new policy little additional benefit will be found. Long-term performance compared with process behavior before the introduction of the scheme is the best way to judge whether this is indeed the case. This information is shown in Fig. 3.27.

Compared with historical performance, shown in Fig. 3.18, where deviations outside of the bounds were frequent (56 percent of samples

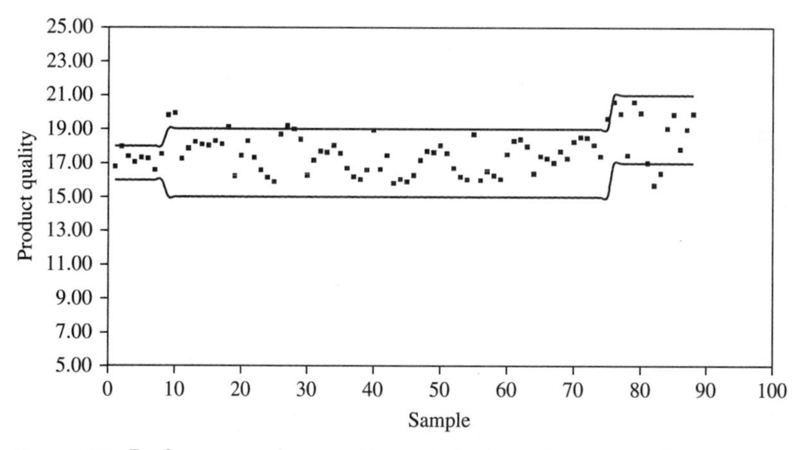

Figure 3.27 Performance subsequent to control scheme implementation.

fall outside of the bounds), with the updated control scheme performance is much improved. Much tighter regulation of the product quality is apparent (10 percent of the samples fall outside of the bounds). Slight oscillatory behavior is observed within the bounds of operation. One of the reasons for this is that raw material loads are not selected at random to go through the production line. The operators make an effort to put raw material loads of a similar quality to the previous load through the process, hence introducing the observed perturbations.

This section has followed through the procedures for assessing performance and improving the control strategy, but there is one fundamental question not answered: is it beneficial from a cost-perspective to carry out the improvements? From a business perspective the question should be broader than this. Not only must a case for improvement be built to justify investment but also the return on this investment in terms of performance improvement must be quantified and verified on project completion. This aspect of the study is considered in Chap. 6.

Concluding Comments

The control of critical process parameters, given appropriate instrumentation and a well-designed and commissioned control system, should be effective in the majority of cases. This is not to say that it always is. There are many pitfalls either with the instrumentation or with the control system that will lead to a degraded process performance. King (2003), in a paper entitled "How to lose money with basic control," describes 21 implementation problems that lead to poor control. While this paper refers specifically to petrochemical plants, there are many generic lessons to be learned. Through understanding of the capabilities and functionality of the elements of the control loop, requirements of process operation, and a maintenance program in place, pitfalls can be largely avoided.

The control procedures discussed earlier continue to be the methods of choice for the majority of processes. There are, however, situations where classical methods could be improved. Among these special conditions are significant interactions between loops, considerable time delays, nonlinearity in response between control actions and outputs, and a high-performance requirement. To tackle these issues advanced control methods have evolved. One of the most popular is continuous chemical manufacture in model-based predictive control (Qin and Badgwell, 1998). Unfortunately, it is too easy to turn to advanced control strategies if a conventional approach appears deficient rather than to consider alternative conventional control strategies. This point is recognized by King 2003 in a companion paper "How to lose money with basic controls." There are situations where more sophisticated control algorithms are justified, and these are discussed in Chap. 5, but they tend to sit above the local control loops using the technology discussed in this chapter.

An appreciation of control procedures delivers more than just improvements to local regulatory control systems. The case study discussed demonstrated that the issue of modification of process operation in light of deviation with humans being an important component of the loop. Such strategies can occur in level process operational tasks as well as low-level regulation, for instance, in modification of batch production strategies in light of observations of previous batch behavior. The effects of ineffective control structures can be more severe in high-level tasks.

Finally, process control should not be seen only as a means of complying with regulatory body specifications. When process information is used to its full potential and control schemes are efficiently designed and maintained, then process control can deliver a real competitive advantage.

References

EEMUA (1999), *Alarms Systems, A Guide to Design, Management and Procurement*, Engineering Equipment and Materials Users'Association, London, UK.

Astrom, K. J., and T. Hagglund (1984), "Automatic tuning of simple regulators with specifications on phase and amplitude margins," *Automatica*, **22**(3):277–286.JH.

Cluley, J. C. (1993), *Reliability in Instrumentation and Control*, Butterworth Heinemann in association with The Institute of Measurement and Control, Burlington, MA.

Dieck, R. H. (1997), "Measurement uncertainty models," *ISA Transactions*, **36**(1):29–35.

Gardner, E. S. and E. McKenzie (1988), "Model identification in exponential smoothing," *J. Oper. Res. Soc.*, **39**(9):863–867.

Health and Safety Executive (1997), *The Cost and Fires at Texaco Refinery, Milford Haven, 24 July 1994*," HSE, 1997, UK.

International Electrotechnical Commission (1999), "Functional safety: safety related systems," Parts 1–7, International Standard, IEC 61508.

Jazwinski, A. H. (1970), *Stochastic Process and Filtering Theory*, Academic Press, New York.

Kessel, W. (2002), "Measurement uncertainty according to ISO/BIPM-GUM," *Thermodynamics Acta*, **382**:1–16.

King, M. (2003), "How to lose money with basic control," *Hydrocarbon Processing*, **82**(10): 51–54.

Kraus, T. W. (1984), "Self-tuning PID controller uses pattern-recognition approach," *Control engineering*, 31(6) (June):106–111.

Leva, A., C. Cox , and A. Ruano (2003), "Hands on PID autotuning: a guide to better utilisation," IFAC professional brief available from http://www.oeaw.ac.at/ifac/.

Luyben, W. L., and M. L. Luyben (1997), *Essentials of Process Control*, McGraw-Hill, New York.

Qin, S. J., and T. J. Badgwell (1998), "An overview of nonlinear model predictive control applications," *Presented at Nonlinear MPC Workshop,* Ascona, Switzerland, June 2–6.

Tham, M. T., and A. Parr (1994), "Succeed at on-line validation and reconstruction of data," *Chem. Eng. Prog.*, May, 46–56.

Thornhill, N. F., M. A. A. Shoukat Choudhury, and S. L. Shah (2004), "The impact of compression on data driven process analyses," *J. Process Control*, **14**: 389–398.

Chapter

4

Knowledge Management

A successful world-class organization has two essential elements: capacity for generating IP and the capability to transform the IP into business reality.

P. MOHAN

Objective

The aim of this chapter is to demonstrate a systematic approach to knowledge management with the focus on building learning and innovative organization.

Learning outcomes of this chapter will include:

- An appreciation of critical need for building learning and innovative organization
- Intellectual property—a key business differentiator
- Knowledge management strategies
- A novel knowledge-extraction technique
- Turning intellectual property into action

The learning is then applied to a comprehensive case study to gain practical understanding of the application of the concepts.

4.1 Introduction

The objective of process development is to generate knowledge, which will become the foundation for manufacturing success (Chaps. 1 to 3). This knowledge generation is one of the most valuable investments of the corporation. Pharmaceutical companies spend 20 to 25 percent of their sales in discovery and development, which for a large size pharmaceutical company can be a few billion dollars. To maximize return from this investment it is vitally important that the knowledge be appropriately captured, integrated, and communicated. The challenge then is to successfully transform the knowledge base (turning knowledge into action) into operational start-up instructions, which is key to the overall manufacturability including control strategy (Chaps. 2 and 3).

In the postindustrial-era economy, intangible assets have become the major sources of competitive advantage, calling for tools that describe knowledge-based assets and the value-creating strategies that these assets make possible. Opportunities for creating value are shifting from managing tangible assets to managing knowledge-based strategies that deploy an organization's intangible assets: customer relationships, innovative products and services, high-quality and responsive operational processes, information technology and databases, and employee capabilities, skills, and motivation. These knowledge-based assets could be summed up as *intellectual property*. Intellectual property is the key differentiator between businesses and is the essence of modern trade. Intellectual property is generated out of learning and innovation within a knowledge-management construct. In the United States, the legal system recognizes four aspects of intellectual property, namely, copyright, patent, trademark, and trade secret.

Copyright applies to original works of authorship as soon as they are fixed in any tangible medium of expression. The author of the work initially owns the copyright, although they may transfer the rights to others. If the work is created by an employee, the author is deemed to be the employer. The copyright term lasts for the life of the author plus 70 years. For the employer or other categories of work, a fixed term of 95 years is used.

The inventor of a product or process has the right to seek a patent on his invention from the U.S. Patent Office. The patent gives the inventor the right to exclude others from making, using, offering to sell, selling, or importing the invention during the term of the patent. The patent term begins when a patent is issued and endures until a date 20 years from the date that the inventor applied for the patent.

Trademark law comprises several related doctrines that serve a common purpose: to allow buyers to reliably distinguish a source of goods or services. A trademark is a symbol used by a person in commerce to indicate the source of that person's goods or services and to distinguish them from those sold or made by others. The trademark owner acquires trademark rights by making bona fide use of the trademark in commerce. There is no time limit on the duration of the trademark.

A *trade secret* is information that has economic value as long as it is not known to or readily ascertainable by those who could gain value from its use or disclosure. A trade secret is the subject of reasonable security

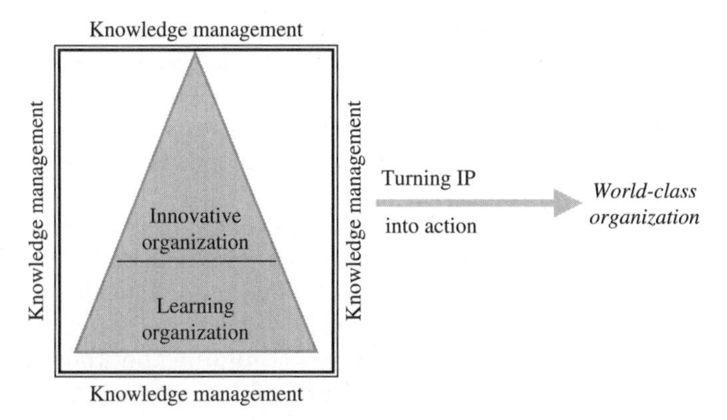

Figure 4.1 Construct for knowledge management.

measures. Typically trade secrets are manufacturing processes, computer programs, new learning, blueprints for equipment/machines and the like.

Thus, there are two key elements leading to a successful world-class organization:

1. Generating the intellectual property within the construct of knowledge management
2. Turning intellectual property into business reality

Figure 4.1 illustrates a construct for knowledge management aimed at developing a world-class organization in this postindustrial era. The building of the knowledge-management strategy requires the creation of a learning organization leading into an innovative organization. Section 4.2 focuses on explaining the concepts around learning and innovation including the concept of the learning curve and its application to manufacturing. Section 4.3 introduces knowledge management including dimensions of knowledge management, knowledge-elicitation methodologies, and the new knowledge acquisition technique (KAT). Section 4.4 highlights the key factors required to turn knowledge into action, and finally Sec. 4.5 demonstrates the application of the knowledge-management concepts for an industrial case study.

4.2 Learning and Innovation

The innovative organization can be developed from the foundation of the learning organization. The five disciplines of the learning organization are fundamental to effective innovation. The spirit of innovation is an exercise in personal mastery. "Thinking out of the box" is an exercise in challenging mental models. To develop a truly innovative organization, people should share the same vision for innovation. They should work closely and find synergy in teams because a team is more powerful than the sum of individuals. Systems thinking allows the organization to innovate in the areas where the highest payoff can be reaped and develop the platform to sustain innovation.

Innovation ensures the longevity of a business. Manifestation of organizational knowledge into innovative (breakthrough) products and services is a very important factor in the growth of a business. Knowledge is the fundamental productivity metric. Knowledge is gathered throughout the evolution of the manufacturing process. Building the knowledge base and the learning curve take years of investment, and preservation of the knowledge for posterity is critical for the survival of the business. A good knowledge-management strategy is becoming a prerequisite for world-class organizations; this includes training as a planned activity for knowledge enhancement.

According to Huber (1991), "An entity learns if, through its processing of information, the range of its potential behaviors is changed." A learning entity can be an individual or a collective. Huber defines learning as follows: "Learning is a knowledge-creation process in which information perception and interpretation lead to a change in the range of an entity's potential behaviors." Kolb (1996) argues that the tendency to define learning by its outcomes may become a definition of nonlearning. This can be understood by the behaviorist axiom, according to which the strength of a habit can be measured by its resistance to extinction. Thus, the better a person has learned a given habit, the longer he or she will persist in behaving that way when it is no longer rewarded.

In examining learning processes Crossan et al. (1999) use a framework that involves three levels of organizational learning processes: individual, group, and organization. The framework links three levels of learning through four subprocesses: intuiting, interpreting, integrating, and institutionalizing. According to Crossan et al. (1999), intuition is clearly an individual phenomenon; ideas come to a person's mind, whether it happens in a group or in an organizational context. Crossan et al. (1999) have two views to intuition: the expert view is a process of past pattern recognition, whereas entrepreneurial intuition is about innovation and change. Individuals use visions and metaphors in explaining their intuition to themselves and others. Individuals build up a picture of one's environment and of one's self as a part of it, based on one's earlier experiences. Interpreting on the group level is a social activity that creates common language, shared meaning, and understanding. Integrating focuses on coherent collective action. Language and storytelling preserve what has been learned. Stories themselves become storage of both individual and collective/organizational knowledge. Institutionalizing is "the process of ensuring that routinized actions occur" (Crossan et al., 1999). What is learned by individuals and groups becomes embedded in organization structures and processes, and hence starts to guide new spontaneous individual and group level learning.

The constructivist perspective on individual learning considers human beings as active, goal-oriented, and feedback-seeking. The individual learning process is determined by one's needs, intentions, expectations, and the feedback perceived by the individual. This does not mean that all learning should be intentional or even conscious, though. Neither does learning always increase the learner's effectiveness, nor does it result in observable changes in behavior. Intuiting is a largely subconscious process and the links between experience and consciousness are complex (e.g., Crossan et al., 1999; Nonaka, 1996; Kolb, 1996). Thus, studying the individual processes of learning in an organizational context is not a simple task. It can be very frustrating, as it was in the beginning of this research. Only a little goal-oriented, "visible" learning seemed to take place.

Conflicts and confusion are efficient initiatives for learning. Individual learning processes are conducted by emotions and feelings, and individuals in an organization create their own interpretation of perceived information. Huber (1991) discusses whether organizational learning should be defined in terms of commonality of interpretation or in terms of the variety of interpretations made by an organization's various units. According to Huber, more organizational learning has occurred when more and more varied interpretations have been developed. Also Kolb (1996) shows that opposing perspectives are essential for optimal learning, and that learning effectiveness is reduced in the long run when one perspective becomes dominating.

Making individual knowledge available for others should be a central activity in the organizational learning process, according to Nonaka (1996). The importance of knowledge sharing must be recognized by all members of the organization—both those who have something to share and those who need the information. Without such recognition the knowledge will remain the individual's private property and never go on to the collective-learning level.

Albert Einstein once said, "The significant problems we have cannot be solved at the same level of thinking with which we created them." To build a lasting learning environment, organizations must begin early by clearly defining what it means to be a learning organization. Growing entrepreneurial companies have a distinct advantage in this regard because their existence hinges on active learning and constant knowledge acquisition. This means that they learn faster and avoid confronting bigger problems later. Though unstructured, these firms have what it takes to build a strong learning culture. What they lack is a clear and functional method of learning, with easy-to-apply management guidelines. Many businesses do not understand how to learn and either fail or remain marginal competitors. Companies that successfully implement learning strategies have the best chance to thrive.

Building a learning organization is also about acquiring knowledge from the external environment and bringing it into the organization to be used to adapt and make changes. This results in a circular process whereby information is constantly fed into organizational processes. Changes are made and monitored with new knowledge continually being fed into the process. This ensures that change and continuous improvement are constants in the firm. No organization today operates in a vacuum. The external environment provides the framework within which the organization conducts its business. The successful firm constantly has its organizational ear to the ground. This organization is reading all the handwriting on the wall outside the organization—recognizing trends in the economy, society, technology, the political-legal arena, and the industry. These trends are monitored to ensure that the organization understands how each of these trends is likely to impact the organization, its industry, and its workforce. A learning organization is capable of aligning its strategic objectives and vision with the capabilities, competencies, and ideas of its employees. Managers within a learning organization seek to create an environment where their employees realize their maximum potential.

4.2.1 Learning curve

The learning curve is the graphical representation of the learning-by-doing phenomenon observed in people performing manual tasks. As the task is repeated, the individual gains experience. The experience stems from becoming familiar with the basic task, becoming familiar with the procedure and interacting with other objects, improving manual dexterity, developing shortcuts to the task at hand, and reducing occurrences of stop and start actions caused by errors in quality. During the 1920s, Wright (1936) gathered manufacturing data to develop the doubling effect theory, which suggests that as the quantity produced doubles, the resources needed are reduced to a percentage of the original requirements.

A composite learning curve is formed when the learning curves of several elements of a single job are summed. There are various strategies for enhancing such a learning curve of an organization. These include:

Step 1: Keep it simple. The first rule in building a learning culture in an organization is to keep things simple. The most effective strategy is one that makes certain that everyone clearly understands what is involved and what is expected from employees and managers.

Step 2: Clearly define strategic objectives. This will ensure that the learning strategy is effective. Bring the management team and the advisors together to clearly define the company's vision and, working backward, outline the critical steps required to attain it.

Step 3: Set up a learning committee and policy. Bring together employees and managers who appropriately represent the scope of the organization. Ideally, the committee should be composed of not more than five to eight individuals and each should have an equal say in the development and implementation of the training policies.

Step 4: Take inventory of existing knowledge and competencies. This is an involved two-step process. First, identify each position in the company and determine the skills and competencies required to effectively complete the responsibilities. Second, document the skills and competencies of the employees currently in these positions.

Step 5: Identify the skills and competency gaps. Once the knowledge inventory is completed, determine the knowledge gaps that exist between the employees' competencies and the skills required to reach the strategic objectives.

Step 6: Determine employee's personal goals. Talk to each employee individually. Determine their personal aspirations within the organization and in life. Help them help you accommodate their needs. This will gain a dedicated and motivated worker. Also, this will help figure out if the individual is a good fit for the company.

Step 7: Develop a training plan (Chap. 2). At this point you know where you want to go and what competencies you need to get there. With this information you are now able to source appropriate training to resolve the identified knowledge and competency gaps. Business people often, mistakenly, equate training with learning. Although these terms are often used interchangeably, they are distinct concepts. Training is instruction for learning a specific task; learning, on the other hand, is a continuous process. Training is one of the many tools used to build a learning environment; learning encompasses an individual's acquisition and assimilation of experiences, information, and daily activities.

Step 8: Integrate learning into daily activities. Implement tactical methods to encourage learning among employees and to support employees with new or more complex responsibilities. Coaching and mentoring are two ways to make the learning process more interesting and an integral part of the company's strategy. Continue to measure, document, and manage the results of these processes.

There is no doubt that a learning organization provides a safe place to take risks and to develop new ideas, behaviors, and the challenge to stretch beyond perceived limits. Everyone's opinions are valued and the amount that people can contribute is not determined by the position they occupy in the organization. Employees at all levels will find it more enjoyable to work in and on the business because it encourages creative ideas and gives people more control on outcomes and the ability to make things better.

4.2.2 Learning curve applied to manufacturing

The knowledge gathered from process development manifests itself in a set of operating instructions contained in process flow description, recipes, manufacturing tickets, *standard operating procedures* (SOPs), and operator training (see Fig. 4.2). These instructions help define the operating space with boundaries and constraints. It is also important that the equipment be maintained in a qualified state to deliver a capable manufacturing process (Chap. 2). A deviation may result from both operational and/or equipment error. Any deviation from the set operating boundaries of process and equipment will lead to an inability to give assurance of the final quality of the product. Quality regulatory (government) agencies ensure that the manufacturing industries have appropriate safeguards in place to contain and investigate such deviations. For example, in the pharmaceutical industry, a paperwork trail is required to investigate deviations, and more importantly appropriate investigation is required to guarantee that the product quality is not impacted.

4.2.3 Information technology application and knowledge flow

The objective of process development is to generate knowledge, which will become the foundation for manufacturing success. This knowledge

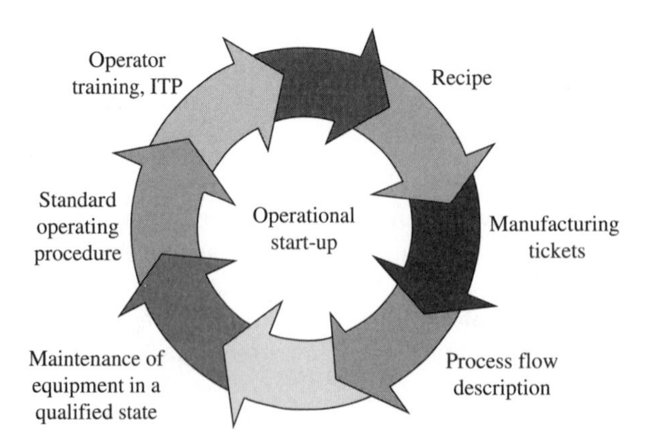

Figure 4.2 Manifestation of process development knowledge.

generation is one of the most valuable investments of the corporation. To maximize return from this investment it is vitally important that the knowledge be appropriately captured, integrated, and communicated. The advancement in the areas of information technology must be engaged in advancing the learning curve of the organization. An IT strategy could be developed to maximize the learning opportunities; one such strategy is illustrated in Fig. 4.3. A simple IT strategy could consist of three layers: the data layer, the analysis layer, and the knowledge-based decision-making layer.

The data layer could be server-based documents, files, and common data. The repository must provide interfaces to document management systems and ERP systems (e.g., SAP), as well as data on the physical properties of materials, equipment specifications, process descriptions, models, expert systems rules, experimental data, historical, and real-time process data . There could be five avenues (complementary) for capturing and sustaining the valuable knowledge base (Fig. 4.3) namely, laboratory data, pilot-scale data, document management, process models, and knowledge extraction. The document management could include key commercialization documents such as the *integrated process development report* (IPDR) and process flow document.

Integrated process development report (IPDR). During process development a lot of activity occurs concurrently in a multifunctional, multisite setting. For example, for case study A in Chap. 1, optimal strain development, media development, and the like will be carried out by scientists (with involvement from engineers); on the other hand developing scale-up/scale-down principles will be carried out by engineers (with involvement from scientists). The formulation of downstream process development (recovery, purification, and packaging) may be happening

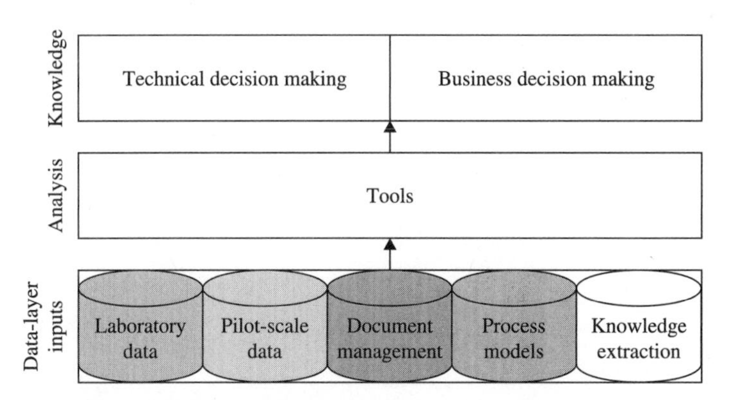

Figure 4.3 IT strategy for knowledge flow.

in a different part of the organization. Numerous detailed technical reports will be generated from these development activities. The IPDR should be a comprehensive summary of the process development of all the process units and should reference all the technical reports. It should also include the technical reports on learning from failures of some experimental programs (though experimentally valid, did not deliver expected outcome) to avoid repeating such experiments in future. The IPDR provides an overall view of the process development including the scale-up/technology transfer aspect spelling out the critical control strategy (Chap. 3) for the manufacturing process. It is important that the IPDR be written for an audience of both technical and nontechnical individuals, including regulatory agencies and management. For example, for case study A, the IPDR would include the summary of Fig. 4.4.

Process flow description (PFD). The PFD is a science and engineering document, which is the foundation for the process and equipment design. It includes process flow charts, the equipment description, the recipe for making the product, and the operating parameters (including critical process parameters and critical operating parameters). The foundation of the PFD is in the IPDR and the technical reports generated by late stage process development activity. The details about the PFD are discussed in Chap. 2.

Modeling. Process modeling offers a structured avenue to capture the knowledge (Chap. 2) on a software platform. It also offers avenues for building the knowledge base to enhance process performance and capacity. One of the major sources of knowledge is data, and developing empirical models from data using a structured platform is key for propagating a sustained process improvement environment. Chapters 6 and 7 of this book highlight some data analysis techniques that can be used in the process development stage.

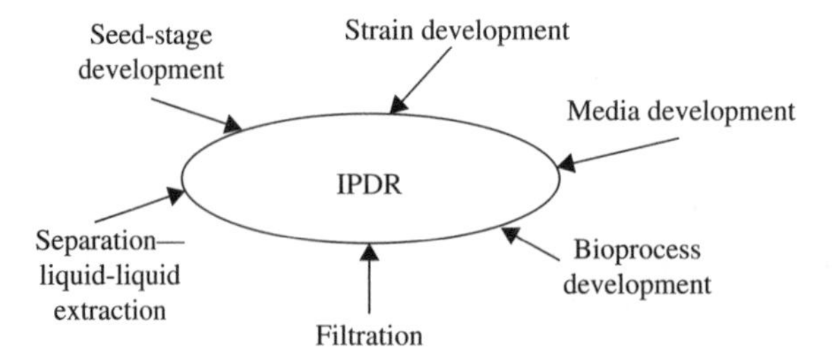

Figure 4.4 IPDR content example.

Knowledge base of new technologies. Exploring new technologies is an integral part of innovation. Implementation of new technologies in process development and manufacturing can lead to competitive advantage, cost reduction, enhanced speed to market, and ability to develop innovative products. A new equipment design aimed at improving process capability, capacity, and product quality can benefit the bottom line of the manufacturing organization. These new technologies can provide the much needed breakthrough improvement for the competitive advantage. Breakthrough improvement is further discussed in Chap. 8.

Knowledge extraction. An important aspect of building the knowledge base is capturing knowledge from a large group of experts around any critical process design or operation issue. The intention of this knowledge base is to generate the operating strategy and development of some "if then" or "if then else" rules to support the manufacturing operation. This aspect will be discussed in detail later in this chapter.

Tools. A broad array of powerful tools is available, tailored to facilitate the different types of analysis and streamline the data handling needed for the chosen work processes. This includes chemical route selection, design of experiments, data collection and analysis, equipment design, production planning and scheduling, process modeling and simulation, materials management, and operating instructions. The "computing engines" in the tools should include expert systems technology to deploy "rules of thumb" for process scale-up and equipment design, check that equipment contents are compatible with materials of construction, examine reactive chemistry issues for mixtures, use heuristics to develop good schedules, and the like. Some of the tools are discussed in Chap. 2, Sec. 2.8.

Knowledge-based decision making. This underlying data and analysis tool results in generating knowledge relevant to the decision-making process. A shared repository for common data ensures that proper information is used and eliminates transcription effort and rework. Collaboration replaces isolated, sequential decision making. Intellectual capital is preserved and leveraged, improving the quality of decision-making processes. Special purpose tools are designed to streamline important tasks and provide decision support. For example, for technical decision making, using modeling simulation (Chaps. 1 and 2) in concert with experiments can reduce the number of runs needed to find the best process parameters. Business tools (e.g., supply-chain models) can assist in making critical business decisions on location of plants, suppliers, and the like.

4.2.4 Enhanced learning curve and innovation—enlightened experimentation

The ability for rapid experimentation is integral to innovation. As researchers conceive of a multitude of diverse ideas, experiments can provide the rapid feedback necessary to shape those ideas by reinforcing, modifying, or complementing existing knowledge. Thomke (2001) introduces the concept of enlightened experimentation—the new imperative for innovation. The essentials for enlightened experimentation include: engage new technologies in experimentation (e.g., rapid methods for testing); organize rapid experimentation; fail early and often, but avoid mistakes; anticipate and exploit early information; and combine new and traditional technologies.

New technologies such as computer simulations and modeling (virtual experimentation) not only make experimentation faster and cheaper, but they also enable companies to be more innovative. There has been a trend to implement rapid screening tools that can enable testing of novel ideas in an aggressive time-line.

Rapid experimentation requires an organization to create small cross-functional teams (scientists, engineers, operators, and so forth) with all the knowledge required to plan, conduct, and analyze the experiments. Parallel experimentation can further assist in rapid experimentation; parallel experiments are most effective if time is more important than money.

Experiments led by out-of-the-box thinking are crucial to innovation. When evaluating numerous such ideas it is important to fail early and often—"to fail often to succeed sooner" (Thomke, 2001). Minimizing variability is an important aspect of experimentation and can help evaluate the ideas with fewer experiments. Chapter 5 discusses variability in greater detail.

Experiments should be carefully planned so as to include gathering critical dataset followed by rigorous data analysis engaging new technologies—fully exploiting early information. New technologies can provide the greatest benefit by identifying and solving problems early in the development cycle, best described as front-loaded development. This is critically important for pharmaceutical sectors as a late failure of drug development could be very expensive.

The true potential of emerging technologies lies in their successful integration with the existing technology platform. For example, in the pharmaceutical sector the new technology of combinatorial chemistry has significantly enhanced the capability for drug screening; however, it must be integrated with the existing technologies for further experimentation of successful leads.

For enlightened experiments in addition to the earlier discussion, it is critically important to engage rigorous statistical methods for generating reliable dataset. Engagement of *design of experiments* (DOE) to study critical variables is an important aspect of enlightened experimentation. DOE assists in optimizing the experimental space followed by statistical analysis to maximize generation of new knowledge. Chapter 8 discusses DOE in greater detail.

Knowledge (leading to intellectual property) is the cornerstone of a learning and innovative organization. Having the knowledge itself does not provide a competitive advantage. Instead, the source of an organization's competitive advantage stems from the firm's ability to effectively manage knowledge to develop new products and services, or make important improvements in the way business is conducted. Knowledge management is critical to organizational success. The growth of positions such as *chief learning officer* (CLO) and *chief intelligence officer* (CIO) reflect the importance of knowledge management today.

4.3 Knowledge Management

It is the acquisition of knowledge and its subsequent management within the firm that can be the source of an organization's competitive advantage. Both knowledge acquisition and knowledge management have become more important as core competencies for organizations across all industries. The effective management of knowledge ensures that everyone has the information they need, can apply it, and then share it with others.

It is not a question of quantity but rather quality when it comes to knowledge. The amount of information is not as important as whether the organization has the "right" information, that is, quality is more significant than quantity. In reality, large amounts of information can actually be counterproductive for the organization. With more information, an information overload may occur and the sheer volume can become so overwhelming that the relevant information is lost in the maze of irrelevant information that must be sifted through. It is not always feasible to weed through these unwieldy amounts of information and "good" information gets lost.

The dissemination of this knowledge also involves information sharing throughout the organization, across organizational departments and divisions. More and more in today's environment, knowledge is transmitted farther down the organizational pyramid and across department lines. Learning that occurred in one division should be shared with other divisions so that they can benefit. With effective emphasis on knowledge management this sharing occurs.

Information technology has further changed the way information is shared and disseminated in organizations today. Traditionally, information flowed along the chain of command in most organizations. Today, with e-mail and modern technology, information is more likely to be shared with those who most need it, regardless of organizational position.

The Internet enables organizational members to learn what other organizations are doing and to benchmark best practices. Then company intranets enable employees to share that information with others inside and outside the organization who may find it useful. Organizational members can even communicate better with and solicit information from customers and suppliers in a two-way exchange of information.

The effective acquisition and management of knowledge enables an organization to be proactive versus reactive. If the organization acquires early signals of shifts in customer tastes, this information can be used to alter product offerings. In some cases, early warning systems can enable the firm to enjoy first mover advantages.

An integral part of the acquisition of knowledge in building a learning organization is benchmarking the best practices. Firms must be committed to continuous improvement. A large number of now defunct companies learned too late that the seeds of their success often grow the seeds of their downfall. That is, when a company learns to do something extremely well and is able to develop a core competency in that area, the firm fails to make changes, believing their future success depends on continuing to do things in the same way.

4.3.1 Dimensions of knowledge management

It is important to identify the knowledge management engine that drives the overall learning dimensions within the organization. Figure 4.5 illustrates such knowledge management dimensions for

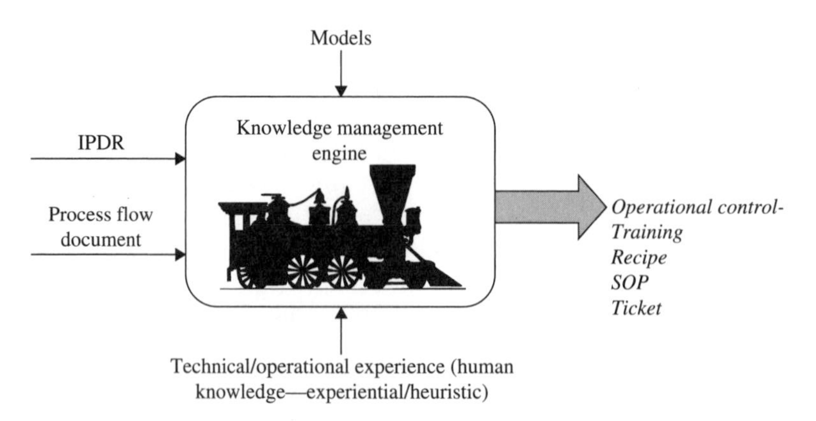

Figure 4.5 Knowledge management dimensions.

the manufacturing process. Operational control sets the boundary of a capable process. The control includes: training for individuals associated with the manufacturing operation, recipes (to transform raw materials into product), standard operating procedures, and manufacturing tickets. A key aspect of the knowledge management capability is to successfully transfer the process development knowledge, which includes the IPDR, models, process flow description, and the technical/operational experience into operational control. This section focuses on the most valuable yet less-understood aspect of the knowledge management engine, which is an extraction of experiential/heuristic knowledge. While there could be numerous ways to achieve this, a new technique is discussed with a case study.

Previous chapters have described the process development and the design stages leading to the specification of a process operating procedure. Throughout this process, data and information about the process have been collected and knowledge formed about the fundamental relationships between the process variables as well as the interactions between the system and its environment. This information and knowledge about the process are typically collected in a variety of forms, such as IPDRs and databases containing process measurements during the development runs. Most of these are in the form of documents and can thus be directly translated into specification sheets, such as standard operating procedures and manufacturing tickets, required for consistent large-scale manufacturing. However, an important source of knowledge is often either overlooked or not taken full advantage of during the SOP specification, that is, the knowledge of people involved in the process development up to this point. While it is common practice that technical operations personnel responsible for scaling up the process oversee its start-up at a large scale, it is important to capture the knowledge of all the staff involved in the process development.

Compiling and sharing best practices of the manufacturing operation is critical to operator effectiveness. This is especially important for manufacturing where there is a high turnover rate and a constant influx of new operators. Typically, one of the major causes of process deviation is "operator error." In a typical manufacturing operation it is generally observed that a shift having experienced operators leads to lower process deviation than a shift with less-experienced operators. Acquiring knowledge from expert operators and making this available to less-experienced operators is important for enhancing operator effectiveness, thus reducing deviation as well as variability in operations. This chapter focuses on techniques to capture process knowledge to develop an effective operational start-up package. An industrial example further illustrates the value of such an approach.

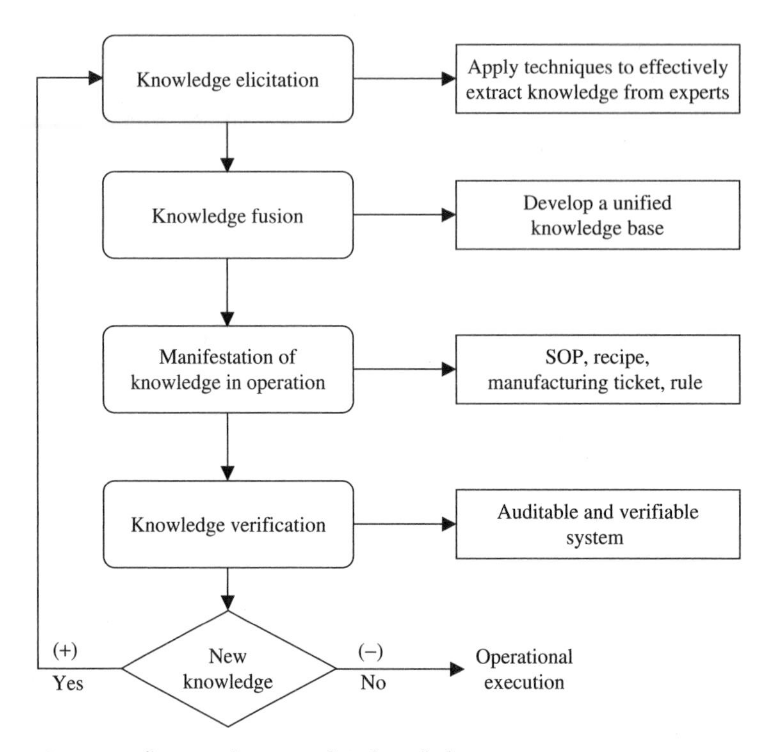

Figure 4.6 Systematic approach to knowledge management strategy.

A structured approach to knowledge management maximizes business return from investment during late-stage commercialization. A knowledge engineer could play an important role in transforming the historic knowledge (IPDR, PFD, and so forth) along with the human experts' knowledge into an effective operating strategy.

The first step is knowledge elicitation (see Fig. 4.6), which is a structured methodology of extracting knowledge from experienced individuals. The knowledge elicited from numerous individuals is then fused into a unified knowledge base. The next step is to manifest the knowledge base into manufacturing operations in the form of SOP, manufacturing recipes, instructions, and improvement opportunities. Following knowledge verification this knowledge base is transformed into action. The knowledge base must be periodically updated with new knowledge so that the operating practices remain current.

4.4 Knowledge Elicitation

Traditionally, capturing knowledge from humans (often referred to as knowledge elicitation) was considered the major bottleneck in developing any system relying mainly on human knowledge. Indeed, it has become almost a part of folklore to describe the experts who are unable to verbalize their knowledge (Berry, 1987) completely and accurately. This could be due to the fact that, in many cases, experts seem to perform almost automatically or intuitively and thus find it difficult to explain their line of thinking. They may simplify or distort their knowledge when conveying it to a nonexpert with limited understanding of the domain. In some instances, when the knowledge of experts is tacit they tend to offer a textbook account during elicitation. The examples of difficulties in knowledge elicitation could be elaborated on several pages or more. Although some of these may be considered less relevant for any particular elicitation method, without doubt that the greatest challenge at this stage of "passing on the acquired process knowledge" is the elicitation of required knowledge from the most appropriate experts and its subsequent representation. This inevitably means that developing and improving methods of knowledge elicitation has attracted much attention and is the subject of significant research activity. In this section an outline of the major techniques is presented to point out their potential and limitations from a business perspective. There are a number of books dealing only with knowledge elicitation and acquisition, for example, Schmalhofer (1998) and Ursino (2002).

Elicitation techniques are classified (Fig. 4.7) on the principle of obtaining the knowledge in two broad categories:

1. Interview-based elicitation
2. Alternative methods of elicitation

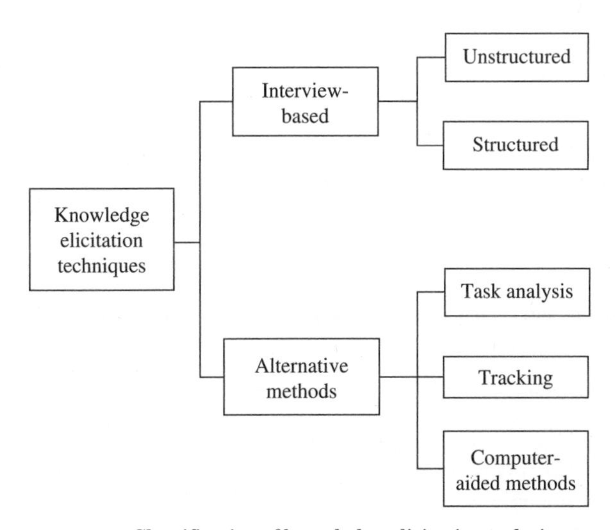

Figure 4.7 Classification of knowledge-elicitation techniques.

The interview-based technique could be either structured or unstructured, while the alternative techniques include task analysis, tracking, and computer-aided methods. This classification is unlike those offered in most of the expert system textbooks and papers (Cooke, 1994).

Before delving into individual elicitation methods, it is worthwhile to establish some goals for judging their suitability for an industrial application. The reader should note that a number of these techniques have evolved from research into human thought processes and have predominantly been developed by psychologists. Thus, they are geared toward collecting knowledge rather than representing it in a usable form from a computational point of view. Ross et al. (1998) suggested the following key criteria that an ideal elicitation technique should meet (see Fig 4.8).

This list was compiled with an objective to maximize the efficiency of knowledge elicitation for computer applications, which is also often highly relevant in later stages of process operation.

4.4.1 Interview-based elicitation methods

The most commonly used form of elicitation is the interview. It involves a direct interaction between the expert and the knowledge engineer in the form of questions and answers. Interviews can be performed without any additional equipment, or with the aid of a variety of instruments

Key objectives to be satisfied by an ideal knowledge-elicitation technique as suggested by Ross et al. (1998):

1. The technique provides a scrutable and auditable knowledge structure.
2. The elicitation session output is comprehensible and communicable between nonknowledge engineers.
3. The technique must provide a complete and exhaustive capture of the expert's knowledge.
4. The session output must be represented in a machine interpretable form.
5. The original output must be provided in a form that can be updated and maintained.
6. The technique must be domain independent.
7. The technique must enable the knowledge engineer to be domain independent.
8. The expert should need no experience of knowledge elicitation.
9. The technique should enable the identification and aid the removal of ambiguity and redundancy.
10. The technique should enable the identification of a lack of deep knowledge or understanding and areas where gaps may be present.

Although not exhaustive, this list provides a sufficient starting point for the assessment of the industrial applicability of these techniques.

Figure 4.8 Key criteria for knowledge elicitation.

ranging from conventional tape recorders and questionnaires, to sorting cards or repertory grids implemented either as physical objects, or using computer technology. In line with the classification proposed in Fig. 4.6, any technique requiring a question-and-answer type of interview with the expert for the first stage of elicitation (i.e., the construction of the first attempt of the domain knowledge map) will be included as a part of this category of methods.

Unstructured interviews. Unstructured interviews are a free form of interviews of experts by knowledge engineers, in which neither the content nor the sequencing of the interview topic is predetermined. These interviews do not require or assume domain knowledge on the part of the knowledge engineer, so they are helpful in getting a broad overview of the domain. Unstructured interviews give more control to the experts who can cover the domain as they see fit. However, on the negative side the unstructured interview does require some training at drawing out or facilitating the conversation and conducting a successful interview. In addition, this method is typically inefficient, in that it produces unwieldy data. The acquired data are often unrelated, exist at varying levels of complexity, and are difficult for the knowledge engineer to review, interpret, and integrate. Perhaps the most important shortcoming is that unstructured interviews do not facilitate the acquisition of *specific* information from experts.

If the knowledge engineer is not skilled in unstructured interviewing techniques, the knowledge base may not be complete or consistent. Typically, there is little training available to knowledge engineers and they often learn from observing others during eliciting. This lack of training often leads to knowledge engineers appearing disorganized during the interview and may cause the experts to lose confidence in them.

With regard to the criteria introduced in Fig. 4.8, this technique would not be the preferred elicitation method for industrial applications. One could argue that only the domain independence of the technique and the knowledge engineer (criteria 6, 7) and the fact that the expert needs no knowledge of the elicitation technique (criterion 8) are fulfilled by this elicitation technique. Indeed, unstructured interviews are mainly used to quickly assess the basic structure of the domain. The full elicitation is subsequently performed through a more formal technique.

Structured interviews. In a structured interview, the interviewing process follows a predetermined format (Schweikert, 1987). Therefore, the knowledge engineer, the questions, and the expert's answers are more restricted than in less structured interviews. Structured interviews are thought to have a more complete coverage of the domain than unstructured interviews. Also, since the information requested is more explicitly defined, this seems to be a more comfortable method not only for the knowledge engineers but also for the experts. The structure reduces the interpretation problems inherent in unstructured interviews and it allows the knowledge engineer to prevent the distortion caused by the subjectivity of the domain expert. For instance, gaps in knowledge can be better identified when systematic questions are asked. On the other hand, the main disadvantage of the structured over the unstructured interview is that the former requires more preparation time and domain knowledge on the part of the knowledge engineer, who determines the interview format.

A range of techniques is traditionally classed as structured interviews. We extend this to include others in which the principal method of eliciting knowledge still remains a structured interview although it is conducted with additional aids. A short overview of some of these techniques is as follows:

1. *Forward scenario simulation.* This technique makes use of simulation to focus on a case. The knowledge engineer presents the expert with an initial situation or a problem and the expert in turn verbalizes the steps he uses for problem solving, describing the relevant concepts and the steps involved in that case. The selection of a case is critical to the success of this technique. The greatest limitation of this technique is the amount of knowledge required of the elicitor to describe the situation in explicit terms familiar to the expert (Cooke, 1994). This method is further limited by the ability of the experts to verbalize their knowledge.
2. *Case study analysis.* This technique focuses on specific experiences. Case study analysis involves simulation in which the knowledge engineer goes through the case step-by-step with the expert. The limitation of this technique is that many of the cases involve old memories and the expert may not remember the exact solution and may tend to reconstruct it in some way. Further, this technique does not guarantee a complete sampling of the cases in the domain (Cooke, 1994). It should be used in conjunction with other techniques, since this incomplete sampling can lead to the most routine cases being missed.
3. *Goal decomposition.* In this technique, the domain is considered as a hierarchy of goals. The knowledge engineer starts with one objective (described by the expert), and by questioning the expert determines how this objective relates to others and finally to the primary goal. In effect, it involves the expert working backward from a single goal, to the evidence that leads to that goal. In this way, the interview is guided toward the construction of specific rules associated with each goal. The technique is not very useful in domains that are

not well structured. The knowledge engineer is left to determine how objectives relate to the primary goal, that leaves the elicitation open to significant bias from the interviewer's point of view.

Cooke (1994) describes a variation of the goal decomposition, *laddering*, which differs by not restricting the concepts to goals and subgoals. Using this technique, a hierarchy of domain superordinates, subordinates, and attributes is constructed in a graphical form. The result of this technique is taxonomy of domain concepts. This technique has common origin with repertory grid analysis (discussed later in this section) and is often used in conjunction with it to expand on either the objects or the attributes in the grid. Corbridge et al. (1994) describe three successful applications of laddering in knowledge elicitation and claim that it was more productive than other techniques investigated (protocol analysis, interviews, and card sort—all described in this chapter). However, the taxonomic focus of laddering makes it inappropriate for domains that are not hierarchical.

4. *Twenty questions.* The elicitor selects a situation, diagnosis, fault, problem solution, or state and the expert has to try to ascertain the specific concept by asking "yes/no" questions (Welbank, 1990). Information about problem-solving sequences and strategies is revealed through the expert's questions. The limitations include the requirement for the knowledge engineer to know enough about the domain to be able to answer questions correctly (the alternative is to replace the knowledge engineer by another expert). It can be a more time-consuming technique and may result in more interpretation disagreements than other techniques (Schweikert et al., 1987).

5. *Likert scale technique.* In this knowledge-elicitation technique statements (Likert scale items) about the task are presented to the expert who rates them along a scale. The points along the scale are associated with categories in some implied order. A typical example could be statements in customer survey questionnaires with which the respondent agrees/disagrees to varying degrees.

6. *Cognitive interviews.* This technique has been developed and extensively tested for use in criminal and medical investigations, specifically to capture episodic knowledge (Fisher and Geiselman, 1992). It involves interviews conducted to assist the expert in recalling events that utilized their expertise, by encouraging the expert to think back to the original event, recalling the physical and physiological surroundings. Moody et al. (1996) highlighted the major advantages of this technique compared to card sort, repertory grid, and other elicitation methods that may leave out important aspects of the knowledge domain under review due to predetermining what will be useful. They believe this technique is not only useful for case-based

knowledge elicitation, but may prove useful as the initial survey tool of the domain before other elicitation techniques are applied. However, one of the major drawbacks is the inability to ascertain the accuracy of the elicited information.

7. *Repertory grid analysis.* Repertory grid is a method of capturing declarative knowledge, and it attempts to capture the underlying constructs that support the decision-making process of humans. The foundation of this method is the personal construct theory developed by the psychologist George Kelly (Turban, 1992). The process of elicitation is summarized as follows:
   - The expert identifies the important *objects* (*elements*) in the domain of expertise.
   - The expert identifies the important *attributes* (*constructs*) that are considered in making decisions in the domain.
   - For each attribute the expert establishes distinguishable characteristics and their opposites.
   - The expert is asked to identify the attributes and characteristics that distinguish any two out of three presented objects from the third.
   - The answers are recorded in a grid for all the objects.

This technique helps the knowledge engineer understand how the expert looks at the domain, but does not lead to the production of clear rules in any direct manner. The attributes may not be equally applicable to all objects and thus judgments may seem awkward or forced. Likewise, the objects and attributes can vary in the level of abstraction and may be difficult to compare. Additionally, the resulting grids are often hard to analyze and interpret, requiring the use of a complex mathematical technique, and this is after the elicitation technique has taken some considerable time to conduct.

8. *Card sort.* Card sorting is a technique aimed toward understanding how experts view element relationships in the given problem domain. Rugg and McGeorge (1997) claim that sorting techniques are useful both for identifying relevant categorization and for investigating common features and differences between experts in the use of that categorization. They are also quick, systematic, and easy to use. To apply this technique paper cards are printed with objects of the domain, in the form of words, phrases, diagrams, or pictures. The expert is repeatedly asked to sort these cards according to different dimensions. The knowledge engineer then derives rules and relationships from these sorts. Rugg and McGeorge (1997) describe briefly the main types of sorting techniques, including *Q sorts, hierarchical sorts, "all in one" sorts*, and offer a detailed tutorial on the *repeated single-criterion sorts.*

However, the sorting techniques only address static, flat, explicit knowledge (Rugg and McGeorge, 1997). They cannot conveniently access knowledge about sequencing procedures, trade-offs, knowledge structures such as hierarchies, or much tacit knowledge. They are not very useful in domains that are not well structured. Overall, they are not very rich in the amount of information produced and, perhaps more significantly, the experts do not generally favor them. Representation of the resulting knowledge base for the purposes of future computer application is also not straightforward, and thus these techniques would not be generally preferred for an industrial application.

9. *Group interviews.* This technique may be adopted where more than one expert is available, although this can introduce additional complexities of multiple expert knowledge elicitation (Beerel, 1993). Some researchers (Grabowski et al., 1992) however, argue that focus groups generated responses that, though similar in quality, were more original than those produced by an individual. A range of variations exists in group interviews including brainstorming, consensus decision making, the nominal group technique, and focus groups; and lately some of these have been computer facilitated. Some of the techniques (brainstorming) emphasize the quantity of responses; others (consensus decision making) concentrate on the quality in that members of the group extensively evaluate each idea.

10. *The Knowledge Acquisition Technique* (*KAT*). This new technique, developed by CKDesign, is derived from the exception structure of Popperian epistemology (a branch of philosophy concerned with the theory of knowledge) and has been shown to allow a rapid and accurate elicitation of knowledge (Duke, 1992). This technique allows the knowledge engineer to build up a directly computable representation of the elements that experts use to make decisions in their domain, by the application of "elicitation by exception." The most important advantage of KAT over other existing methodologies is that it is extremely fast to extract knowledge from the experts and convert it into rules, which can be directly represented in a computer. This speed of elicitation is partly due to the highly structured nature of questioning and the "falsificationist" approach to it (see Fig. 4.9). Furthermore, the knowledge engineer is not required to be aware of the domain or any computer software. Also, the knowledge is directly recorded in graphical form (e-graph) that is easily transformed into "if-then" production rules.

    This technique appears to be the most appropriate for rapid elicitation of knowledge to be used in practical implementation of knowledge-based systems, and its effectiveness is discussed in detail in this chapter, including an industrial case study.

The human nature is in principle *falsificationist*, in that when we look at the world to try and discover information that will provide us with an ability to make predictions about it, we progress through several stages. We observe some specific datum or a data set and then perform a generalization from the specific observances to a general rule or hypothesis. Such a hypothesis is then supported by further, incidental or intentional observation to a satisfactory degree of evidential support for the hypothesis (or *belief state*). However, regardless of how much evidence we accrue in support of a particular belief state, we only require one verifiable and repeatable counter instance in order for us to be forced to review our position. Thus, what we are actually doing in testing our hypotheses is trying to force their boundaries, that is, we are trying to falsify them *not* trying to verify them. In an empirical setting, no matter how much evidential support we gather in favor of a particular hypothesis, we can never collate sufficient evidential support to categorically establish a particular belief state. On the other hand, we need only one true piece of counterevidence to destroy that belief state.

A natural tendency of an expert to justify their decisions by supporting evidence can potentially lead to a "never-ending" elicitation process. On the contrary, the task of identifying exceptions is empirical, explicit, and definitive with respect to the boundaries of the expert's knowledge of the domain.

Figure 4.9 The principle and procedure of KAT.

4.4.2 Alternative elicitation methods

The underlining characteristics of this class of elicitation methods are varied, but none of the methods involves a direct interview with a domain expert in the initial stages of elicitation, hence they are grouped together here.

Task analysis. Task analytic techniques are commonly used in industrial practice (at operational levels), however, their application in knowledge engineering is less common. The introduction to this chapter discussed in detail the characteristics of the knowledge held by human experts and the traditional means of communicating this knowledge across the company. The current strict regulatory pressure on some businesses forces them to hold great repositories of standard operating procedures, manuals, best practices, process flow diagrams, and so forth. Compiling these documents can actually be viewed as a process of eliciting the knowledge of perceived experts in a particular area, whether it is a result of an introspective activity of an individual expert, or an outcome of a consultation process involving a group of experts from that area.

The process of compiling the knowledge would have been performed for other purposes and in its own time allocation. For the purposes of knowledge elicitation, the knowledge engineer can (depending on the quality of the information) construct the first version of the mediating knowledge representation solely on the basis of these documents. This means that valuable expert time is not wasted in explaining the principal characteristics of the domain, which are already captured in the documentation. To obtain the "working" knowledge of the domain, however, this first version of mediating representation will have to be analyzed and critiqued by the expert. This would be performed most likely in a guided interview process, where the structure is shown to the expert who is asked to comment on it, making any additions, deletions, or modifications as necessary. The guided interview process should be more rapid than an interview "from scratch," since the documentation is often based on the experience of the expert and covers the fundamental features of their knowledge.

A number of variations of task analytic techniques exist. *Functional flow analysis* involves the creation of diagrams that display the primary system functions and subfunctions in sequence, thus relating to the sequencing, timing, and flexibility of the various functions. *Operational sequence analysis* reveals sequence information at a much finer level of actions and decisions, as opposed to functions. *Information flow analysis* exposes the flow of information necessary to carry out the system functions. *Interaction analysis* consists of the identification of constraints imposed on the system.

Decision trees. This technique attracted widespread interest from a range of communities due to its ability to extract rules and discriminating factors from data while capturing the human decision-making process in a graphical form. We will concentrate on the methods of identifying critical factors for categorizing processes into high and low performing. The decision trees easily handle such categorization. There are a wide variety of algorithms for tree construction (Cao-Van and De Baets, 2003) although most of them are based on the ID3 classification algorithm by Quinlan (1993). Examples of C5.0 and *chi-squared automatic interaction detection* (CHAID) inducers are described in literature for a range of subject areas (Moshkovich et al., 2002; Chae et al., 2003). In general, the algorithms quantify the ability of a variable to discriminate between the defined classes and select the variable with the highest explanatory power as the highest-level node in the tree. On completion of the tree the decision process is unveiled from the data and this can be compared to the human mental causal relationship model of the process at this stage of process development.

Case base reasoning (CBR). This approach is characterized by its ability to capture past experience and knowledge for case matching in various applications. It is popular especially in areas where problem solution is not clearly defined. CBR is based on the human mental model of searching for similar situations occurring in the past and applying the experience learned in those situations to the current circumstances (Leake, 1996; Lau et al., 2003). Theoretical investigations have proven that experts employ effective indexing schemes (Dahr and Stein, 1997), which allow them to locate cases similar to the current problem (Aadmodt and Plaza, 1994). This ability to mimic experts' reasoning allows this technique to be applied in a wide variety of areas, for example, medicine (Lopez and Plaza, 1997), offshore well design (Mendes et al., 2003), product design (Cheng, 2003), and design of chemical absorption plants (Arcos, 2001).

Tracking methods. In principle, similar to the task analytic techniques, tracking methods are associated with specific tasks, but the elicitation technique is performed concurrently with task performance. Two principal tracking methods will be discussed later and the benefits and limitations of each will be highlighted. In all tracking methods the selection of tasks to be tracked is a critical issue and can become one of their main limitations.

Observations. In this technique the expert is observed as he or she performs a domain-related task or solves a domain-related problem (Hoffman, 1987). Observation involves recording the actions of the experts, while the expert is performing a task in the domain. Observations can be used, for instance, to identify problem-solving strategies that are not consciously accessible; study the physical movements of hands, eyes, and so forth; identify the task involved in the domain; and identify the information for the task. The methods of observation have the advantage of minimal interference with the expert's task and environment. However, in situations like real-time process control detailed observation is often problematic if not impossible (Clarke, 1987). In addition, it is difficult to extract rules from the information available from observations, and the influence of the elicitor's presence on the expert's behavior is difficult to ascertain.

Protocol analysis. Protocol analysis is one way of capturing procedural knowledge from the experts. The concept behind this technique is the fact that, when the expert encounters decision points in the task, certain conditions are perceived by him, resulting in the expert taking an action, thereby performing in an "if-then" manner. The experts are asked to perform their task and talk about their thoughts while doing the task. All of these are recorded. The transcript of this thinking aloud is analyzed and coded by the knowledge engineer. Unfortunately, it is difficult to go through all the cases in the domain while the knowledge engineer is present. In addition, the knowledge engineer has to derive rules from the experts' statements, which is not easy. Problems arise because of the sheer amount of data generated, the qualitative nature, the complexity and the nonorderliness of the data, and the subjective nature of its interpretation (Cooke, 1994).

Computer-aided elicitation tools. We include a range of tools into this class, from tools assisting either the knowledge engineer or the expert during the elicitation process to tools aiding the analysis of the results from previously mentioned elicitation methods. This is by no means an exhaustive list of tools currently available, and is intended only to give the reader an idea about the wide-ranging support for elicitation.

The benefits of enabling the domain expert to have more control and participation in the elicitation process by providing him or her with elicitation tools are summarized in Sandahl (1994). In general, such tools tend to elicit knowledge of specific types. For example, *Cognosys* (Woodward, 1990) is a semiautomatic system that characterizes a domain through the identification of a concept and its organization into a graph structure. Other tools may be more suited to the development of models of diagnostic knowledge. For example, *MORE* (Kahn et al., 1985) asks the expert to distinguish between hypotheses and symptoms and test the relations between them. Corbridge et al. (1994) describe a *knowledge engineering workbench* (KEW) tool, constructed as a part of an ESPRIT project to provide computerized support for all stages of the knowledge acquisition process. Several tools have been built on the basis of the repertory grid methodology [e.g., KITTEN, KSS0, and KSSn (Eriksson, 1994)]. *Expertise transfer system* (ETS) interviews experts to uncover vocabulary conclusions, problem-solving attributes, their structures, weights, and inconsistencies (Turban, 1995). It also takes the resulting rating grid and, by making use of logical entailments in the grid, transforms it into a set of production rules associated with strength values.

Another example of automated knowledge elicitation is the use of OPAL (OPAL is a high-order, pure functional software language) for eliciting knowledge in the cancer chemotherapy domain (Musen et al., 1987). OPAL helps oncologists build cancer treatment protocols that are used as a knowledge base in the expert system, *Oncocin*. It requires users to fill out slots in forms using a predefined vocabulary of representation primitives. OPAL does not require users to concern themselves with details of implementations, such as which medical parameters are referenced by such actions in the internal workings of Oncocin. A visual programming language facilitates the acquisition of procedural knowledge. The graphical interface allows the user to create icons for plan elements, and arrange them into a graph structure. By positioning these elements and drawing connections between them, the user can create charts that mimic the control flow of conventional programming languages (Jackson, 1999).

Automated protocol analysis tools. Several automated protocol analysis tools have been developed for analyzing data in the form of verbal and nonverbal protocols. For example, PAS-I and PAS-II (Waterman and Newell, 1971) are systems that use natural language parsing of verbal input to generate problem behavior graphs. While PAS-I was developed within the specific domain of crypt arithmetic problems and its performance compares favorably to manual encoding, PAS-II is context-free and involves more extensive human interaction (Cooke, 1994). There are also tools combining several knowledge-elicitation approaches. *KRITON* (Diederich et al., 1987) is a technique that combines methods of laddering, forward scenario simulation, and repertory grid to elicit a set of domain concepts. It also involves the collection and analysis of protocol and the automatic generation of rules for a knowledge base suited for analysis problems. The system segments the protocol using pauses in speech, searches the transcribed protocol for domain concepts, and carries out a semantic analysis to transform the protocol segments into propositions (Cooke, 1994). The most pronounced drawback of this class of automated techniques stems from the limitations of the underlining elicitation approaches upon which these are based. Namely, the large amount of often-unmanageable data that is difficult to interpret.

Ripple-down rules (RDRs). RDRs are a knowledge acquisition technique as well as a knowledge representation scheme that nonmonotically reason over a list of ordered rules with exceptions. Knowledge acquisition is achieved through RDRs with minimal analysis on the basis that the experts use their knowledge to "make up" a solution to fit the situation. In a sense the capture of behavioral knowledge based on cases appeared to offer a solution that avoided the attempt to understand the thinking process of the experts (Richards and Compton, 1999). The utility of RDRs has been demonstrated by the *pathology expert interpretive reporting system* (PEIRS, Edwards et al., 1993) routinely used in a large Sydney hospital (Richards and Compton, 1999). Current evaluations of commercial systems have shown that experts can build a system with 3,000 to 4,000 rules in about one-person week.

Once the knowledge is extracted and a knowledge base is developed, it becomes a part of the intellectual capital for the organization. This intellectual property must be turned into action for business results. The next section focuses on this aspect.

4.5 Turning Intellectual Property into Action

Knowledge leading into intellectual property must be appropriately manifested into action to deliver organizational performance. However, the knowing-doing gap exists in almost every organization (sometimes called *performance paradox*). Pfeffer and Sutton (2000) have conducted extensive research exploring the performance paradox. Their research indicates that the knowing-doing gap arises from a constellation of factors and that it is important for the organizational leaders to understand them all and how they interrelate. These factors could be summarized as:

1. Explain the philosophy first—why a certain action is being pursued before plunging into the how piece?
2. Knowing comes from doing and teaching others how.
3. Action counts more than elegant plans and concepts.
4. There is no doing without mistakes and a tolerant attitude to risk taking is critical.
5. Fear fosters knowing-doing gaps, so drive out fear.
6. Foster teamwork and avoid internal competition.
7. Measure what matters and what can help turn knowledge into action.
8. Conducive management environment.

In Chap. 8, a comprehensive strategy for transforming knowledge into productive actions is discussed in detail. Chapter 8 discuses the need for transformational leadership and a culture of siloless synergy required to transform the knowledge (and hence the intellectual property) into business results (optimal productivity). The concepts discussed in this chapter will be applied to an industrial case study to demonstrate the value of a structured approach to knowledge management.

4.6 Industrial Case Study D—The Knowledge Acquisition Technique (KAT)

Bioprocesses are generally considered highly complex often leading to undesired levels of process variability. An important process stage in a large-scale bioprocess is fermentation of a seed stage, which serves as an inoculum for the final product fermentation. The quality of the seed stage is critical to the success of the process, while operating instructions exist in forms of standard operating procedure, manufacturing tickets, and the like. However, the best human operators can collect extensive knowledge over years of operating a process and demonstrate their expertise by achieving tight control leading to higher productivity. From a management point of view it is critical to capture such knowledge and expertise and build it into standard operating procedures, manufacturing tickets, and the like, to be implemented by all operators. This case study demonstrates how KAT can elicit knowledge about the seed stage of a bioprocess to define the best practice leading to good quality seed. The "state" of a fermentation seed is difficult to quantify, but without this information prediction of the behavior of a production vessel can become imprecise. Additionally, the identification of seed quality prior to its transfer to the production vessel may have added benefits in terms of scheduling and process economics.

Selection of experts. One of the most critical steps in knowledge elicitation is the identification of appropriate experts. Thus, a discussion was held with the industrial project leader (the knowledge-base owner) and other company staff, and as a result a list of experienced scientists, shift leaders, and operators was drawn up. From this list, the 12 most experienced people whose expertise covered the domain of interest were short-listed. Following the identification of the experts, the next step was to meet them, inform them about the project objectives and, in a sense, convince them about the worth of the project. The cooperation of the experts is critical to the success of the project. Therefore, before starting the interview process it is necessary to have a general consensus of the worthiness of the undertaking among all the experts. In other words, the experts should be able to see how the results of the project are going to help them and the company.

Knowledge-elicitation process. Before starting the interviews, it was necessary for all the experts, the project owner, and the knowledge-base owner to agree upon a key question. This exercise was conducted in a brainstorming session. Because this project was being implemented on the seed stage, the key question agreed upon was "Is the seed good?" The seed stage consists of three phases, and for a seed to be good, all three have to be standard. Therefore, the key question was further subdivided into three subcategories: Is phase-1 OK? Is phase-2 OK? Is phase-3 OK?

It is difficult to predict the length of an interview process with a particular expert as the depth and scope of knowledge involved in any given task are rarely obvious. However, previous studies have suggested one to five days per expert (Duke, 1998). At the outset, two months were allotted for the entire interview process and three knowledge engineers conducted it. Most of the interviews took two man-days, with the most experienced expert's interview extending to four man-days.

According to best practice, the person nominated by the project manager as the most experienced expert should be interviewed first. This was the policy adopted as far as practicable, although some deviations from it are described later. Theoretically, KAT does not require the knowledge engineer to have domain experience. Domain stands for the knowledge boundary of the expertise. Practically this also holds true, but experience suggests that, if the knowledge engineer has some domain appreciation, the process of knowledge elicitation becomes faster. However, the knowledge engineer (with domain experience) should be careful that he or she:

- Simply records the knowledge of the experts and does not preempt or optimize it with (possibly subconscious) personal attitudes toward the domain. In particular it is important to avoid rationalizing the knowledge into the framework with which the interviewer is familiar. This is important especially since the knowledge elicited from the experts is already in an interpreted form, that is, it is in a sense prefiltered by the experts from their knowledge and experience of their domains; and it is thus very likely to be optimized in a manner suitable for them and their working practices in those domains.
- Does not intimidate the experts by virtue of their domain experience. For example, if the knowledge engineer happens to have a similar level of knowledge as the experts then it is more likely that the experts might feel awkward while being interviewed in that they might feel that the gap in their knowledge (if any) might get exposed to the knowledge engineer or experts' colleague.

In this study, due to constraints on the experts' and the knowledge engineers' time, the interviews of different experts were conducted simultaneously by different knowledge engineers. For example, say on a particular day "expert A" was interviewed by "knowledge engineer A" and "expert B" was interviewed by "knowledge engineer B." The key expert's interview was spread over several days (due to his time constraints) therefore, to meet the project deadline the knowledge engineer started interviewing other experts even before the interview of the key expert was complete.

Because the knowledge engineers had basic understanding of the domain, they at times prompted the experts toward a particular aspect, which the expert had not volunteered. For example, say after interviewing "expert A," "knowledge engineer A" started interviewing "expert C," and if "expert C" missed some point that was covered by "expert A" the knowledge engineer could point "expert C" toward identifying it. However, if "expert C" believed that the point is not particularly important for his decision making, then it was not recorded and the difference was sorted out in the knowledge fusion session as discussed later. In addition, since the aim of the project was to monitor the process in real time and advise the operators in case of an anomaly, the elicitation was kept focused toward this aspect.

Due to constraints on the time of the knowledge engineer, the situation arose that "knowledge engineer B" completed the elicitation of an expert following initial interview sessions with "knowledge engineer A." This type of change is very easily and effectively handled in KAT because of the structure and representation of KAT. Because, a practical industrial environment is full of these time constraints, this flexibility provided by KAT helps in saving the time and effort required for knowledge elicitation.

Knowledge fusion. The interview with 12 experts resulted in 12 different knowledge bases (in the form of eGraphs). Although the basic concepts in all the knowledge bases elicited from different experts were the same, they still differed slightly because of the varied experience of the experts. The next step was to merge these into a single knowledge base, which combined the knowledge of all the experts. This was done in a joint session involving the project owner, the knowledge-base owner, and the knowledge engineers. The knowledge base of the most experienced expert was taken as the basis and all the additional features that were found in the knowledge bases of the other experts were incorporated into it. At each step, a consensus was reached among all the participants. The whole exercise took 3 sessions of 3 h each. The project owner and the knowledge-base owner sanctioned the final eGraph.

Implementation. For the purposes of SOP specification, the final eGraph was used and transcribed in a straightforward manner. This is demonstrated on the example of an eGraph for sterilization procedure assessment shown in Fig. 4.7. Sterilization was identified as one of the key phases in seed fermentation and it is important that all operators adhere to a standard sterilization procedure. However, if the expert knowledge reveals any legitimate reasons for deviations, this would serve as a foundation for change of procedure.

Figure 4.10 indicates the decision points that experts consider when assessing whether the sterilization procedure has been followed in any particular case. As a simple illustration for node A14 the knowledge extracted indicates that, if temperature and back pressure on the sterilizing tank do not correlate, this may mean that the sterilization is suspect and further action must be taken. The further action within node A14 could be to check the temperature and back-pressure measurement. If they are OK, then it might mean that the tank has not been sufficiently deaerated prior to sterilization, which will lead to a failed sterilization.

For the purpose of demonstrating the applicability of this procedure, it is not necessary to discuss in detail each of the nodes and the meaning of the symbols within each node. It is sufficient to note that each node is linked to a set of methods to determine its "truth state" and often a list of actions is attached to a node (indicated by Axx). Once the eGraph

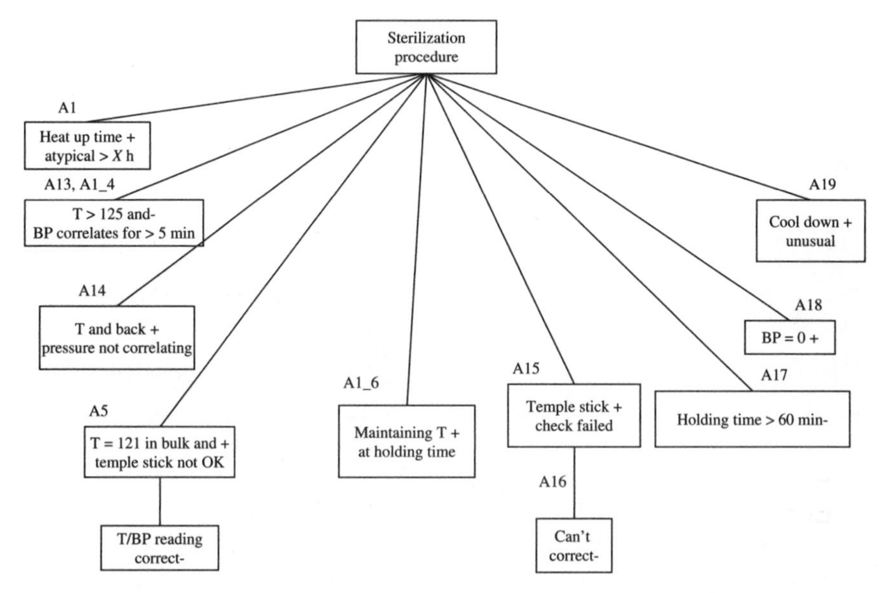

Figure 4.10 Example of an eGraph for sterilization phase of seed fermentation process.

is completed, these methods and actions are compared to the current sterilization procedure, and if any changes are required, they can be implemented in a straightforward manner.

4.6.1 Comprehensive knowledge base for SOP and manufacturing ticket

One of the advantages of such a knowledge elicitation is that an integrated (human- and data-based) control strategy could evolve that could form the basis of operating practices (SOP and tickets). This approach will help reduce variability by making the operating processes knowledge based, thus eliminating the need for subjectivity.

The first step of converting the knowledge base into instruction is to distill the major nodes into control points. The next step is to document the demonstrated best practices for optimum results. Finally, the action plan to troubleshoot should be formulated and institutionalized. From the earlier eGraph the instructions listed in Table 4.1 for the operating processes could be extracted.

The operating rules generated by experts could be translated into SOP and troubleshooting instructions. For example (see Table 4.1):

1. Heat-up time for the sterilization should be less than 2 h, if this is not the case then check steam temperature, pressure, and valve setting.

TABLE 4.1 Operating Rules

Control point	Instructions	Troubleshooting action
Heat-up time	The heating-up time should be less than 2 h	Check steam temperature, pressure, and valve setting
Sterilization temperature	121–124°C	Check back pressure
Sterilization check	Templestick should melt on contact with sterilizing surfaces	Check temperature and back pressure reading
Hold time	40–45 min	Check temperature and back pressure
Cool down	3 h cooldown	Check valve, cooling water temperature
Maintain back pressure	0.3 bar-back pressure	Check valve and air pressure

2. Sterilization temperature should be within 121 to124°C, if this is not the case check the back pressure.
3. If the templestick (an indicator of temperature) does not meltty when in contact with sterilizing surface, then check temperature and back pressure.
4. If the sterilization hold-time is more than 45 min, check for temperature and back pressure.
5. If the cooldown takes longer than 3 h, check for valve setting and temperature of cooling water.
6. If the back pressure in the tank after sterilization is not maintained at 0.3 bar, check for valve and air pressure.

However, the transcription of an eGraph into an SOP is just one means of utilizing the captured expert knowledge about the process. Some "if then" or "if then rules" are also generated from this knowledge base, this could be used in process control strategy. Chapter 6 will describe the application of this knowledge in a real-time knowledge-based system for intelligent alarming, warning about potential process deviations, and providing expert advice to the operators about the best course of action in any particular situation. Thus, one knowledge-elicitation exercise provides benefits to the company on a number of levels and should be seriously considered as a good practice in knowledge management.

Concluding Comments

Building a learning organization is an integral component of success in business. While the bottom line lesson for organizations may remain the same, the consequences of ignoring the message and the way in which it is achieved are somewhat different in today's competitive landscape. A failure to learn carries a higher price tag today than ever before. The very survival of the organization may be threatened. Businesses have been warned to be fast, agile, and responsive. Yet some have still not heeded the message. In a rapidly changing world, those organizations that fail to meet the challenge will not survive. Building a learning organization will assist companies in more effectively meeting this challenge—to change, to adapt, and to learn.

Knowledge leading to intellectual property is the ultimate source of competitive advantage of an organization. The knowledge must be captured appropriately. New techniques of knowledge extraction, like eGraph, are instrumental in maximizing the knowledge base and storing it in a simple, easy-to-understand system. This also makes it easy to convert the knowledge base into computer-based programs for control of the process including intelligent alarming.

The valuable knowledge gained from various sources must be retained, as this is the primary weapon for competitive advantage. The knowledge in the form of documents, databases, and the like, must be stored in a secure location and should be easily searchable and accessible to the employees (and only to the employees) of the company. Knowledge that is inaccessible is knowledge lost. For a knowledge-based-learning organization, it is critically important that the knowledge base is in front of the employees and that the new knowledge gained is appropriately communicated. This communication can be in the form of seminars, presentations, news flashes in emails, and the like. It is not unheard of that the same knowledge is developed again and again in large organizations leading to inefficiencies, opportunities lost, and financial losses. Further, the knowledge base must be maintained to keep it current.

References

Aadmodt, A., and E. Plaza (1994), "Case-based reasoning: foundational issues. Methodological variations and system approaches," *AI Com*, **7**(1):39–59.

Arcos, J. L. (2001), "T-air: A case-based reasoning system for designing chemical absorption plants," *Lect. Notes Artif. Int.*, **2080**:576–588.

Beerel, A. (1993), *Expert Systems in Business: Real World Applications*. Ellis Horwood, Chichester, UK.

Berry, D.C. (1987), "The combination of explicit and implicit learning-processes in task control," *Psychol. Res.*, **49**(1):7–15.

Cao-Van, K., and B. De Baets (2003), "Growing decision trees in an ordinal setting," *Int. J. Intel. Syst.*, **18**:733–750.

Chae, Y.M., H. S. Kim, K. C. Tark, H. J. Park, and S. H. Ho (2003), "Analysis of healthcare quality indicator using data mining and decision support system," *Expert Syst. Appl.*, **24**:167–172.

Cheng, C. B. (2003), "A fuzzy inference system for similarity assessment in case-based reasoning systems: An application to product design," *Math. Comput. Model*, **38**(3–4): 385–394.

Clarke, B. (1987), "Knowledge acquisition for real-time knowledge based systems." *Proceedings of the 1st European Workshop on Knowledge Acquisition for Knowledge Based Systems*. University of Reading, UK.

Cooke, N. J. (1994), "Varieties of knowledge elicitation techniques," *Int. J. Human-Computer Studies*, **41**:801–849.

Corbridge, C., G. Rugg, N. P. Major, N. R. Shadbolt, and A. M. Burton (1994), "Laddering: Technique and tool use in knowledge acqusition," *Know. Acquis.*, **6**:315–341.

Crossan, M. M., H. W. Lane, and R. E. White (1999), "An organizational learning framework: From intuition to institution," *Acad. Manage. Rev.*, **24**(3):522–537.

Dahr, V., and R. Stein (1997), *Seven Methods for Transforming Corporate Data into Business Intelligence*, Prentice Hall, Englewood Cliffs, NJ.

Diederich, J., I. Ruhmann, and M. May (1987), "KRITON: A knowledge acquisition tool for expert systems." *Int. J of Man-Machine Studies*, **26**:29–40.

Duke, P. (1992), *KAT A Knowledge Acquisition Technique, Methodology manual.* CKDesign, UK.

Edwards, G., P. Compton, R. Malor, A. Srinivasan, and L. Lazarus (1993), "PEIRS: a pathologist maintained expert system for the interpretation of chemical pathology reports." *Pathology*, **25**:27–34.

Eriksson, H. (1994), "Models for knowledge-acquisition tool design," *Knowl. Acquis.*, **6**:47–74.

Fisher, R., and R. Geiselman (1992), *Memory Enhancing Techniques for Investigative Interviewing*," Charles Thomas, Springfield, IL.

Grabowski, M., A. P. Massey, and W. A. Wallace, (1992), "Focus groups as a group knowledge acquisition technique," *Knowl. Acquis.*, **4**:407–425.

Hoffman, R. (1987), "The problem of extracting the knowledge of experts from the perspective of experimental psychology," *AI Mag.*, **8**:53–67.

Huber, G. B. (1991), "Organizational learning: the contributing processes and the literatures," *Organ. Sci.*, **2**(1):88–115.

Jackson, P. (1999), *Introduction to Expert Systems*. Addison-Wesley Longman, Boston, MA.

Kahn, G., S. Nowlan, and J. McCermott, (1985), "Strategies for knowledge acquisition." *IEEE Trans. Pattern Anal. Mach. Intell.*, **7**:511–522.

Kolb, D. A. (1996), "Management and the learning process," in: K. Starkey, (ed.), *How Organizations Learn*, International Thomson Business Press, London, pp. 270–87.

Lau, H. C. W., C. W. Y. Wong, I. K. Hui, and K. F. Pun, (2003), "Design and implementation of an integrated knowledge system," *Knowledge-Based Sys.*, **16**:69–76.

Leake, D. B. (1996), *Case-Based Reasoning Experiences, Lessons and Future Directions*, MIT Press, Cambridge, MA.

Lopez, B., and E. Plaza (1997), "Case-based learning of plans and goal states in medical diagnosis," *AI in Med.*, **9**(1):29–60.

Mendes, J. R. P, C. K. Morooka, and I. R. Guilherme (2003), "Case-based reasoning in offshore well design," *J. Petrol Sci. Eng.*, **40**(1–2):47–60.

Moody, J.W., R. P. Will, and J. E. Blanton (1996), "Enhancing knowledge elicitation using the cognitive interview," *Expert Sys. Appl.*, **10**:127–133.

Moshkovich, H. M., A. I. Mechitov,. and D. L. Olson (2002), "Rule induction in data mining: effect of ordinal scales." *Expert Syst. Appl.*, **22**:303–311.

Musen, M. A., L. M. Fagan, D. M. Combs, and E. H. Shortliffe (1987), "Use of a domain model to drive an interactive knowledge editing pool," *Int. J. Man Mach. Stud.*, **26**: 105–121.

Nonaka, I. (1996), "The knowledge-creating company," in: K. Starkey, (ed.), *How Organizations Learn*, International Thomson Business Press, London, pp. 18–31.

Pfeffer, J., and R. I. Sutton (2000), *The Knowing-Doing Gap: How Smart Companies Turn Knowledge into Action*, Harvard Business School Press, Boston, MA.

Quinlan, J. R. (1993), *C4.5: Programs for Machine Learning,* Morgan Kaufmann, San Mateo, CA.

Richards, D., and P. Compton (1999), "An alternative verification and validation technique for an alternative knowledge representation and acquisition technique," *Knowledge-Based Sys.*, **12**:55–73.

Ross, D., D. I. McBriar, and P. Duke (1998), "Rapid knowledge elicitation and expression graphically in G2," *Gensym User Symposium '1998*, Chicago, USA.

Rugg, G., and P. McGeorge (1997), "The sorting techniques: A tutorial paper on card sorts, picture sorts and item sorts," *Expert Sys.*, **14**:80–93.
Sandahl, K. (1994), "Transferring knowledge from active expert to end-user environment," *Knowl. Acquis.*, **6**:1–22.
Schmalhofer, F. (1998), *Constructive Knowledge Acquisition: A Computational Model and Experimental Evaluation*, Lawrence Erlbaum, Mahwah, NJ.
Schweikert, R., A. M. Burton, N. K. Taylor, E. N. Corlett, N. R. Shadbolt, and A. P. Hedgecock (1987), "Comparing knowledge elicitation techniques: a case study." *Artif. Intell. Rev.*, **1**:245–253.
Thomke, S. (2001), *Enlightened experimentation: The New Imperative for Innovation*, Harvard Business School Publishing Corporation, Boston, MA.
Turban, E. (1992), *Expert systems and applied artificial intelligence,* Maxwell Macmillan, Singapore.
Turban, E. (1995), *Decision support and expert systems: Management support systems,* Prentice-Hall, Hoboken, NJ.
Ursino, D. (2002), *Extraction & exploitation of intentional knowledge from heterogeneous information sources: semiautomatic approaches & tools*, Springer 1st ed., New York.
Waterman, D. A., and A. Newell (1971), "Protocol analysis as a task for artificial intelligence," *Artif. Intell.*, **2**:285–318.
Welbank, M. (1990), "An overview of knowledge acquisition methods," *Interact. Comput.*, **2**:83–91.
Woodward, B. (1990), "Knowledge acquisition at the front end: defining the domain," *Knowl. Acquis.*, **2**:73–94.
Wright, T. P. (1936), *Factors Affecting the Cost of Airplanes, J. Aeronaut. Sci.,* **3**:124–125.

Part 3

Variability Reduction

Variability could render a product out of specification, leading to rejection and re-work. When variability is out of control there can be a significant threat to manufacturing. On the other hand, if variability is controlled so that the mean productivity of the process improves, this will have a significant positive business impact. Besides the cost-benefit issue variability makes process improvement efforts difficult as any small improvement may be masked by variability. Therefore, reducing variability should be an important priority for a manufacturing operation. This part discusses concepts of variability reduction including some emerging concepts along with industrial case studies.

Chapter

5

Fundamental Strategies for Variability Reduction

"In God we trust, others must use data."

DEMING

Objectives

The previous chapters (Chaps. 1 to 4) have dealt with process design and implementation of a manufacturing process. A key aspect of implementing a robust process and its subsequent sustenance is the ability to reduce variation.

The learning outcomes of this chapter will include:

- Historic perspective on variation
- Understanding types of variation and process capability quadrants
- *Statistical process control* (SPC) concepts
- *Root cause analysis* (RCA) and its role in identifying special cause variation

The learning is then applied to a case study to gain practical understanding of the application of the concepts.

5.1 Introduction

A manufacturing process is impacted by numerous process parameters in the effort to deliver a quality product of a certain specification. The variation in the quality attribute of the product is of primary concern to any operational leadership. Variability could render a product out of specification leading to rejects and rework. When variability is out of control, it can be a significant threat to manufacturing. During the 1920s, Dr. Walter Shewhart made significant contributions on a scientific basis to address variability for economically controlling the quality of manufactured products. He developed the control charting methodology as a fundamental tool to understand and address variability. This forms the foundation of what is now called statistical process control (Shewhart, 1986). Along with Dr. Shewhart, Dr. W. Edwards Deming is regarded as a cofounder of modern quality concepts; he defines quality as "Good quality does not necessarily mean high quality. It means a predictable degree of uniformity and dependability at low cost, with a quality suited to the market." From his definition "a predictable degree of uniformity" could be inferred as "a predictable degree of variability (Deming, 1950)." Thus, a defined and predictable (and acceptable) degree of process variability is a keystone in the success of a manufacturing business.

Dr. Deming (1950, 2000) summarizes variability and its consequences in Fig. 5.1. The illustration indicates that reduced variability leads to improvement in product quality; less variability leads to less rework, fewer mistakes result in improved productivity; a better quality product and lower price leads to a competitive advantage and hence the business grows.

If variability is controlled such that the mean productivity of the process improves, this will have significant business impact too. Besides the cost considerations, variability makes process improvement efforts difficult as any small improvement may be masked by the variability

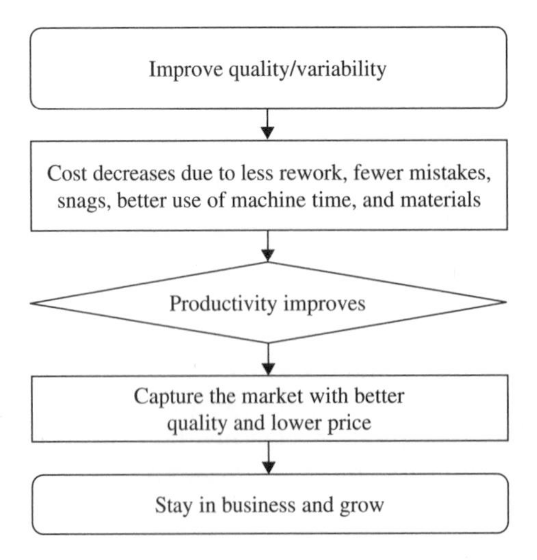

Figure 5.1 Variability diagram.

in the process itself. Therefore, reducing variability should be the first priority of any operation.

Dr. Shewhart (1986) laid the foundation of modern thinking in understanding variation. During the 1920s, Dr. Shewhart studied variation at Bell Telephone Laboratories. He concluded that:

1. All processes display variation.
2. Some display controlled variation.
3. Others display uncontrolled variation.

The controlled variations are uniform and predictable over time and due to "chance causes," they are in a state of statistical control. Uncontrolled variations on the other hand are unpredictable, unstable, and have "assignable causes." The process is then said to be out of statistical control.

Dr. Deming expanded on Dr. Shewhart's logic and suggested that a key deliverable from a manufacturing operation could be "a predictable degree of variation." A process in statistical control would have common cause variation, which according to Dr. Deming is systemic or part of the system. Dr. Deming classified (Table 5.1) process variability into two components: (1) common cause variability and (2) special cause variability. Common cause variability is related to inherent variability of the system—machines, materials, methods, and working environment. Generally, common cause variability is built into the specification of the product, and a process modification/improvement is required to reduce the variability further. On the other hand, special cause variability, as the name suggests, has associated causes, which could be identified and fixed (variability shift due to new batches of raw materials, inexperienced operators, and so forth). Typically, special cause variability has profound impact on the quality of product and may lead to rejection of the batch of product incurring significant losses. Such variations must be immediately addressed.

Deming in his famous compilation *Out of Crisis* (1982) emphasized the need for a new culture and management leadership to bring about the quality revolution. He suggested 14 points to enable this metamorphosis, which include:

TABLE 5.1 Classification of Variation

	Shewhart	Deming
Systemic cause	Chance causes	Common causes
Extrinsic cause	Assignable causes	Special causes

- Cease dependence on inspection to achieve quality, instead build quality in the process design
- Minimize total cost
- Make persistent effort toward improvement of products
- Provide training
- Use transformational leadership and teamwork breaking barriers between departments

A successful organizational quality improvement culture is a balance of both top-down and bottom-up approaches with visionary management leadership.

In this chapter fundamental approaches to variability are described. The systematic approach (Fig. 5.2) is in the PDCA form (plan-do-check-act) with control charting, special cause identification, root cause analysis, and the fix states.

Variability defines the capability of a process; an out-of-control process is not capable. The two important aspects of capability are: (1) degree of statistical control and (2) conformance with specification.

Shewhart (1986) defined a minimum criterion to establish statistical control. If for an average and range chart (described later in this chapter) with subgroup of size $n = 4$ at least 25 consecutive subgroups do not indicate any lack of control, then the process may be said to display a reasonable degree of statistical control. It was also stated simply as, if the 100 consecutive observations fail to indicate a lack of control then the process has a reasonable degree of statistical control. Also, Shewhart pointed out that "Data normality is neither a prerequisite nor a consequence of a state of statistical control."

Conformance with specification means that the product is within the critical product parameter defined during the process development.

Process capability could be categorized in quadrants as shown in Fig. 5.3.

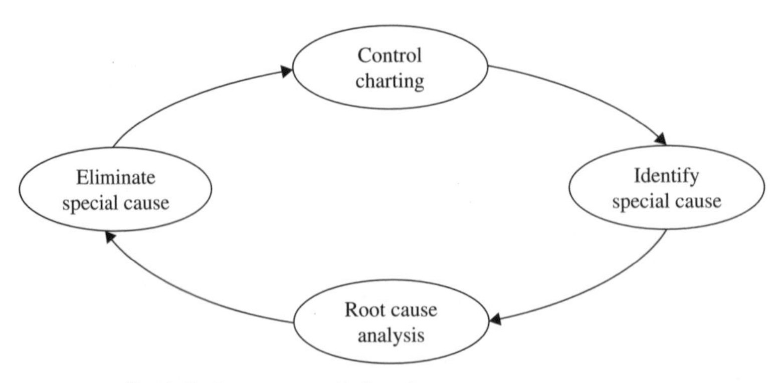

Figure 5.2 Statistical process control cycle.

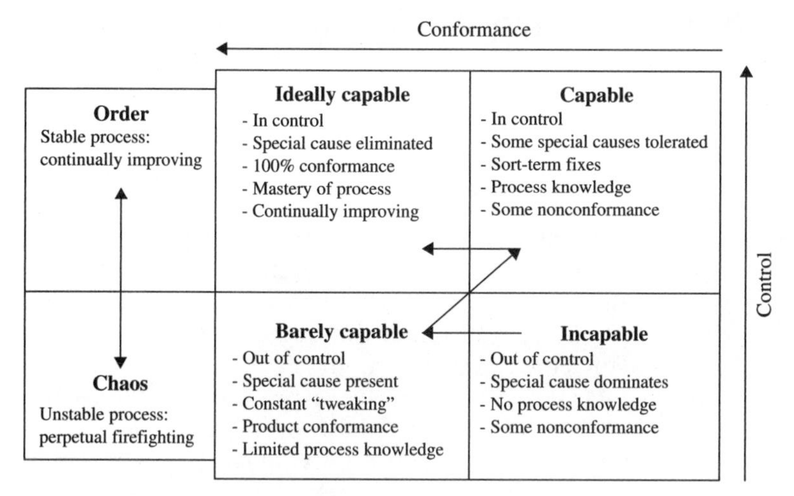

Figure 5.3 Process capability quadrants.

The ideally capable state should be the target of every manufacturing process. A process that is stable, compliant (100 percent conformance to the quality specification), in control, and continually improving, characterizes this state. The foundation of this state is a high level of process knowledge and a persistent "process vigilance" using control charts. Any special cause is analyzed and eliminated following the SPC cycle. Such a state is ideal for process improvement.

Capable process state suggests a reasonable degree of statistical control, that is, the variations in the product attributes are consistent over time. However, some special causes occur from time to time and short-term fixes are put in place. This generates deviation and typically there are systems in place to deal with the deviation backed by documentation. Some deviations lead to nonconformance leading to rejects and rework. The focus of this state should be to enhance process knowledge and migrate the process into the ideal state by eliminating the special cause (long term) thereby enhancing process robustness.

Barely capable process state is statistically out of control, but is still producing mostly conforming products. The variations are not consistent over time, and due to limited process knowledge the special causes are addressed by constant "tweaking" of the process variables. Most of the tweaking is based on dominant opinion at worst or empiricism at best. This state lends itself to perpetual firefighting that consumes the energy of the manufacturing organization. There is inadequate time for process understanding, which could lead to long-term solutions.

Incapable processes are characterized by a statistically out-of-control process dominated by special causes producing some nonconforming products. There is total lack of process knowledge, and the operation is consumed in perpetual firefighting. This is ultimate chaos and a manufacturer's nightmare.

5.2 Control Charting for Variables

Variability visibility is central to variability reduction; it signifies the data collection and its representation using appropriate techniques. There are several statistical tools that can provide visibility to variability, which include cause and effect diagram, flow chart, Pareto chart, histogram, run chart, scatter diagram, and control chart. Histograms, flow charts, run charts, and scatter diagrams show the overall picture, while Pareto charts show problem areas. These methodologies are primarily designed to provide snapshots in time. These methods do not indicate limits under which the process has to operate, and the variability depicted cannot be clearly identified. The modern SPC methodology has its foundation in control charts. These were developed as long ago as 1924 with the first method for charting data developed by Walter Shewhart. Dr. Shewhart's work was followed by the work of Dr. Deming in the 1950s with the successful implementation of control charting in Japan that established SPC as a useful industrial tool. The fundamental SPC tool, the control chart, will be discussed in detail in this section.

A chart with limit lines (or controls) is known as a *control chart*, and the lines are called *control lines*. There are three kinds of control lines: the upper control limit is known as UCL, the lower control limit is known as LCL, and the mean is denoted as $\overline{X}$ or $\overline{R}$. Control charts have several benefits, which include:

1. Improve productivity
2. Prevent defects—*Do it right the first time*
3. Prevent unnecessary process adjustments—"If it isn't broken, don't fix it"
4. Supply diagnostic information—information about change or step improvement
5. Supply process capability information—information about the value of important process parameters

However, there are also risks associated with the application of control charts. In general these risks are divided into:

Type I risk. The risk that an extreme sample (caused by explainable influences) will lead to a decision to take action although no process change has occurred.

Type II risk. The risk that a sample will fall within the control limits although there has been a change in the process and thus, the change is not detected.

The design of a control chart will compromise between these two risks and the choice of the control limits will influence the effectiveness of a scheme based on this type of chart.

5.2.1 Choice of control limits

Widening control limits. Decrease the risk of Type-I error and increase the risk of Type-II error.

Narrowing control limit. Increase the risk of Type-I error and decrease the risk of Type-II error.

It is also important to determine how quickly a control chart will indicate that the process is out of control, generating a false alarm (Type-I error). For this purpose, the concept of *run length* is introduced. The run length represents a number of observations plotted on the charts before an "out-of-control" signal is given. It is useful to calculate a long run average over a very large number of trials. This is referred to as the *average run length* (ARL) and it reflects the average number of points that must be plotted before a point indicates an out-of-control condition.

For example, to calculate ARL for a process within the conventional 3σ bounds of "in-control" area of the control chart and assuming normal distribution (thus, $p = 0.0027$), the following result is obtained:

$$\text{ARL} = \frac{1}{0.0027} = 370$$

This means that if the process remains in control, an out-of-control signal will be generated every 370 samples on average. Obviously, the frequency of the out-of-control call will depend on the frequency of sampling. To illustrate this, let us assume that a sample is taken every hour. In this case a false alarm will be generated every 370 h on average . For $\overline{X}$ with 3σ control limits for a normal distribution the value of $P = 0.0027$. This means that the probability of an incorrect out-of-control signal or false alarm will be generated in only 27 out of 10,000 data points.

Traditionally, the selection of control limits for $\overline{X}$ (± 3 standard deviations around the process mean) is based on a model with four assumptions (Pitt, 1994):

1. The process is in statistical control (no special causes).
2. The process mean $\overline{X}$ is known.
3. The standard deviations for sample averages $\sigma_{\overline{X}}$ is known.
4. The process standard deviation from individual units σ_X is known.

In practical situations, the process mean and standard deviation are usually not known definitively at the outset of the control chart preparation, and thus they have to be estimated from sample data. A sample average $\overline{X}$ and variability for each subgroup can be calculated from the observations and their range. By taking the average of all the sample averages $\overline{X}$ we obtain $\overline{\overline{X}}$, which is an estimate of the process mean. We can use this value as the center line for the $\overline{X}$ chart.

For the $\overline{X}$ and R and $\overline{X}$ and σ charts, the control limits are based on $\pm 3\sigma_{\overline{X}}$, as suggested by Shewhart. The intention of a control limit is that it is insensitive to special causes or shifts in the process average. For $\overline{X}$ and σ the limits are derived as:

$$\frac{U}{\mathrm{LCL}_{\overline{X}}} = \overline{\overline{X}} \pm 3\sigma_{\overline{X}} \tag{5.1}$$

where $\sigma_{\overline{X}}$ is the standard deviation of the averages of the subgroup, also referred to as standard error of the mean. However, this standard deviation is sensitive to the special causes or drifts in the process mean. On the other hand σ_X, which is the standard distribution of the individual values in the distribution, is less sensitive to special causes or drifts in the process mean.

$$\sigma_{\overline{X}} = \frac{\sigma_x}{\sqrt{n}} \tag{5.2}$$

For a subgroup of two or more, another important form of standard deviation is the average of the standard deviation of each subgroup "$\overline{\sigma}$."

$$\overline{\sigma} = C_2\sigma_X \tag{5.3}$$

where C_2 is a constant, which depends on the subgroup size. For a large subgroup size $\overline{\sigma}$ trends toward σ_X.

Thus, by substitution:

$$\frac{U}{\mathrm{LCL}_{\overline{X}}} = \overline{\overline{X}} \pm \frac{3\overline{\sigma}}{C_2\sqrt{n}} \text{ (for a given } n\text{)} \tag{5.4}$$

For a given subgroup, C_2 and n are constants, hence

$$A_1 = \frac{3}{C_2\sqrt{n}}$$

Hence, for $\overline{X}$ and s chart,

$$\frac{U}{\mathrm{LCL}_{\overline{X}}} = \overline{\overline{X}} \pm A_1\overline{\sigma} \tag{5.5}$$

TABLE 5.2 Factors for Control Charts

Subgroup size, n	A_1	A_2	B_3	B_4	d_2	D_3	D_4
2	3.76	1.88	0	3.267	1.128	0	3.27
3	3.39	1.02	0	2.568	1.693	0	2.57
4	1.88	0.73	0	2.266	2.059	0	2.28
5	1.60	0.58	0	2.089	2.326	0	2.11
6	1.41	0.48	0.030	1.970	2.534	0	2.00
7	1.28	0.42	0.118	1.882	2.704	0.08	1.92
8	1.17	0.37	0.185	1.815	2.847	0.14	1.86
9	1.09	0.34	0.239	1.761	2.970	0.18	1.82
10	1.03	0.31	0.284	1.716	3.078	0.22	1.78

The limit to the ranges are given as (Wheeler and Chambers, 1992):

$$\mathrm{UCL}_\sigma = B_4\overline{\sigma} \tag{5.6}$$

$$\mathrm{LCL}_\sigma = B_3\overline{\sigma} \tag{5.7}$$

The value of A_1,,which is dependent of the subgroup size, is given in Table 5.2.

For $\overline{X}$ and R, the σ_X and $\overline{R}$ are related by a constant (d_2) for a given subgroup size (from distribution theory) as:

$$\sigma_X = \frac{\overline{R}}{d_2} \tag{5.8}$$

Therefore, through substitutions:

$$\frac{U}{\mathrm{LCL}_{\overline{X}}} = \overline{\overline{X}} \pm A_2\,\overline{R} \tag{5.9}$$

where $A_2 = 3/(d_2\sqrt{n})$.

A_2 is a factor influenced by sample size and can be found in Table 5.2 for control charts.

It is important to interpret not only the $\overline{X}$ charts designed to detect changes in the process mean, but also to follow the process variability. Indeed, the ability of the $\overline{X}$ chart to detect changes in the process mean depends on the amount of noise (i.e., the extent of process variability) in the system. Thus, it is important to construct equivalent range charts, where center line, upper control limit, and lower control limit could be represented (Pitt, 1994) as:

$$\text{Central line} = \overline{R}$$

$$\mathrm{UCL}_R = D_4\overline{R} \tag{5.10}$$

$$\mathrm{LCL}_R = D_3\overline{R} \tag{5.11}$$

where D_3 and D_4 are factors influenced by sample size and can be found in Table 5.2 for control charts.

For individual charts the limits for individual values are called *natural process limits*. The three-sigma range for individual charts is calculated as:

$$\text{UCL} = \overline{X} + 3\frac{\overline{R}}{d_2} \tag{5.12}$$

$$\text{LCL} = \overline{X} - 3\frac{\overline{R}}{d_2} \tag{5.13}$$

For individual chart $n = 2$ and the value of $d_2 = 1.128$

$$\text{UCL} = \overline{X} + 2.66\overline{R} \tag{5.14}$$

$$\text{LCL} = \overline{X} - 2.66\overline{R} \tag{5.15}$$

Control charts monitor both the location and dispersion of the data.

5.3 Special Cause Identification

One of the key functionalities of control charting is the ability to identify special causes to initiate root cause analysis and the subsequent elimination of the special causes. Table 5.3 captures some rules-of-thumb in analyzing control charts for special cause identification.

5.4 Understanding Control Charting Using Case Study E

Crystallization is a very important unit operation in the pharmaceutical and process industry. Particle size, distribution, and morphology are key attributes of crystallization. Typically, particle size is tightly controlled during the manufacturing process. In most cases, mean particle size is closely monitored and specified as a critical process parameter and quality attribute. In this case study the data from the mean particle size as measured by image analysis (advanced microscopy) is used to understand the control charting methodology.

5.4.1 Organize the data

List the data in a spreadsheet; the data for the case study include production batch number and measurement of five samples from the batch. The data analysis is very basic and includes averages of the samples in each batch, range, moving ranges, and the subgroup's standard deviation. The following simple mathematics were used for this calculation:

TABLE 5.3 Special Cause Identification

Alarm	Example	Description
Outside limits	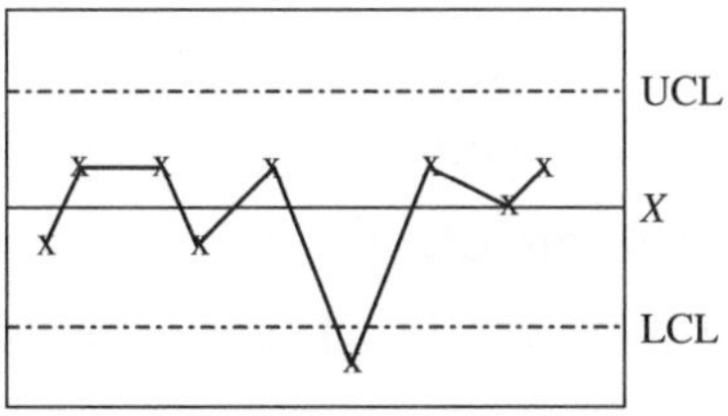 UCL X LCL	Special cause. 1 or more points outside the control limit
Run of 7	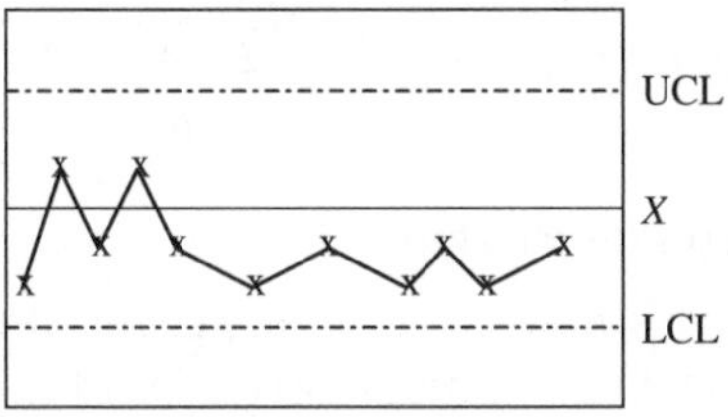 UCL X LCL	Special cause. 7 or more points above or below the center line
7 of 7 trending	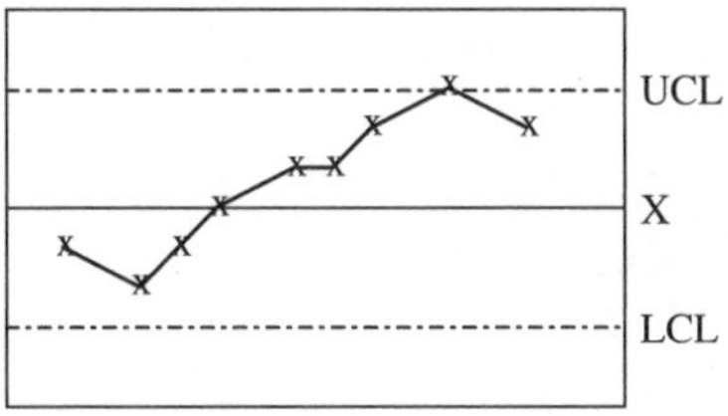 UCL X LCL	Special cause. 7 or more points trending up or down
14 or more up and down	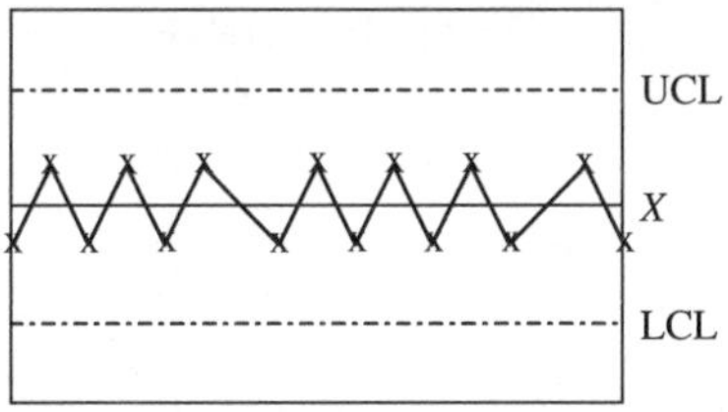 UCL X LCL	Special cause. 14 or more points alternating up and down
1 sigma unit	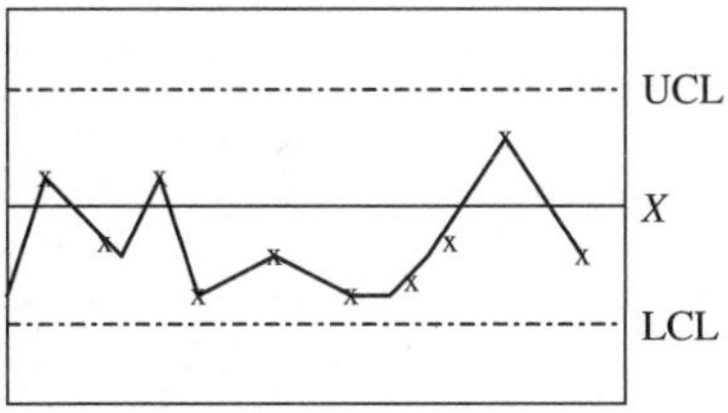 UCL X LCL	At least 4 out of 5 successive values fall on the same side at more than 1 sigma unit distance from the centerline

Average,

$$\overline{X} = \frac{(\text{sample1} + \text{sample2} + \text{sample3} + \text{sample4} + \text{sample 5})}{5}$$

$$\text{Range, } R = \text{maximum} - \text{minimum}$$

$$\text{Moving range} = \text{abs}\,|A_{n+1} - A_n|$$

Moving range is the absolute difference between the consecutive average values.

$$\text{Standard deviation, } \sigma = \sqrt{\frac{\sum (X - \overline{X})^2}{n - 1}}$$

5.4.2 Select the type of control chart to use

Selection of a control chart is an important aspect of SPC. There are several types of control charts: individual chart, $\overline{X}$ and R chart (most popular), $\overline{X}$ and s chart, moving average chart, exponential moving average chart, and *cumulative sum* (CUSUM) chart. Figure 5.4 recommends a

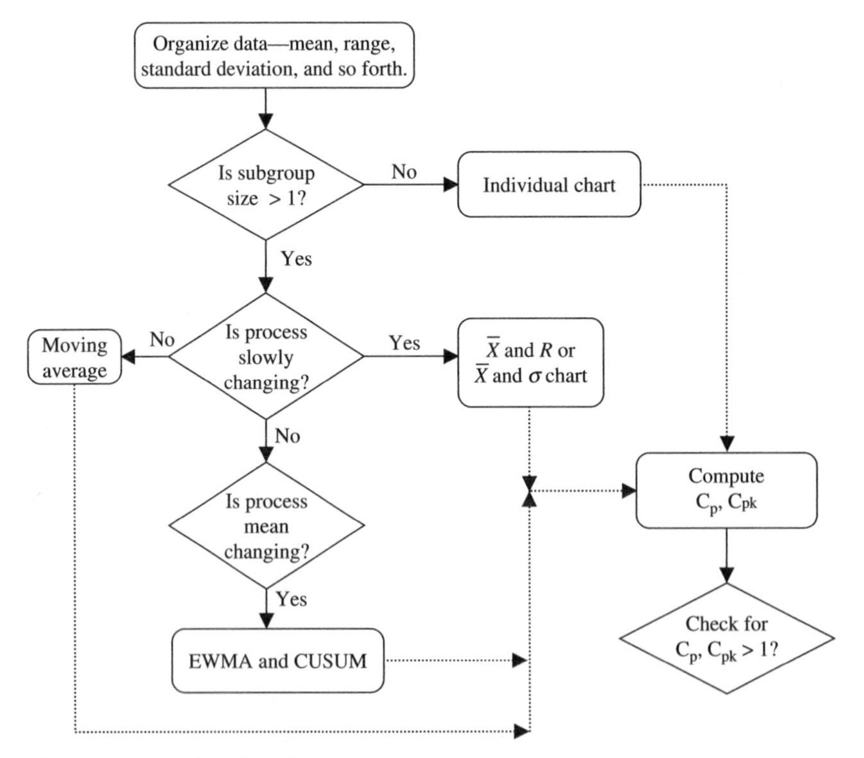

Figure 5.4 Type of control chart.

logical way to choose the right charting methodology. These charts will be discussed with the case study to highlight their features and suitability of use.

5.4.3 Individual charts

Individual control charts are used with a limited dataset. For example, for the case study, suppose that only one sample (for the case study assume it to be the average value of the subgroups) was available for each batch in which case the data set is limited. Due to the limited dataset moving range is used, which is the difference between the consecutive values of the variable (in this case averages of the mean particle size).

Here are the averages of the mean particle size as listed in Table 5.4:

$$\text{The grand average, } \overline{\overline{X}} = 195.46$$

$$\text{Average of moving range} = \overline{R} = 8.74$$

$$\text{UCL} = \overline{X} + 2.66\overline{R} = 195.46 + 2.66 \times 8.74 = 218.7$$

$$\text{LCL} = \overline{X} - 2.66\overline{R} = 195.46 - 2.66 \times 8.74 = 172.2$$

$$\text{UCL}_R = D_4\,\overline{R} \qquad \text{LCL}_R = D_3$$

For $n = 2$, as the moving range is a difference of two consecutive numbers

$$D_3 = 0 \quad \text{and} \quad D_4 = 3.27 \text{ (Table 5.2)}$$

$$\text{UCL}_R = 8.74 \times 3.27 = 28.5$$

$$\text{LCL}_R = D_3\overline{R} = 0 \times 8.74 = 0$$

The averages control chart indicates that the process is out of control. There are indications of three instances of special causes as highlighted in Fig. 5.5. The first special cause is indicated as 7 consecutive points are above the average line (Table 5.3), the second special cause is indicated as there are 14 consecutive alternating up and down points, and the third special cause is indicated by a trend of 7 consecutive points. The moving average chart shows the variation of ranges; it also reflects that it is an unstable process.

5.4.4 Control charts for subgroups

While individual charts are used for a limited dataset, typically in a manufacturing process the quality attributes are measured in subgroups of more than two for each batch. Though changes in process average level tend to be more frequent than changes in process spread,

TABLE 5.4 Dataset for Mean Particle Size Distribution

Batch #	Sample 1	Sample 2	Sample 3	Sample 4	Sample 5	Average	Range	Moving range	Standard deviation
1	185	205	175	180	190	187	30		11.51
2	185	190	215	205	200	199	30	12	11.94
3	180	190	195	190	195	190	15	9	6.12
4	195	190	180	180	195	188	15	2	7.58
5	155	185	165	185	175	173	30	15	13.04
6	210	200	195	200	195	200	15	27	6.12
7	195	180	195	205	185	192	25	8	9.75
8	200	190	175	180	180	185	25	7	10.00
9	175	190	195	190	200	190	25	5	9.35
10	200	210	205	210	180	201	30	11	12.45
11	180	190	205	200	175	190	30	11	12.75
12	185	220	225	190	200	204	40	14	17.82
13	190	180	210	190	175	189	35	15	13.42
14	205	190	195	210	230	206	40	17	15.57
15	210	205	195	195	210	203	15	3	7.58
16	205	205	210	195	200	203	15	0	5.70
17	220	200	210	195	200	205	25	2	10.00
18	200	225	195	220	195	207	30	2	14.40
19	195	205	205	215	205	205	20	2	7.07
20	210	200	215	195	215	207	20	2	9.08
21	205	180	195	200	185	193	25	14	10.37
22	210	190	185	200	190	195	25	2	10.00
23	195	210	190	205	200	200	20	5	7.91
24	195	210	190	215	195	201	25	1	10.84
25	185	190	185	195	195	190	10	11	5.00
26	200	205	200	210	175	198	35	8	13.51
27	185	200	180	190	195	190	20	8	7.91
28	205	195	210	200	185	199	25	9	9.62
29	200	190	205	175	180	190	30	9	12.75

30	205	190	185	210	200	198	25	8	10.37
31	205	175	180	205	180	189	30	9	14.75
32	205	210	195	200	185	199	25	10	9.62
33	195	190	170	185	200	188	30	11	11.51
34	210	210	215	190	175	200	40	12	16.96
35	185	190	195	185	190	189	10	11	4.18
36	210	190	185	200	210	199	25	10	11.40
37	195	175	190	190	180	186	20	13	8.22
38	195	200	190	220	200	201	30	15	11.40
39	185	195	185	170	195	186	25	15	10.25
40	200	190	200	175	185	190	25	4	10.61
41	185	190	220	220	195	202	35	12	16.81
42	205	185	200	170	170	186	35	16	16.36
43	200	205	170	185	180	188	35	2	14.40
44	205	200	185	176	189	191	29	3	11.64
45	215	195	195	210	190	201	25	10	10.84
46	210	205	195	215	195	204	20	3	8.94
47	210	220	190	205	205	206	30	2	10.84
48	206	220	205	214	195	208	25	2	9.51
49	190	170	185	210	175	186	40	22	15.57
50	200	190	215	220	195	204	30	18	12.94
51	210	190	185	210	205	200	25	4	11.73
52	189	195	185	191	205	193	20	7	7.62
53	194	180	175	196	190	187	21	6	9.11
54	190	210	180	185	205	194	30	7	12.94
Total						10555	1410	463	587.67
Average						195.46	26.11	8.74	10.88

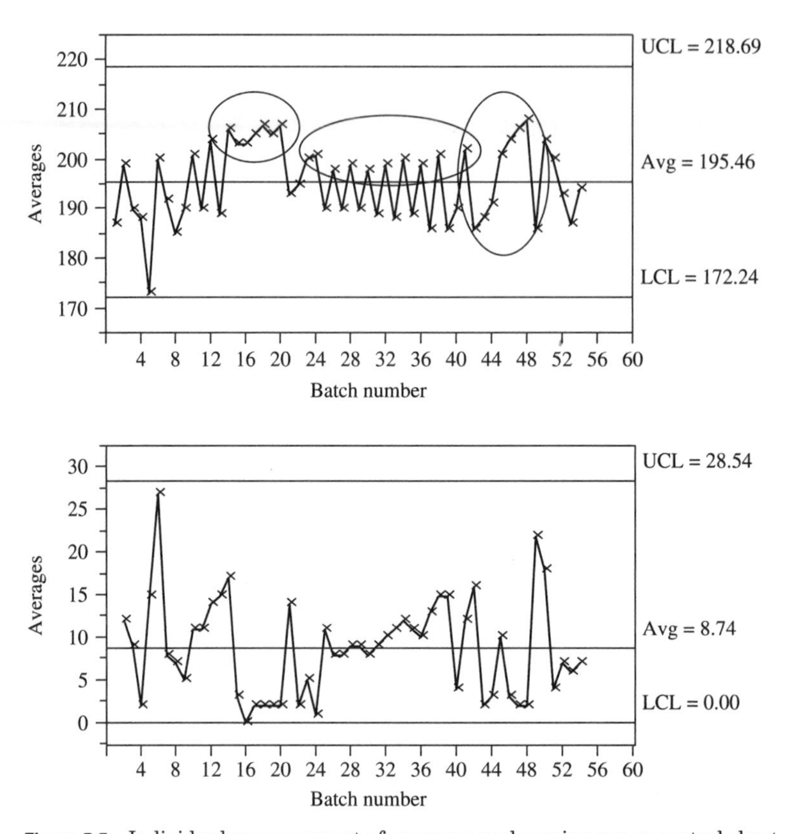

Figure 5.5 Individual measurement of averages and moving range control chart.

it is still necessary to control the variability or the spread of the process to achieve real process improvements. Clearly, even if the process is controlled to the required average level, if the variability is not controlled, the specification bounds can easily be violated leading to an off-specification product. Moreover, tight control of (and reduction in) the variability allows the process mean performance to be moved closer to the higher specification level, thus leading to increased process effectiveness.

Two basic methods of controlling variability for subgroups are:

$\overline{X}$ and $\overline{R}$ chart. Control chart for the range (more widely used).

$\overline{X}$ and σ chart. Control chart for the standard deviation.

$\overline{X}$ and $\overline{R}$ chart. The method described here is based on a number of basic assumptions (Wetherill, 1991):

1. The group sizes are equal.
2. All groups will be weighted equally.
3. The groups are independent of each other.
4. All between-group variation is due to special causes.

To construct an $\overline{X}$ and $\overline{R}$ chart it is necessary to determine the average range either from process capability studies or from at least 20 groups of data using the following calculations:

$$R_i = X_{i,\max} - X_{i,\min}, \overline{R} = \frac{R_1 + R_2 + \cdots + R_n}{n} \tag{5.16}$$

Calculate averages of the samples:

$$\overline{X} = \frac{(\text{sample 1} + \text{sample 2} + \text{sample 3} + \text{sample 4} + \text{sample 5})}{5}$$

Calculate ranges:

$$R = \text{maximum value} - \text{minimum value}$$

$$\text{Average of the averages} = \overline{\overline{X}} = 195.46$$

$$\text{Average of the range} = \overline{R} = 26.11$$

$$\text{UCL}_{\overline{X}} = \overline{\overline{X}} + A_2\overline{R} \qquad \text{LCL}_{\overline{X}} = \overline{\overline{X}} - A_2\overline{R}$$

$$A_2 = 0.58$$

$$\text{UCL} = 195.46 + (0.58 \times 26.11) = 211$$

$$\text{LCL} = 195.46 - (0.58 \times 26.11) = 180$$

$$\text{UCL}_R = D_4\overline{R} = 2.11 \times 26.11 = 55.1 \qquad \text{LCL}_R = D_3\overline{R} = 0$$

Once a chart is constructed, its interpretation is fairly straightforward. As with an individual control chart, the averages control chart indicates that the process is out of control. Besides the three special cause indications similar to that of the individual chart, there is an additional special cause indication—an out-of-control point that was not picked up by the individual chart. This indicates that obtaining data for subgroups is important for a robust control charting methodology. There are four instances of special causes as highlighted in Fig. 5.6. The first special cause is an outside the control limit, the second special cause is the 7 points above the average line (Table 5.3), the third special cause is the 14 alternating up and down points, and the fourth special cause is indicated by a trend of 7 points. Unlike the individual moving range chart, the range chart shows that the variation within the subgroup is controlled.

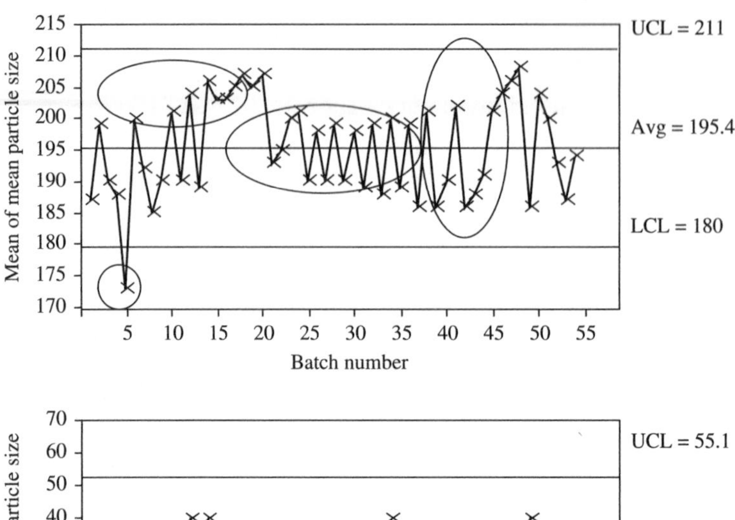

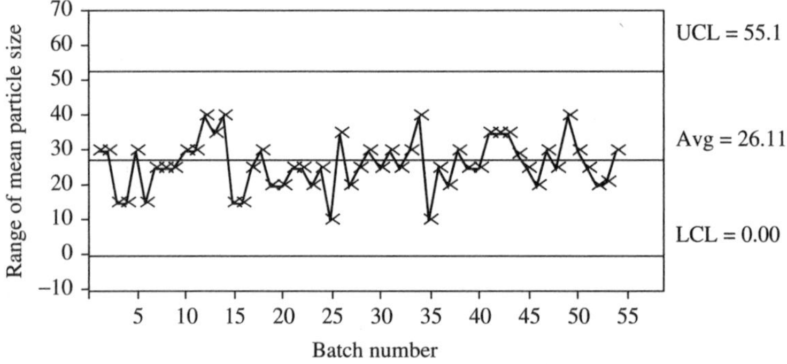

Figure 5.6 X bar and range of mean particle size.

$\overline{X}$ and σ chart

$$\text{U/LCL}_{\overline{X}} = \overline{\overline{X}} \pm A_1\overline{\sigma}$$

where $\overline{\sigma} = 10.88$
$A_1 = 1.60$
$\overline{\overline{X}} = 195.4$

$$\text{UCL}_{\overline{X}} = 195.4 + 10.88 \times 1.6 = 212.8$$

For the ranges:

$$\text{UCL}_{\sigma} = B_4\overline{\sigma} = 2.089 \times 10.88 = 22.73$$

$$\text{LCL}_{\sigma} = B_3\overline{\sigma} = 0 \times 10.88 = 0$$

The $\overline{X}$ and σ chart (Fig. 5.7) and the $\overline{X}$ and $\overline{R}$ chart are very similar in their approach and methodology giving similar output. The $\overline{X}$ and $\overline{R}$

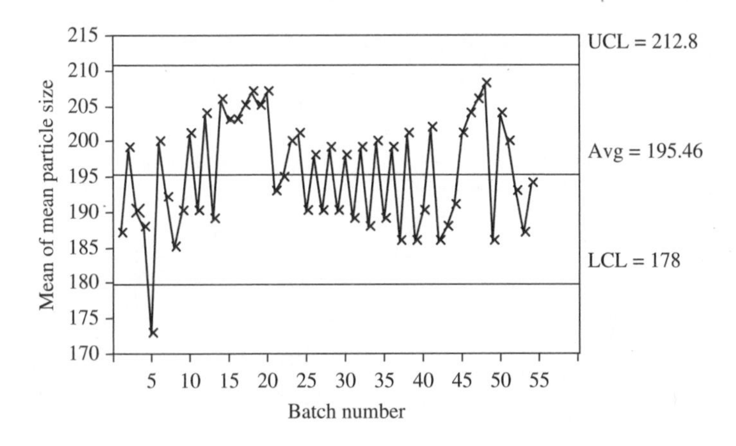

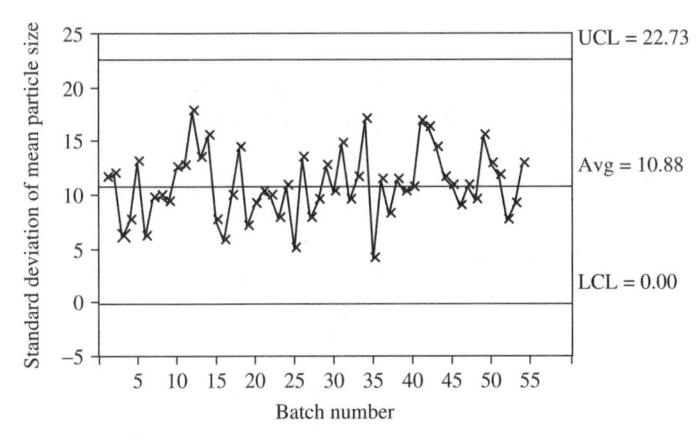

Figure 5.7 $\overline{X}$ and σ of mean particle size.

chart is more popular as there is no need to compute standard deviation of the subgroup, therefore computationally it is easier.

5.5 Advanced Charting Methods

The major disadvantage of the Shewhart control charts described in the preceding section is that they only use the information about the process contained in the last plotted point, and ignore any information given by the entire sequence of points. This feature makes the Shewhart control charts relatively insensitive to small shifts in the process, say on the order of about 1.5 or less.

Three effective alternatives described in the following sections are:

1. Moving average charts
2. *Exponentially weighted moving average* (EWMA) control charts
3. CUSUM control charts

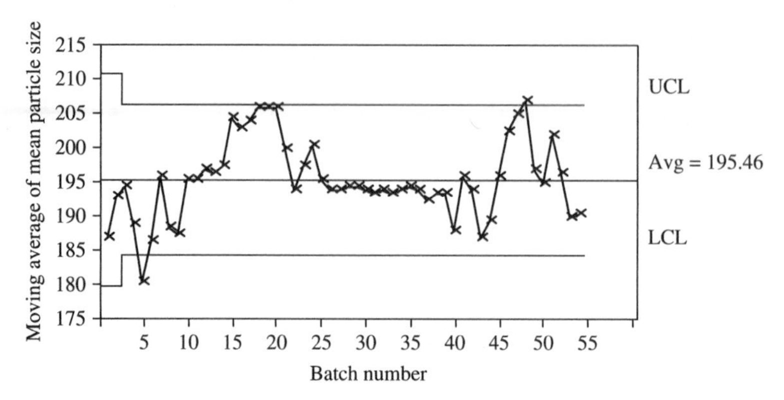

Figure 5.8 Moving average of mean particle size.

5.5.1 Moving average control charts

The moving average control chart, as shown in Fig. 5.8, is based on a simple moving average of the last w groups of size n. Such charts can be used successfully if the group size is limited for some reason, the true mean changes rather slowly, and the process spread is relatively stable. However, since the averages are taken over a period of time, a significant time delay can arise before an out-of-control signal is recorded. Also, if a large number of groups are averaged and only one of them is out of control, it can easily be masked by averaging with other in-control groups.

The moving average chart is fairly simple to construct using the following equations:

$$M_i = \frac{n_i + n_{i-1} + \cdots + n_{i-\omega+1}}{w}$$
$$\text{L/UCL} = \overline{\overline{X}} \pm \frac{3\sigma}{\sqrt{w}} \tag{5.17}$$

This type of a control chart is more effective than the Shewhart chart in detailing small process shifts. However, it is generally not as effective against small shifts as either the EWMA or the CUSUM described.

5.5.2 Exponentially weighted moving average control chart

The EWMA control chart is a modification of the moving average chart in which the past data has some effect on the current value, but this influence rapidly diminishes. This type of chart is particularly useful for monitoring processes with slowly drifting means rather than those liable to sudden jumps. By choice of weighting factor, λ, the EWMA control procedure, can be made sensitive to a small or gradual drift in the process, whereas the Shewhart control procedure can only react when the last data point is outside a control limit. To construct an EWMA chart, it is necessary to calculate the moving weighted average of the past group means, $EWMA_0$, which is the mean of historical data.

$$EWMA_t = \lambda Y_t + (1 - \lambda)\, EWMA_{t-1} \tag{5.18}$$

where $t = 1, 2, 3, \ldots, n$
Y_t = observation at time t
n = number of observations to be monitored including $EWMA_0$
$0 < \lambda \leq 1$ = constant that determines the influence of past observations

A value of $\lambda = 1$ implies that only the most recent measurement influences the EWMA (like a Shewhart chart). A smaller value of λ gives more weightage to historic data. In general λ in the interval $0.05 \leq \lambda \leq 0.25$ works well in practice, and the values $\lambda = 0.05$, $\lambda = 0.10$, and $\lambda = 0.20$ are popular. Lucas and Saccucci (1990) give tables that help the user select λ. The control limits are then calculated as follows:

$$\begin{aligned} UCL &= \mu_o + L\sigma\sqrt{\frac{\lambda}{(2 - \lambda)}} \\ LCL &= \mu_o - L\sigma\sqrt{\frac{\lambda}{(2 - \lambda)}} \end{aligned} \tag{5.19}$$

where L represents the width of the control limit usually tabulated in relevant literature (e.g., see Wetherill and Brown, 1991, Lucas and Saccucci, 1990) for different group sizes and λ values. L, equal to 3 (Lucas and Saccucci, 1990), works reasonably well, particularly with the larger value of λ.

A good way to improve the sensitivity of the control procedure to larger shifts without sacrificing the ability to detect small shifts is to combine a Shewhart chart with the EWMA.

When using such schemes—slightly wider than usual limits on the Shewhart chart (3.25σ or 3.5σ) are used—it is also possible to plot both n_i (or $\overline{n}_i$) and the EWMA statistic on the same control chart along with both Shewhart and EWMA limits.

5.5.3 CUSUM

The CUSUM chart is another method for detecting changes in the operating mean. The concept is extremely simple, but it is a powerful method that indicates changes in process performance and provides clues to assist in assigning causes. The chart is constructed by subtracting the overall mean from a data set and then cumulating the differences from the mean. A section from a typical CUSUM chart is shown in Fig. 5.9.

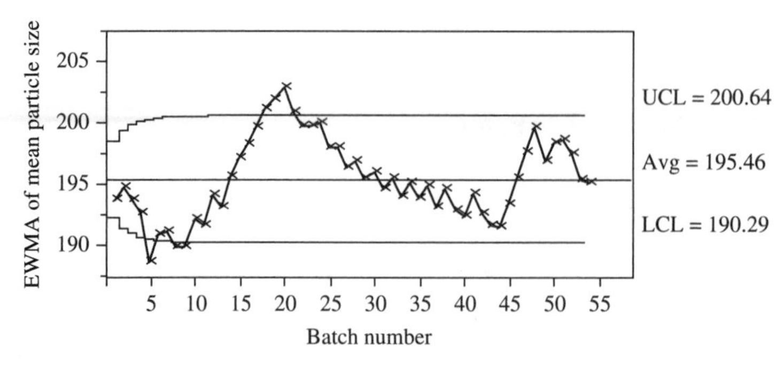

Figure 5.9 EWMA of mean particle size.

The change of slope in the chart indicates some change in the process. For example, for case study A for fermentation operation, this chart may correspond to the final fermenter production rate over a number of batches (300 in this case). Significant changes in the slope occur at around batches 70, 160, and 220, thus indicating a change in the process conditions. Although this does not indicate the cause of the deviation, pinpointing the batches at which changes occurred is useful as operating records may show whether any routine operation deviations occurred. So, for example, it may be found that a different feed supplier was used for batch 160 onward—such subtleties can be difficult to detect by looking at traditional time trends.

To construct a CUMSUM control chart the following parameter is calculated:

$$\subset_i = \sum_{j=i}^{i} (\overline{u}_j - \mu_o) \tag{5.20}$$

where μ_o = target for the process mean

$\overline{u}_j$ = average of jth sample

$\subset_i$ = cumulative sum up to and including the ith sample

If the process remains in control at the target value μ_o, the cumulative sum defined in equation is a random walk with mean zero. For example, if $\mu_o = 10$

$$\begin{aligned}\subset_i &= \sum_{j=1}^{i} (n_i - 10) \\ &= (n_i - 10) + \sum_{j=1}^{i=1} (\lambda_j - 10) \\ &= (n_i - 10) + \subset_{i-1}\end{aligned} \tag{5.21}$$

Two ways of interpreting the CUSUM control charts are used in industry. They are as follows:

Tabular (or algorithmic) CUSUM. Tabular or algorithmic CUSUM to monitor the process mean works by accumulative derivation from μ_o that is above target with one statistic $\subset^+$ and accumulating durations from μ_o that are below target with another statistic $\subset^-$. $\subset^+ + \subset^-$ are called one-sided upper and lower CUSUMS and are calculated as follows:

$$\subset_i^+ = \max\,[o, n - (\mu_o + u) + \subset_{i-1}^+] \tag{5.22}$$

$$\subset_i^- = \max\,[o, (\mu_o - u) - n_i + \subset_{i-1}^-] \tag{5.23}$$

where the starting values are $\subset_o^+ = \subset_o^- = o$K, which is the reference value or slack, often chosen about halfway between the target and the out-of-control value.

Mask forms (V-shape, semiparabolic, or snub-nosed V-masks) of CUSUM. Similar to Shewhart charts, which use simple decision rules (the action and warning lines) to determine when the process is out of control, it is possible to construct such decision rules for CUSUM charts using masks. A typical V-mask is shown in Fig. 5.10, where the decision lines and the means of constructing them (using h and f values) are indicated.

In the standard V-mask, $h = 5$ and $f = 1/2$, so that the decision interval is 5 standard deviations. This standard V-mask can be modified (semiparabolic or snub-nosed V-mask) to improve the behavior of

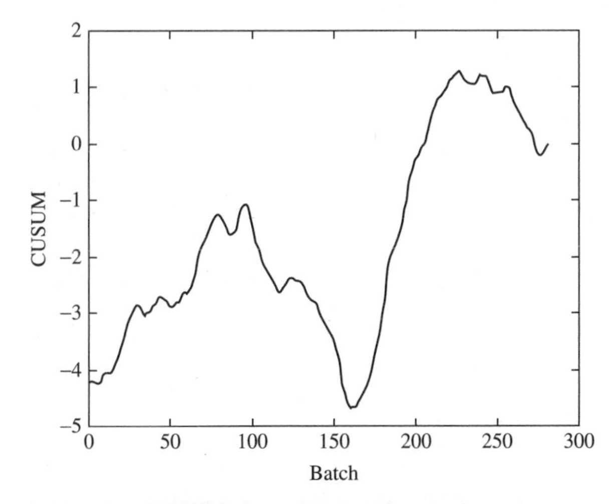

Figure 5.10 CUSUM chart showing changes in mean operating level.

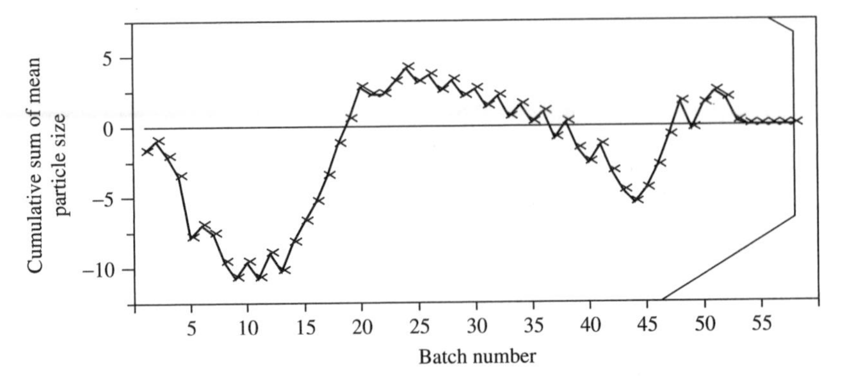

Figure 5.11 CUSUM of mean particle size.

CUSUM charts in the case of very large changes in process mean. The case study CUSUM is shown in Fig. 5.11 and is a standard V-mask.

Overall, the CUSUM charts show improvement in efficiency over Shewhart charts in the region of 0.5 to 2.0 standard deviations. They provide a useful visual indication of change in mean by change in slope and allow the point of change to be pinpointed. They are also useful where the measurements are fairly expensive and the observations are available only singly. However, they are more complex to use than Shewhart charts and more difficult to interpret when process mean changes with process improvements.

5.6 Capability Analysis of a Stable Process

The previous sections focus on understanding and quantifying variation. It is equally important to understand the capability of a stable process—a kind of a robustness scale. Dr. Taguchi's definition of process capability as "on target with minimum variance," will result in minimum nonconformance and low cost of production. Along with the variance it is also equally important to understand the positioning of the process in relation to the specification limits. This aspect of positioning is described by the parameter, *distance to nearest specification* (DNS).

$$\text{DNS} = \text{minimum of } \{Z_U, Z_L\} \tag{5.24}$$

where $Z_U = (\text{upper specification limit} - \text{average})/\sigma_x$
$Z_L = (\text{average} - \text{lower specification limit})/\sigma_x$

DNS value output is in the units of sigma; it is generally convenient to represent this as a ratio of 3 sigma; this ratio is called the *capability ratio* C_{pk}.

$$C_{pk} = \frac{DNS}{3} \tag{5.25}$$

Another capability ratio computes the spread of the specification limits highlighting the elbow room. This is depicted by:

$$C_p = \frac{(\text{upper specification} - \text{lower specification})}{6\sigma_x} \tag{5.26}$$

As a rule-of-thumb, C_{pk} and C_p should be greater than 1 for a stable process. Further, in comparing different processes a comparatively higher value of capability ratio may indicate a greater centrality and stability.

Assuming that the special causes are eliminated for the crystallization case study leading to a stable process, the above concepts could be as follows:

Average = 195.46

Standard deviation = 7.689

Upper specification = 280

Lower specification = 110

Z_U = (upper specification limit − average)/σ_x = (280 − 195.46)/7.689 = 11 σ

Z_L = (average − lower specification limit)/σ_x = (195.46 − 110)/7.689 = 11.1 σ

DNS = 11 σ

C_{pk} = DNS/3 = 11.1/3 = 3.7

C_p= (upper specification − lower specification)/6σ_x = (300 − 110)/ 46.1 = 4.1

The given numbers imply that the process is centered and is capable. Taguchi has indicated that departure from centrality would lead to more losses (Taguchi loss function). It is important to strive to keep the process centered on the target even when the capability ratios are well above 1.

The earlier sections cover the concepts of SPC and the related charting methodologies. SPC is generally confused with *engineering process control* (EPC); the next section focuses on the distinction between the two.

5.7 EPC and SPC

EPC (Chap. 3 covers EPC in detail) schemes only react to process upsets. The control system reacts in a feedback mode when there are error signals generated by a difference between the set point and the measured value. EPC schemes do not make any effort to remove the assignable causes. Consequently, in processes where feedback control is used, there may be substantial improvement if control charts are also used for statistical process measuring.

Successive deviations beyond control limits or single deviations beyond action limits indicate process changes or problems. Warning and action limits are set at three standard deviations from the mean operating level. A typical $\overline{X}$ chart is shown in Fig. 5.12.

In Fig. 5.12 the signal mean is displayed, along with upper and lower control limits (UCL and LCL) and upper and lower action limits (UAL and LAL). It is possible to set additional rules, for example, if there are seven successive points increasing or decreasing then take action, but such rules can increase the risk of taking action when none is required.

The concept of taking action only when required is different from that generally adopted by control engineers. For example, in the case of PID-type algorithms (the method adopted in all standard controllers) for process regulation, control action is taken when any deviation from the set point is observed. Using the SPC methodology, action is taken only when significant deviations are indicated. The difference primarily arises at the level of control hierarchy at which the methods operate.

There should be an integrated approach between EPC and SPC to sustain a robust process. In an ideal scenario, SPC should focus on identifying special cause variations and addressing those, and EPC should focus on adjusting the set point to the common cause variation holding

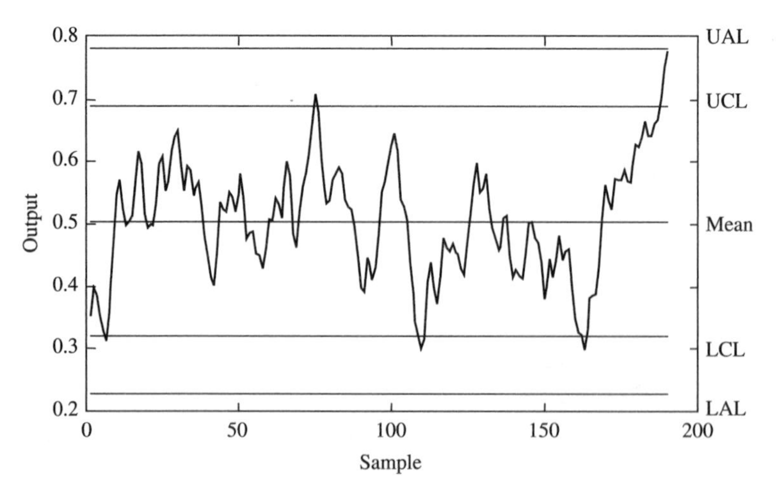

Figure 5.12 Shewhart $\overline{X}$ chart.

the process steady. An important aspect in SPC is identifying the special cause; RCA is the technique used for special cause identification.

5.8 Root Cause Analysis (RCA)

It is important to learn a structured technique to understand the causes of variation especially special cause variation, so that corrective action can be taken. RCA is a structured investigation into a problem to identify the underlying true cause and the actions required to eliminate it (see Fig. 5.13). This is a problem-solving process with roots in the quality methodology and strong linkages to the PDCA approach.

Problem identification and understanding are the first steps of RCA to ensure that efforts are directed at the right problem. Tools for problem identification include flowcharts, critical incident charts, spider charts, performance matrices, and so forth. The most widely used tool is flowcharting and it will be used in this book. Appropriate data collection distinguishes between haphazard problem solving and structured RCA. Data could be obtained from a variety of sources—people, data historians, surveys, checksheets, fishbone diagrams, and so forth. The next step is root cause identification, which includes understanding causal relationship by using analysis tools such as histograms, Pareto charts, scatter diagrams, and regression analysis. The intention is to establish causal relationship using a cause-and-effect chart, matrix diagram, and five "whys" (Andersen and Fagerhaug, 2000). A technique introduced in Chap. 4 of this book is knowledge extraction, and it could

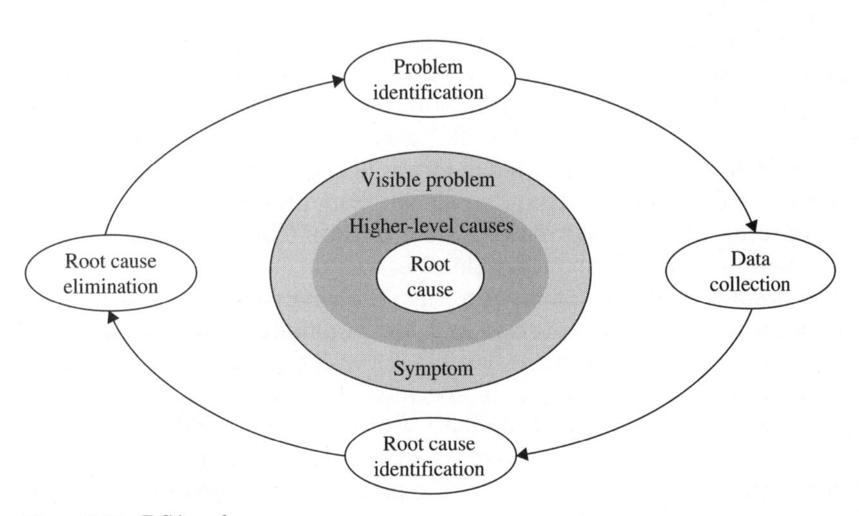

Figure 5.13 RCA cycle.

be used as a comprehensive causal analysis tool. Finally, the identified root cause is eliminated.

The RCA methodology could be illustrated using an example. For case study E on crystallization, the root cause methodology could be utilized to address the first special cause identified as outside the control limit (see Fig. 5.6). This out-of-control batch led to deviation, and the batch was held until a full investigation indicated suitability for release.

5.8.1 Problem identification

A root cause analysis could be effective to eliminate special cause, as identified by the control chart. The first step in problem identification is a rigorous analysis of the possible reasons to generate a hypothesis, which could then be investigated further with appropriate data collection. A flowchart can be used to map a process that illustrates where problems occur and which problems should be solved. Also, a flowchart could be used to help generate a hypothesis that could lead to a focused data collection state.

Figure 5.14 shows the flowchart for the case study stating the problem. The mean particle size was found to be within the conformance limit though the process was out of control. The possible causes were identified as raw material and environment; people and operation; and equipment and maintenance. A structured knowledge extraction was conducted to further explore the possible causes to ascertain the root cause.

It is critical to extract knowledge in a structured fashion such that appropriate focus and direction can be achieved, rather than to jump into immediate firefighting with no clear direction. A clear focus can lead to a targeted data acquisition effort to assist understanding and development of causal relationship to resolve the problem. The knowledge extraction technique described in Chap. 4 will be applied to this case study.

In this case study a knowledge extraction team was created with a knowledge engineer eliciting knowledge from an engineer, a scientist, an operator, and one maintenance technician. The key question was: what will lead to different particle sizes?

The knowledge extraction led to the following possible causes under the categories of raw material and environment (color coded gray); people and operation (color coded black); and equipment and maintenance (color coded white). The likely cause in the category of raw material and environment included different reactant and antisolvent concentration. The likely cause under the category of people and operation included the standard operating procedure not followed; the temperature not maintained; and the manual valve operation not done due to an untrained operator. The final category of equipment and maintenance

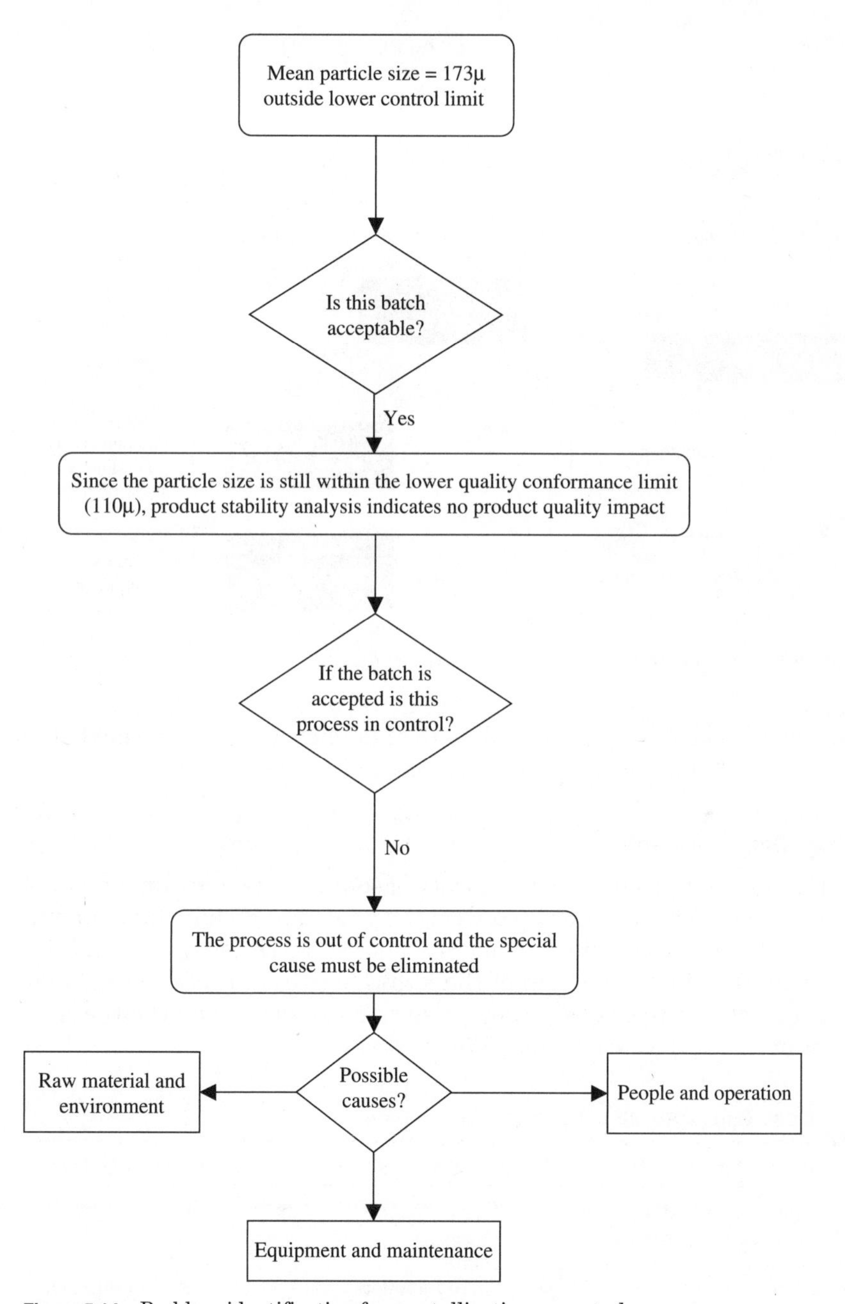

Figure 5.14 Problem identification for crystallization case study.

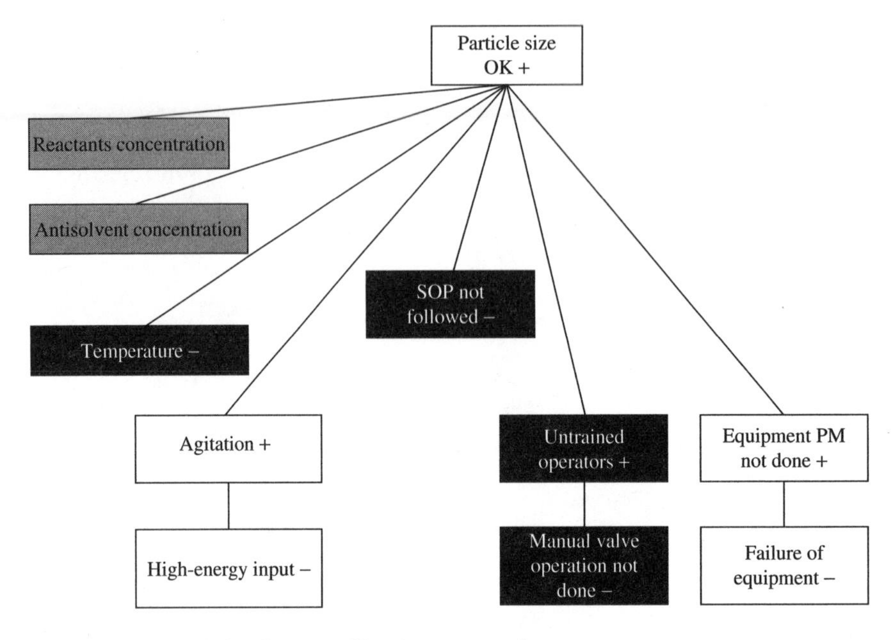

Figure 5.15 eGraph for the crystallization case study.

included change in agitator leading to high-energy input. The next step was to collect data for the above categories.

5.8.2 Data collection

Data could be collected from a variety of sources, the most important of which are the people associated with the process, which includes operators, engineers, scientists, and maintenance staff. The eGraph (Fig. 5.15) presents the elicited version of the knowledge from those individuals. The next step is to use this knowledge to construct causal relation verification using data (see Table 5.5).

TABLE 5.5 Causal Relation Verification Using Data

Control points	Prior to special cause	During special cause
Reactant concentration	Same	Same
Antisolvent concentration	Same	Same
Temperature	Same	Same
Agitation	A310 impeller Power = 4984 W	Rushton impeller Power = 11604 W
SOP followed	Yes	Yes
Untrained operators	No	No
Equipment PM	Done	Done

5.8.3 Root cause identification

From knowledge extraction and data collection from the data historian "change of impeller from a low shear (A310) to a high shear impeller (Rushton) operating at the same impeller speed" was identified as the root cause. A scientific causal relationship (empirical or mechanistic) must be established prior to troubleshooting. For the case study a review of literature suggested that the impeller shear could be related to the power input to the tank or power per unit mass. An analysis was conducted with the data, as shown in Table 5.6.

Thus, the symptom was modification in the mean particle size, and the cause was increased power per unit mass from 0.36 to 0.84 W with the modified impeller setup. This increase in power per unit mass has created a higher shear leading to the reduction in particle size from 195.4 to 173 μm.

5.8.4 Root cause elimination

Once the root cause is identified and the causal relationship established using scientific rationale, then the next and final step is to eliminate the root cause. For the case study the root cause of impeller modification could be eliminated in two ways: (1) retrofit the original impeller or (2) optimize the operation of the new impeller to provide the same power per unit mass. Optimization of the operation of the new impeller system was preferred for the case study mainly for the speed and ease of root cause elimination. The strategy employed was to manipulate the impeller speed for the new impeller system (Rushton impeller) to give the same power per unit mass as the previous impeller system (A310).

TABLE 5.6 Root Cause Identification

Parameters*	Old setup/axial flow impeller – A310	New setup/radial flow impeller – Rushton
Tank diameter, T(m)	2	2
Volume, V(m^3)	12.56	12.56
Impeller diameter, D(m)	1	0.67
Impeller speed, N(rpm)	150	150
Power number, N_p	0.29	5
Power (W)	4984	11602
Power per unit mass (W/Kg)	0.36	0.84
Mean particle size (μm)	195.4	173

*Parameter definitions: P = power input, W; $P = N_p \rho N^3 D^5$; N_p = power number of the impeller; D = diameter of the impeller, m; ρ = density of process fluid = 1100 kg/m^3; and power per unit mass = $P/V\rho$, W/kg.

Following calculations, it was established that at the impeller speed of 113 rpm the new impeller system delivered the same power per unit mass as the old system. This was an easy fix as it required only a change in the impeller speed and the subsequent batch had a mean particle size of 200 μm, thus ensuring that the root cause was eliminated.

Typically, in an industrial environment one of the frequent root causes for increased variability is manufacturing system decay, especially important is the decay associated with the physical assets (namely equipment, facilities, and the like). The next section covers system decay related to equipment.

5.9 System Decay

A manufacturing system like any universal system will decay over time. The second law of thermodynamics implies that systems will deteriorate on their own, unless appropriately worked upon. System decay is one of the major causes of variability; for example, old equipment without proper maintenance will lead to stoppages and/or deviations. Maintenance of the manufacturing system is an important factor in sustaining a capable manufacturing system, and a good maintenance strategy is an important element of world-class manufacturing organizations.

There have been various approaches to equipment maintenance. Corrective maintenance is a reactive approach—fixing things either when they are found to be failing or when they have failed. A reactive approach involves "fixing after it is broke;" this used to be the old maintenance model. The maintenance strategy has evolved over time leading into the modern *reliability centered maintenance* (RCM) strategy.

RCM is defined as a process used to determine what must be done to ensure that any physical asset continues to do what its users want it to do in its present operating context (Moubray, 1997). RCM focuses on functions and performance standards, functional failures, failure modes, failure effects, failure consequences, proactive tasks, and default actions.

The first step in the RCM process is to define the functions of each asset in its operating context, together with the associated desired standards of performance. The next step is to understand the functional failures that will make the asset unable to fulfill a function to a standard of performance, which is acceptable to the user. Once each functional failure has been identified, the next step is to identify all the events which are likely to cause failure or failure modes. The fourth step in the RCM process entails listing failure effects, which describe what happens when each failure mode occurs. The failure effects lead to an understanding of failure consequences. *Failure modes and effects analysis*

(FMEA) could be used as a structured methodology for failure analysis of mode and effects.

The RCM divides failure management techniques into two categories, proactive maintenance and default actions.

Proactive maintenance includes what is traditionally known as preventative and predictive maintenance. Preventative maintenance involves taking action at or before a specified age limit. This action can take one of two forms:

1. Scheduled restoration, which entails remanufacturing a component or overhauling an assembly at or before a specific age limit, regardless of its condition at the time.
2. Scheduled discard, which entails discarding an item or component at or before a specified life limit, regardless of its condition at the time.

This is a widely used technique, therefore it is important to note here that this technique should be carefully engaged otherwise it could be a waste of resources. This caution stems from the traditional understanding of age-related equipment failure model (*bathtub model*). The bathtub model assumes that the probability of failure is higher during the initial states of equipment life (due to issues of equipment integration, tuning, and the like). This is followed by a low probability of failure (the useful life of the equipment), and then followed by an increased probability of failure due to wear and tear, fatigue, and corrosion. This model holds good typically where the equipment comes in direct contact with the product, for example, pumps, mixers, tanks, and so forth. However, there are five other failure models described by Moubray (1997). As an example, a study performed on several hundred mechanical, electrical, and structural components of civil aircraft showed that only 4 percent of the failure followed the bathtub model (Moubray, 1997). Therefore, the bathtub model is not the only failure model, and a careful evaluation should be done on age-related failure mechanism (following the RCM process) before engaging in preventative maintenance.

Predictive maintenance is applied when identifiable physical conditions indicate that a functional failure is about to occur or is in the process of occurring. This technique if not directed properly could be a waste of resources. The RCM process of failure mode identification and its prioritization helps direct the resources to appropriate tasks.

Default actions include failure finding, redesign, and no scheduled maintenance. Failure finding focuses on checking the function of the equipment periodically to ascertain any failure. Redesign involves making one-off changes to make the design more reliable. No scheduled maintenance implies that the equipment is run-to-failure.

The RCM methodology determines the appropriate maintenance strategy by applying structured methodology, namely FMEA. This methodology helps make the decision which (if any) of the proactive tasks is technically feasible in any context. It also incorporates criteria for evaluating whether the task is worth doing with regard to the consequences of failure. Further suggested reading on RCM and maintenance strategy includes Bloch and Geinter (1990) and Smith and Hinchcliffe (2004).

Appropriate engagement of RCM will significantly impact physical-asset-related manufacturing systems decay. RCM can lead to greater safety and environment integrity, improved operating performance, greater maintenance cost-effectiveness, greater teamwork, and a comprehensive maintenance database.

Concluding Comments

Variability in a manufacturing operation must be controlled and minimized. The key aspects of variability reduction are

- Making variability visible by control charting the key variables and quality attributes
- Maintaining a persistent watch for out-of-control signals
- Acting immediately to ascertain the root cause for special causes as indicated by the control chart
- Using structured methodology like root cause analysis to identify the root cause
- Moving and maintaining the manufacturing process in the ideally capable quadrant
- Managing system decay

References

Andersen, B., and T. Fagerhaug (2000), *Root Cause Analysis*, ASQ Quality Press, Milwaukee, WI.

Bloch, H. P., and F. K. Geintner (1990), *An Introduction to Machinery Reliability Assessment*, 2d ed., Gulf Publishing Company, Houston, TX.

Deming, W. E. (1950), *Some Theory of Sampling*, Dover Publications, New York. Deming, W. E (2000), *Out of Crisis*, MIT Press, Cambridge, MA.

Lucas, J. M., and M. S. Saccucci (1990), "Exponentially weighted moving average control schemes: Properties and enhancements," *Technometrics* vol 32, No. 1, pp. 1–12.

Moubray, J. (1997), *RCM II*, 2d ed., Butterworth Heinemann, Oxford, UK.

Pitt, H. (1994), *SPC for the Rest of Us*, Addison-Wesley, Reading, MA.

Shewhart, W. A. (1931), *Economic Control of Quality of a Manufactured Product*, Reinhold, Princeton, NJ.

Shewhart, W. A. (1986), *Statistical Method from the Viewpoint of Quality Control*, General Publishing Company, Toronto, Ont.

Smith, A. M., and G. R. Hinchcliffe (2004), *RCM – Gateway to World Class Maintenance*, Elsevier Butterworth-Heinemann, Burlington, MA.
Wetherill, B., and D. Brown (1991), *Statistical Process Control. Theory and Practice*, Chapman and Hall, UK.
Wheeler, D. J, and D. S. Chambers (1992), *Understanding Statistical Process Control*, 2d ed., SPC Press, Knoxville, TN.

Chapter

6

Emerging Monitoring and Control Strategies

Control makes the product, advanced control makes the profits!
JIM ANDERSON

Objective

This chapter introduces new and emerging monitoring and control strategies aimed at further reducing variability and enhancing process performance.

Learning outcomes of this chapter will include:

- Techniques for advanced process control including artificial neural network and model predictive control
- Introduction to *multivariate statistical process control* (MSPC)
- Introduction to artificial neural network pattern recognition
- Introduction to knowledge-based systems

The learning is then applied to a comprehensive case study to gain practical understanding of the application of the concepts.

6.1 Introduction

New technologies have brought new opportunities for productivity enhancement in process monitoring and control. This chapter considers methods that have been proven to be industrially relevant but are yet to find general application across all sectors of the process industry. It is not the technology that limits their general applicability, but simply that confidence in use and an appreciation of benefits take time to permeate throughout the industry. The four strategies considered are:

- Artificial neural networks can be used to build models that relate on-line process measurements to off-line assays, with the models being used to provide on-line estimates of critical quality variables.
- Model-based predictive control can be used to overcome process challenges that conventional control fails to deal with. Interactions between loops, delays, and complexity lead to deterioration in quality control, but through the use of models cast in a predictive algorithm it is possible to maintain control within tight operating limits.
- Interpretation of numerous process measurements to extract information pertaining to process quality deviation is not a simple task for the process operator. Univariate SPC is a useful tool but its multivariate counterpart (multivariate SPC) compresses process data to produce operating signatures that are more straightforward to interpret.
- Knowledge-based systems provide a means within which to implement human expertise to act as a computer-based advisory system. With links established to process databases, real-time advice can be generated for process operators to act upon. The rapid detection of deviations due to process abnormalities will lead to greater consistency in quality.

The common characteristic of the four strategies is that the technology is not developed to such an extent that a nonexpert can implement it. This chapter reviews the methods with the emphasis being to provide an appreciation of potential rather than attempting to be a guide to implementation.

6.2 Advanced Process Control

Advanced process control can mean different things to different people. Some control engineering textbooks refer to "advanced control" as "optimal control" or "$H\infty$ control." Other texts focus more toward the process sector and use the term "advanced control" to mean some of the approaches discussed in Chap. 3. Clearly what constitutes advanced control is sector specific. In the process industries basic regulatory controls, whether they be feedback, cascade, or feedforward systems, are commonplace and in this sense are not considered here to be "advanced." But advanced control strategies found in other sectors such as aerospace are hardly used in the process sector. So what constitutes advanced control for the process sector? The answer without doubt is *model-based predictive control* (MPC). Its popularity may not have "filtered through" to all process industry sectors, but from its conception in the oil industry, it is now starting to impact across the process sector. Continuous chemical operations have followed the oil industry lead with batch plant application now starting to be realized.

To understand the popularity of MPC it is necessary to consider what makes process control difficult:

- *Background.* Process engineers find it difficult to express process dynamic characteristics in the form required by some algorithms. A process engineer will generally be able to specify response times and gains but would struggle to translate this to a Laplace domain process model.
- *Constraints.* Both hard and soft limits exist for plant variables as a result of physical limitations (e.g., valves open from 0 to 100 percent) or the operating policy (e.g., the dissolved oxygen must always be above 10 percent).
- *Process delays.* A delay or dead time between control actions being implemented and process output changing makes determining appropriate modifications difficult as observation of their effect takes a considerable time to propagate around the control loop. Significant delay is normally considered to be larger than the process time constant.
- *Multivariable structure.* Controlling a single process variable is seldom sufficient. Typically, process plants will have multiple loops that interact causing degradation of process performance. In addition, it is necessary to be aware of the operational objectives of the plant in total, not just individual units. Inevitably the resulting control problem is multivariable.
- *Process nonlinearities and difficult dynamics.* Process dynamics that change as operating levels vary can cause a well-tuned controller to become less effective. Certain types of linear dynamics can also be difficult, such as inverse response where the initial dynamic is in the opposite direction to the final response or when the time delay varies, typically as a consequence of changes in flow-related transport.
- *Process uncertainty.* Exact models can never describe the process plant. Process changes, natural deviations, and complexity of behavior result in a significant level of imprecision.

Classical control strategies described in Chap. 3 are limited in effectiveness when faced with these problems. Furthermore, the more advanced strategies of control such as optimal control approaches fail to address the practical application issues. This motivated those involved in the process industry to formulate an algorithm that was both suited to the problems faced in the industry and at the same time was comprehendible by the plant engineers. Cutler and Ramaker, while working for Shell Oil in the early 1970s instigated the development of MPC with their algorithm *dynamic matrix control* (DMC). Although the initial algorithm considered unconstrained systems, subsequent developments brought in constraints. Numerous other commercial packages have followed with subtle variations but the basis of the algorithm is predominantly the same. Further details of the historical development of MPC can be found in a review paper by Qin and Badgwell (2003). The paper also provides information on the more common commercial MPC products that are available.

6.2.1 The concept of model-based predictive control

The basic philosophy for the control system is that if there is a model of the process it can be used to predict how the process will respond in the future. MPC is a discrete controller as it determines control changes at fixed sample intervals and holds the signal constant between the samples. Thus, the process models in the controller are specified as discrete time series or continuous time models converted to discrete time series. Further details of modeling processes using discrete time models can be found in Ljung (1999).

When a model is available, future response can be predicted, and then current control action can be chosen to tailor future conditions and thus satisfy a reference trajectory. However, from a practical perspective it is important for the process output to be forced toward the set point, but it is also essential to take account of the control effort to get the process output to set point (desired value). This leads to a simplistic profit function of the form:

$$J = \sum (r - y)^2 + \sum \lambda \Delta u^2 \tag{6.1}$$

where r = reference trajectory (set point)
y = model predicted output
Δu = change in manipulated variable (control effort)

This cost function balances deviation from set point with changes in control effort where λ sets the balance. An algorithm that minimizes this cost function will determine Δu that minimizes J. As an example, r could be a temperature set point trajectory in a batch reactor, y the future temperature of the reactor as predicted by a model, and Δu changes in the cooling water valve position.

There are several issues around using this cost function. First, the deviation from set point is best assessed over a period of time rather than at a single value, hence the summation. The period of time over which the deviation is calculated would sensibly start from the point when the calculated update in control action could influence the process output. The purpose of the controller is to drive the process output to be equal

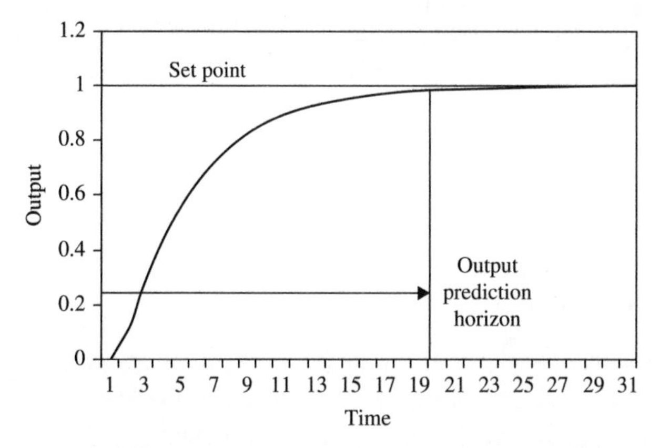

Figure 6.1 Future process behavior and the output prediction horizon.

to the set point, thus after a period of time there will be no benefit to calculating the summation of errors as they should be zero. The limit over which errors from set point are calculated is termed the *output prediction horizon*. This concept is demonstrated in Fig. 6.1.

If the future response of the process is to be predicted then it is necessary to know how the manipulated variable moves. This requires the determination of future manipulated variable moves. The concept is demonstrated in Fig. 6.2.

In this case the changes in manipulated variables up to time = 4 are postulated and after that time control actions are assumed to be fixed. In the future the limit over which manipulated control move changes are allowed is termed the *control action horizon*.

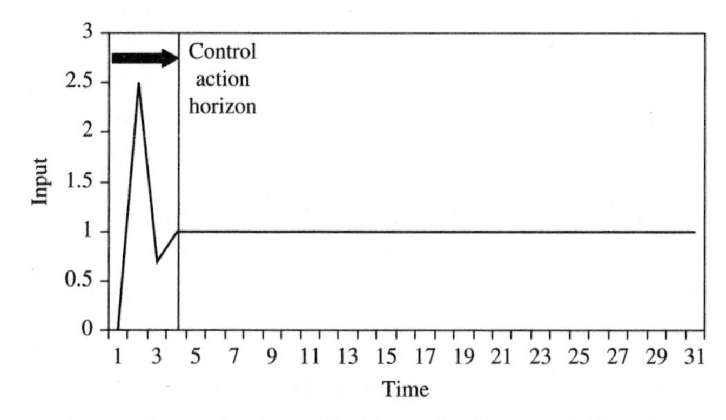

Figure 6.2 Determination of future manipulated variable values.

Given the concept of horizons, the controller profit function can now be modified and extended:

$$J = \sum_{j=k}^{N_2} Q(r(t + j) - \hat{y}(t + j))^2 + \sum_{j=1}^{N_u} \lambda \Delta u(t + j)^2 \qquad (6.2)$$

where N_2 = output prediction horizon
N_u = control action prediction horizon
λ = control action weighting

and ^ is used on $y(t + j)$ to refer to the fact that it is estimated from the process model. The introduction of the term Q into the first term of the cost function raises an important aspect of the controller. In Chap. 3 the control problem was primarily considered as a single-input/single-output problem. This may be satisfactory but in many cases interactions between control loops can cause deterioration in the performance. For example, a temperature control loop in an exothermic reactor, which modifies temperature by changing cooling water flow, would interact with a loop modifying reactant additions to control product concentrations. The cost function could consider deviations in both concentration and temperature from respective reference trajectories and account for the control effort to achieve them. In such a case, errors in temperature control would have to be balanced against errors in product concentration. This is achieved by the term Q that serves to weight the output errors based on the relative importance of achieving the objective. In practice, application of such methods to the plant considers multistage processes and control problems are typically of high complexity (i.e., many tens of inputs and outputs).

Minimization of the cost function allows the manipulated variable changes to be determined. But how can this function be minimized? In the absence of process constraints it is possible to derive an analytical solution and determine the manipulated moves directly. Unfortunately, satisfying process constraints is usually a fundamental operating objective. Maximum, minimum, and rate-of-change limits on manipulated variables and maximum and minimum limits on output variables would be typical, and consequently the solution of the cost function is more complex. Quadratic programming (QP) is the most common solution strategy as the cost function with constraints is of the appropriate form (Luenberger, 2003). Essentially, QP is an optimization search procedure looking for manipulated variable changes that minimize the function without violating a constraint. There may be circumstances that arise when it is not possible to solve the cost function without violating constraints. In these cases constraints need to be relaxed in a systematic manner to achieve a feasible solution.

An important characteristic of the algorithm has yet to be mentioned. The earlier discussion considers a single sample instant in time. At this point a vector of future control moves is calculated and the first move in the sequence is implemented. Further elements in the vector are discarded. At the next sample instant the *whole* procedure is repeated with a new set of increments determined.

The theory behind MPC and constrained MPC in particular is quite considerable. All that has been discussed is a brief summary of the strategy with reference books such as those by Maciejowski (2001) or Soeterboek (1992) providing a comprehensive and detailed theoretical basis for the MPC algorithm.

6.2.2 Implementation issues

It is first important to recognize that the implementation of MPC is far more involved than the simple feedback control loops discussed in Chap. 3. Applications generally involve multivariate problems where the MPC controller sits above local feedback loops such as flow control loops. The complexity of the problems tends to mean that implementation costs are not insignificant, but the benefits to be achieved can far outweigh the initial outlay, resulting in payback times measured in terms of months rather than years. The first step in the implementation of an MPC is to ensure that the problem that it is to be applied to is financially viable. Here the cost-benefit-risk assessment as discussed in Sec. 7.5.1 should be followed. With the financial case justified, this section considers the practical implementation steps that must be followed.

Familiarization with plant operation. The implementation of an MPC strategy is not a job for the nonexpert in control systems. It is therefore probable that the team responsible for the implementation is brought in to carry out the project and is therefore not familiar with the intimate details of the process operation. Not only must they understand the process dynamic characteristics and operating policy, but they also need to be acquainted with the financial aspects of the process. The MPC in essence controls the process to satisfy a cost function, the basis of which is likely to be financially based. Operating policies and constraints may change as market conditions vary, and the MPC must accommodate this. Furthermore, in the early stages of an implementation project it is essential to build up the confidence of the plant operating staff because their assistance in determining the plant models is paramount.

Model determination. At the heart of the MPC scheme are models that describe the behavior of the process. The algorithm uses the models to forecast the effects of changing the process-manipulated variables on the output variables. It is therefore important to derive models that adequately describe process dynamics. These models are not based on mechanistic differential equations but identified from process responses. Normal plant operation does not usually provide sufficiently rich information to identify the models, and thus it is common to carry out a series of plant trials. Disturbing the process manipulated variables with identification sequences and recording the output variable responses provide the data for model identification. Undertaking a series of plant trials obviously upsets the process and may compromise plant profitability to an extent. These costs need to be factored into the initial cost-benefit analysis study. The identification sequence may involve the use of *pseudo random binary sequences* (PRBS), steps, or other disturbance forms. A typical response from a process step test is shown in Fig. 6.3.

Process operation is under closed-loop control until around sample number 130. The oscillatory nature of process operation is characteristic of the process prior to the implementation of the MPC strategy. At around sample 130, a step down is introduced into one of the manipulated variables and the process output responds. A step up in the manipulated variable follows at sample number 215. The process settles out around a steady state and closed-loop feedback control is reapplied at sample number 350. Note that the steps are not immediate as they are constrained by rate limits on manipulated variables. Fitting a model describing such a response is straightforward using system identification procedures (Ljung, 1999). The precise structure of the model depends on the commercial MPC package being used. Many such steps or similar disturbances are required to achieve a complete description of the whole process under consideration.

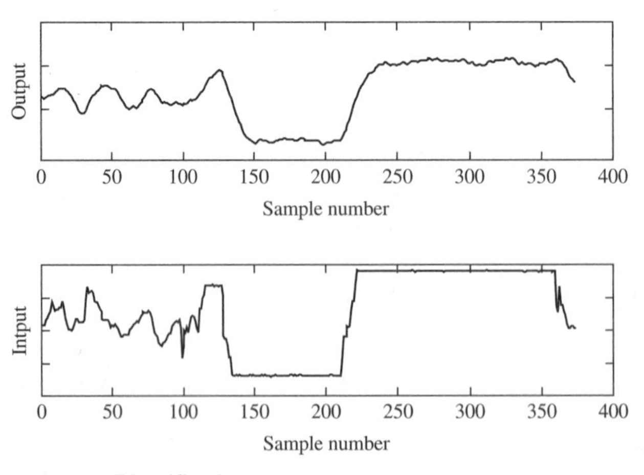

Figure 6.3 Identification step test.

Control system design. With a plant model available, the control strategy can be devised. Process simulation using the identified plant gives some prior indication of likely behavior so that it can assist in controller initial tuning. One of the major benefits of MPC over PID controllers is that the settings of the controller have more physical meaning. Constraints in operation can be coded and strategies for their relaxation considered from process operating objectives. Controller parameters can be specified with relative ease with, for instance, the output prediction horizon being chosen to cover the vast majority of the closed-loop process response. The control action horizon is chosen to be small. With a value of 1 the response is overdamped and sluggish, while 2 or 3 provides a more responsive controller. Here it is important to balance the controller speed of response against stability. Changing the weightings in the cost function provides for fine-tuning of the controller. Problems of stability arise mainly from tuning the controller for fast response and applying it in conditions when the controller models are not accurate. It is thus common to sacrifice some degree of performance for robustness to process model inaccuracy. Process plant response is seldom known with precision and can change over time, for instance, as catalysts degrade. Long-term drift is manageable, but one of the major problems that can be faced is if the plant has significant nonlinear response. In practical terms this means that the dynamic response is different as operating levels vary. In such situations switching between a suite of models or a nonlinear predictive control strategy, such as Apollo from Aspen Tech may be required.

Commissioning and long-term maintenance. Commissioning of the MPC control strategy follows control system design and involves modification of the controller parameters to fine-tune the response and train the operators in system use. During the early stages of the use of the MPC system it is also advisable to attempt to obtain a measure of plant performance compared to that prior to implementation. MPC control system sanction was obtained on the basis of a cost-benefits analysis, and for the control team it is vital to show that the benefits have been obtained if not exceeded. It could be argued that performance improvement will always exist immediately after a project just from increasing awareness of the operation, but if this were the sole source of benefit, then performance improvement would not persist. It is essential therefore to also verify that benefits continue to be achieved. With this consideration, it may be necessary to upgrade models on an infrequent basis if process plant changes are obvious. In this sense MPC and conventional control share a characteristic—implementation and commissioning is not the "end of the story." Long-term maintenance is essential if quality performance is to be maintained.

6.2.3 Examples of performance—distillation case study F

An example of the ease of implementation and tuning can be gained from a relatively straightforward application. Consider a distillation column where the objective is to control both the top and bottom product composition using reflux flow and steam flow to the reboiler. Simple step tests were used to identify a process model, an MPC controller implemented with an output prediction horizon covering the closed-loop response, and a control action prediction horizon of two. Fine-tuning of the response was obtained by changing the controller weighting. The first results shown in Fig. 6.4*a* demonstrate a change in bottom product concentration set point while wanting to maintain a fixed top product concentration with the manipulated variable moves required shown in Fig. 6.4*b*.

It can be seen that tight regulation of the top composition is achieved with a rapid rise of the bottom composition to the new set point. Reflux and steam flow modifications by the controller have achieved this desirable behavior. It could be argued by some that such performance could be attained using classical control strategies and feedforward to decouple loop interactions. Much more challenging for a classical control system is the satisfaction of constraints. Take the example shown in Fig. 6.5.

Here, a set point change in the bottom composition is implemented and the continuous line shows the response obtained. If operational objectives now dictate a constraint on the top composition (as shown by the thick line), then the unconstrained response is no longer satisfactory (as shown by the dashed line). Asking the controller to attempt to satisfy the constraint and using QP to solve the cost function result in

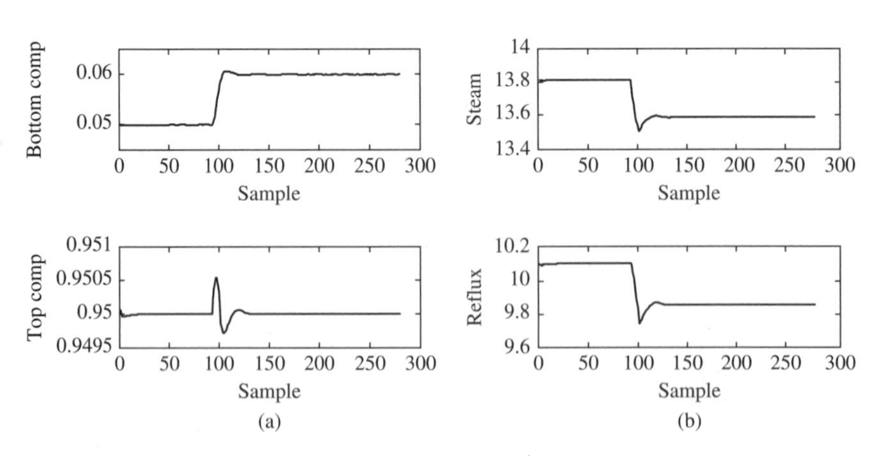

Figure 6.4 (*a*) Distillation output responses and (*b*) manipulated variable moves.

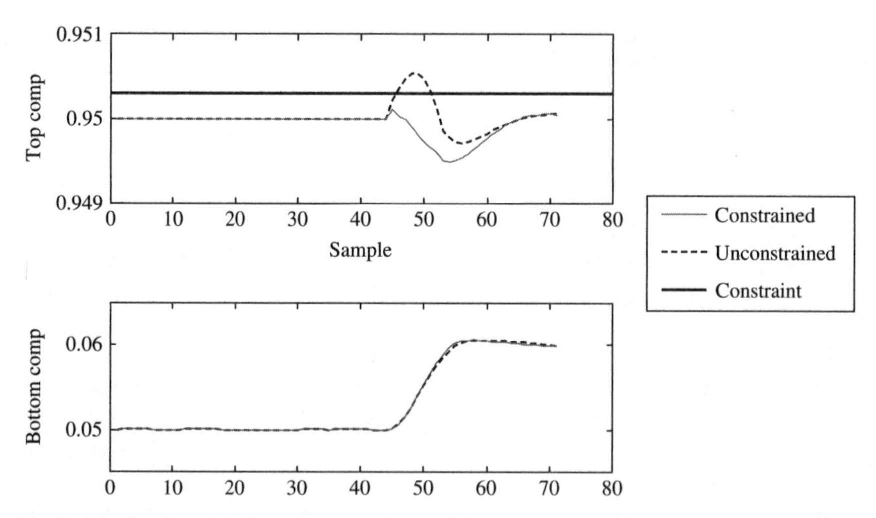

Figure 6.5 Implementation of output constraints.

the response shown by the continuous line. At around sample 45 in the top composition response, the controller predicts that the top composition constraint will be violated. An alternative control profile is implemented that avoids violation but is more "costly" than the unconstrained response. The difference in the bottom composition response is difficult to distinguish.

6.2.4 Benefits of model predictive control

At the start of this section the problems that make process control difficult were considered. These partly motivated the development of MPC, and as Qin and Badgwell (2003) report, applications have flourished. MPC technology by no means solves all control problems but contributes to the solution of many. If implemented correctly, classical control strategies (discussed in Chap. 3) can prove satisfactory in the majority of cases, but may cause struggle with long dead times, difficult dynamics, multivariable systems, process constraints, and process uncertainty. MPC offers improvement because the models in the controller enable it to:

- Forecast the effect of control actions rather than waiting for them to propagate around a loop with a large delay in it.
- Predict long-term performance to steady state thus overcoming problems such as inverse response.

- Predict the consequences of changes in manipulated variables in a multi-input/multi-output system, thus enabling loops to be decoupled.
- Limit changes in process-manipulated variables and predict the consequences of their implementation on process outputs, thus allowing outputs to be tailored to avoid violating output constraints.

The consequence of limiting increments in control action to a short horizon gives robustness to the controller that enables it to cope more effectively with process uncertainty than a conventional control system can.

Finally, although MPC has many attractive features, it should be implemented in situations where classical techniques have proved ineffective. A sensible approach is to consider why classical control strategies fail, and if persuasive arguments can be put forward, then MPC may be considered. Note that all the difficulties listed at the start of this section are not overcome with MPC. Although MPC algorithms that allow the use of nonlinear models are starting to become available, they are by no means common. Also, while developed by process engineers to solve process control problems, MPC is not a strategy that can be implemented by the nonexpert in control systems. In the foreseeable future, algorithms that could be implemented by nonexperts are unlikely to be developed, particularly when control systems suppliers make a portion of their profits by selling implementation expertise.

6.3 Multivariate Statistical Process Control

Chapter 5 discussed how process monitoring using univariate statistical process control techniques could assist in the detection of deviations and through interpretation of the information "pinpoint" causes of deviation. While univariate SPC can be very effective and has been used widely, it does suffer from a number of practical drawbacks:

- Interactions between process variables cause univariate SPC to fail to recognize off-specification behavior.
- Univarariate charts indicate off-specification behavior, but fault conditions are indicated on multiple charts making interpretation of the cause difficult.
- Non–steady state behavior, process dynamics, time delays, and the like cause univariate charts to be inappropriate.

To compound the problems further, most industries now collect large amounts of data, the extent of which obscures the detection of deviations.

The application of multivariate statistical process control procedures is now growing in popularity as an approach to overcome these problems and to achieve variability reduction. Before extensions to standard univariate SPC are considered, the limitations outlined are considered in greater depth.

6.3.1 Limitations of univariate SPC

What has now become a standard justification for the adoption of multivariate as opposed to univariate SPC is presented in Fig. 6.6. The figure shows the time course of two variables (1 and 2) on $\overline{X}$ charts and the individual limits on each. Both variables remain within the limit, so normal production might be expected at the time that corresponds to the point labeled X on both graphs; however, the combined effect of variables 1 and 2 can result in unacceptable operation—point X falls outside the area of acceptable operation when both variables are considered together. The interactions between variables 1 and 2 are such that an unacceptable product is produced. From an SPC perspective, the conclusion from this simple example is that univariate control charts are not necessarily capable of extracting the key information on unacceptable production and multivariate approaches must be adopted.

When considering data from a multivariate process, it may indeed be the case that univariate charts indicate that deviations in operation are occurring. This information may be provided to the operators in the form of charts or text warnings as alarm or action limits are violated. The problem that arises is that a single problem cause can

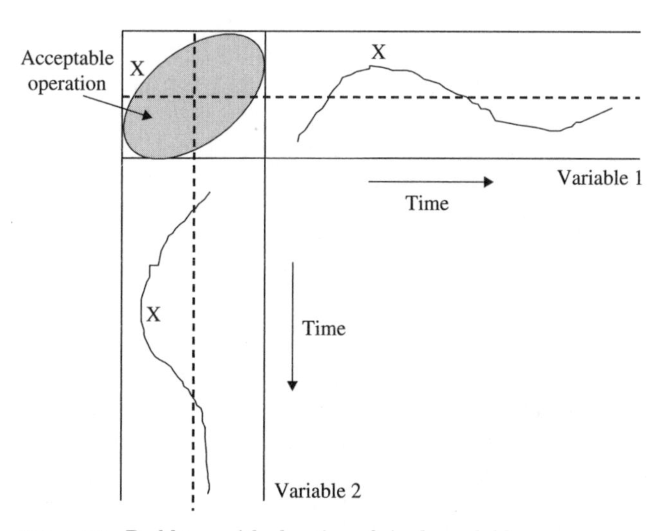

Figure 6.6 Problems with charting of single variables.

result in a multitude of univariate alarms occurring over a period of time as the problem propagates through the process. Interpretation of the causes of a problem from observing the effects in terms of the pattern of alarms can be quite a challenge to the operator. A more sophisticated strategy is required to look at the alarm sequence pattern to pinpoint a cause.

As a strategy is formulated, a key consideration is that a process plant can seldom be considered to be operating at steady state. A continuous process plant is constantly subject to disturbances due to external influences, operator actions, raw material changes, and so forth, so non–steady state operation is inevitable. Batch processes, in addition to suffering from disturbances in the same manner as continuous counterparts, are naturally non–steady state. So as well as taking into account time delays that cause temporal alarm sequences to occur, it is also necessary to make allowance for the dynamic progression of variables. This can result in a very different picture of plant behavior than that arising from simple steady state analysis.

One potential solution to this is discussed further in Chap. 4, where human expertise is coded within an expert system in an attempt to detangle a complex pattern of signals to determine the cause of a process problem and offer a solution. An alternative strategy is to determine a "fingerprint" of process operation and interpret information using a limited number of charts—this is the basis of multivariate statistical process control.

6.3.2 Data compression methods for multivariate statistical process control

To overcome the problems associated with univariate SPC, multivariate SPC (MSPC) techniques have been developed (Nomikos and MacGregor, 1995; Martin and Morris, 1996). Multivariate SPC developments have drawn heavily on the basic concepts of *principal component analysis* (PCA) and *partial least squares* (PLS), also known as projection to latent structures. The algorithms were originally formulated for application to continuous systems; however, simple data preprocessing procedures can be employed to enable their application to batch processes.

Principal component analysis. PCA is a method used to analyze the covariance of a set of plant variables. The approach transforms a matrix $[\mathbf{Z}]$ containing measurements from n process variables, into a matrix of mutually uncorrelated variables $\mathbf{t}_k$ (where $k = 1$ to n). These variables, called *principal components* (PCs), are transforms of the original data into a new basis defined by a set of orthogonal *loading* vectors $\mathbf{p}_k$. The individual values of the principal components are called *scores*. The transformation is defined by

$$[\mathbf{Z}] = \sum_{k=1}^{np<n} \mathbf{t}_k \mathbf{p}_k^T + \mathbf{E} \tag{6.3}$$

The loadings are defined here as being orthonormal, and so they become the eigenvectors of the data covariance matrix $\mathbf{Z}^T\mathbf{Z}$. The $\mathbf{t}_k$ and $\mathbf{p}_k$ pairs are ordered so that the first pair captures the largest amount of variation in the data and the last pair captures the least. In this way it is generally found that a small number of PCs (*np*) can account for much of the variation in the data. The remaining variation constitutes the error term E. When Eq. 6.3 is applied to a single vector of new process measurements $\mathbf{z}^T$, the resulting term $\mathbf{E}$ is called the *prediction error*. There are several methods for determining the suitable value for *np*. One method is to continue to add PCs until the variation explained in the retained PCs exceeds a particular value; however a more suitable approach, and the technique generally preferred, is to use cross-validation (Wold, 1978).

Why is PCA a useful technique for MSPC? The ability to describe data variation in a small number of principal components allows a considerable reduction in the number of monitoring charts. Furthermore, the data transformation resulting in orthonormal principal components effectively deals with the problems brought about through data correlations from the single-cause/multiple–effect problems described.

Partial least squares. PCA is suited to the analysis of interacting process variables that are the cause of process deviations. PLS (Geladi and Kowalski, 1986) is a tool suitable whenever plant variables can be partitioned into cause ($\mathbf{X}$) and effect ($\mathbf{Y}$) values. The method may be used for regression or similarly to PCA, with reduction of the effective dimensionality of data. The approach works by selecting factors of cause variables in a sequence that successively maximizes the explained covariance between the cause and effect variables. Given a matrix of cause data $\mathbf{X}$ and effect data $\mathbf{Y}$, a factor of the cause data $\mathbf{t}_1$ and effect data $\mathbf{u}_1$ is evaluated, such that:

$$\mathbf{X} = \sum_{k=1}^{np<nx} \mathbf{t}_k \mathbf{p}_k^T + \mathbf{E} \quad \text{and} \quad \mathbf{Y} = \sum_{k=1}^{np<nx} \mathbf{u}_k \mathbf{q}_k^T + \mathbf{F}^* \tag{6.4}$$

where $\mathbf{E}$, $\mathbf{F}$ = residual matrices

np = number of inner components that are used in the model

nx = number of causal variables

These equations are referred to as the *outer relationships*. The vectors $\mathbf{t}_k$ are mutually orthogonal. These vectors and the $\mathbf{u}_k$ are selected to maximize the covariance between each pair, $(\mathbf{t}_k, \mathbf{u}_k)$. Linear regression is performed between the $\mathbf{t}_k$ and the $\mathbf{u}_k$, to produce the *inner relationship*, such that:

$$\mathbf{u}_k = b_k \mathbf{t}_k + \varepsilon_k \tag{6.5}$$

where b_k is a regression coefficient and ε_k refers to the prediction error. The PLS method provides the potential for a regularized model through selecting an appropriate number of latent variables $\mathbf{u}_k$ in the model (np). The number of latent variables is again typically chosen through the use of cross-validation.

Multiway MSPC. Conventional PCA and PLS are linear procedures and therefore limited in their effectiveness when applied to batch problems that typically tend to be nonlinear in form. Two options exist for improving the capabilities of the techniques when applied to batch systems. The first is to develop nonlinear counterparts to PCA and PLS, and the second is to transform the batch data to remove the major nonlinear characteristics. Although nonlinear MSPC techniques exist and have been applied successfully (Dong et al., 1996), the transformation of batch data has proved to be a more effective option. The most common form of data transformation, termed multiway PCA and PLS, was initially proposed by Nomikos and MacGregor (1994). Since then other researchers have adopted the approach and applied it to a variety of processes. For example, Gallagher et al. (1996) applied the technique to monitor nuclear waste storage vessels, and Gregersen and Jorgensen (1999) and Lennox (1999) investigated the detection of faults in a fed-batch fermentation processes.

The concept of multiway PCA and PLS is a relatively straightforward extension to the approach taken for continuous systems, but deviations from mean trajectories rather than steady state values are considered. The following description, together with Fig. 6.7, explains the multiway approach to PCA. [Further details can be found in Kourti and MacGregor (1995).]

First, m historical batches are selected that have exhibited "good" behavior. These batches are referred to as *nominal* batches and will be used for comparison. The duration of each batch is likely to differ and therefore the data from each batch are considered usually only until the shortest run length. Techniques exist that allow all the measured process data to be used regardless of the shortest run length (Weitz and Lakshminarayanan, 1996). The next step is to identify the n process variables that are to be monitored. For each variable, the

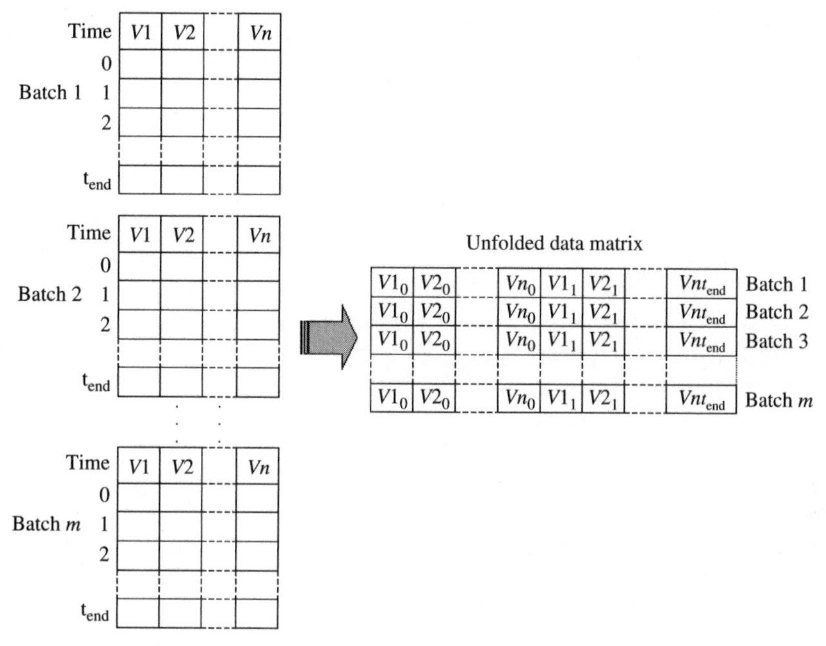

Figure 6.7 Unfolding batch data for multiway MSPC analysis.

mean trajectory over all the nominal batches is calculated and the mean value at the appropriate time point is removed. This effectively removes the major nonlinearity from the data and leaves a zero mean trajectory for each variable.

The individual data matrices from each batch are then unfolded into a single unfolded data matrix, as depicted in Fig. 6.7. In this figure $V1$ refers to variable 1, $V1_t$ refers to the value of variable 1 at time t, and t_{end} refers to the time at which the batch finished. Finally, all columns of the unfolded data matrix are standardized to unit variance. PCA can now be applied to this unfolded data matrix using two techniques:

1. *Projection PCA.* A PCA model is developed using the entire unfolded data matrix, i.e., there is a loading variable calculated for every measured process variable at each sampling instant. The subsequent use of this model to monitor a batch on-line poses the problem that it is necessary to know the values of all process measurements through to the end of the batch. This means that with the exception of the end point of the batch, it is necessary to estimate the future values of all the measured variables.

   The prediction of future process values is normally referred to as *filling up* the matrix and is typically achieved by assuming that either all future scaled process values remain at the mean value of zero (*filling method* 1) or the future process values remain at the current offset from the mean value (*filling method* 2). An alternative technique is to use the capability of PCA and PLS to handle missing data (Nelson et al., 1996). Research work in this area has tended to concentrate on using PCA and PLS to infer process measurements at the current sampling instant when a particular measurement is unavailable. However, the same techniques can be used to fill up the data matrix (*filling method* 3).

2. *Moving window PCA*. A PCA model is developed on a moving window of data. For example, at the 50th sampling instant, a PCA model may be developed using data collected from the 41st to the 50th sampling instants. In this case the moving window length would be said to be equal to 10. Such a model effectively compares operating conditions over the last few sample points with those experienced at the same time in the nominal batches. The advantage of this technique is that it does not require the values of future measurements to be predicted. However, it does require that the PCA model be reevaluated at each sampling point.

The concept of multiway PLS is very similar to multiway PCA. The unfolded data matrix represents the cause matrix and is created in exactly the same way. The construction of the effect matrix is problem dependent, and an example is given in the case study later.

6.3.3 Process monitoring using PCA and PLS

Process monitoring is achieved with PCA by developing the PCA model on a selected set of data from the process. The data chosen should be from a successful, high-performance operation and are referred to as *nominal data*. The model developed with this data is later recalled to interpret on-line data and assess its consistency with nominal data. Various approaches for assessing this consistency have been suggested in the literature. It is commonly found that for PCA, the analysis should be applied separately to data variation in the space of the principal components and in the prediction errors.

Variation in the space of the principal components is provided through the T^2 statistic that is defined as:

$$T^2 = \sum_{k=1}^{np} \mathbf{t}_k \sigma_k^2 \mathbf{t}_k^T \tag{6.6}$$

where σ_k^2 is the variance of the kth score.

Variation in the prediction error is expressed as the *squared prediction error* (SPE) and is defined as $\mathbf{E}^2$. Control limits can be applied to the T^2 and SPE charts based on the assumption that the statistics follow a normal and chi-squared distribution, respectively. Details of these confidence limits can be found in Lennox et al. (1999).

While a process is running, the T^2 and SPE values can be plotted. Any violations of the confidence limits should indicate that the process is deviating from nominal conditions. However, the manner in which it is deviating from normality differs depending on which chart is violating the confidence limits. Zhang et al. (2000) demonstrated that changes in the relationships between variables, such as would be experienced if a sensor failed, tended to be detected on the SPE chart, while changes in the operating conditions, for example a grade change, were typically identified on the T^2 chart. There will be exceptions to this, for example, a high-impact fault that significantly affects a number of variables is likely to be detected on both the T^2 and SPE charts.

An important consideration when constructing the SPE and T^2 charts is setting the confidence limits. Most reported applications tend to place 95 or 99 percent confidence limits on the charts. However, experience shows that the correct setting of the confidence limits is critical to the success of the monitoring system. Process operators will quickly lose confidence in a system that gives many false alarms, and therefore the limits should be set as a compromise so that such alarms are minimized but sensitivity to fault detection maintained.

While highlighting abnormal conditions is very useful, assigning the cause of the abnormalities is even more important. Technology in this aspect of MSPC is limited and is an area of significant research (Gertler et al., 1999), but currently the predominant approach is using simple contribution charts to indicate cause.

Contributions to the T^2 statistic are obtained by taking the gradient of T^2 with respect to each variable. The contribution to the SPE statistic from a given variable is simply the squared prediction error on that variable.

Previous work has demonstrated that PLS can be used for process monitoring in two different ways. The first possibility is to use PLS as a prediction tool to estimate the end concentration of product in a batch for example. The second possibility is similar to PCA and involves monitoring the inner latent variables produced by the algorithm. Previous work (Lennox et al., 1999) has reported that the second approach offered no tangible benefits over PCA. Using a PLS model to predict the final biomass concentration was, however, found to be beneficial. The PLS model acted as a classification system and indicated if the current batch is more consistent with high- or low-yield batches. Since it is used as a classifier, it is important that the model is developed using historical data from both high and low-yield batches. The construction of the cause and effect matrices is relatively straightforward. The cause matrix is identical to that used in multiway PCA and the effect matrix is a column vector containing the product concentration from each of the historical data.

Confidence limits can be placed around the PLS prediction (Nomikos and MacGregor, 1995), and if the PLS model predicts that the required production is unlikely to be achieved then a warning message can be relayed to the operator.

6.3.4 Can bioprocesses benefit from MSPC?

When operating bioprocesses it is vitally important to maintain operation within strict limits. There are two main reasons for this. First, biological based systems by nature are highly sensitive to abnormal changes in operating conditions. To ensure that the maximum possible yield of product is obtained from the system it is necessary to make sure that conditions remain closely fixed around a prespecified "ideal" trajectory. Second, for many compounds, as part of the procedures to guarantee product chemical consistency, regulatory authorities (such as the FDA in the United States) demand proof that consistent operation is adhered to. Without this the product cannot be sold. In industrial bioprocess systems, process operators are typically employed to achieve consistent operation through manual monitoring and control. The operator uses experience and knowledge to detect potential problems and make modifications when necessary. The importance of effective operator control cannot be underestimated as the performance is largely dependent on the ability to keep the system operating smoothly. A process that is free from major upsets is likely to be more productive than one subject to significant disturbances. Therefore, the earlier a potential problem to the system can be detected, the less severe its influence will be, and the resulting corrective action will consequently be more restrained.

Although operator control may be adequate in certain situations it makes little direct use of the historical data that is routinely gathered and logged. Historical data will typically contain information on high- and low-productivity batches, as well as information on the consequences of performing particular actions in response to problem situations. It is possible therefore to develop simple rule-based structures that compare individual variables with historical records. Any deviations that indicate reduced productivity from the current batch can then be brought to the attention of the operators. Univariate SPC advice in the form of rules can be formulated, but the sheer complexity of the system, together with the need to account for temporal patterns, can challenge rule-based strategies.

Traditional univariate SPC leading to operational rules may be suitable for selected bioprocess systems, but in many cases a multivariate approach to SPC is likely to prove beneficial.

6.3.5 Case study in process monitoring with MSPC

This case study concerns the development of a process-monitoring system for a fed-batch fermentation system operated by Biochemie in Austria. In developing the system the capabilities of a variety of MSPC techniques have been compared and assessed. In addition, a number of modifications to the existing technologies are described and how they can improve the monitoring capabilities of the system is demonstrated.

Data issues. It is important that any monitoring system that is developed be capable of handling common data problems such as noise and outliers. These procedures involve filtering out noise using low-pass filters and detecting and removing outliers through the identification of trends in the data. These were discussed in Chap. 3.

A common problem in process systems is that process measurements may not always be available. For example, a faulty thermocouple may be withdrawn from service for the duration of a batch. Such a situation would mean that it would not be possible to apply any previously developed PCA or PLS model that used this thermocouple measurement to the current batch. In fact, any blank entries in the unfolded data matrix, which would result from a single missing measurement, would render the PCA or PLS model invalid during an on-line batch. In this case study it was considered likely that one or more process measurements would be unavailable during a typical batch run. It was therefore felt it was essential that this problem be addressed.

Problems associated with missing data values have been researched at length by Nelson et al. (1996). In their studies they identified three methods for estimating missing data values, each of which makes use of the predeveloped PCA or PLS models. Each of these approaches was applied with mixed results. It was found that two of the methods, projection to the model plane and replacement by the conditional mean, created matrix singularity problems and were unsuitable for this application. However, the single-component regression technique was found to provide acceptable results. Complete details of these algorithms can be found in Nelson et al. (1996).

Application of MSPC. A PCA model was developed from historical data from 10 high-yield industrial batches. The ability of this model to monitor a number of subsequent batches was then determined. Some of these subsequent batches operated smoothly while others were affected by disturbances.

Figure 6.8 displays the SPE for each of the 10 reference batches. Based on this information a 97 percent confidence limit has been evaluated and also plotted. In this example, the unfolded data matrix has been filled using filling method 2. A 97 percent confidence limit (Q limit) has been specified in this application because at this level none of the nominal batches violate the limit, thus reducing the potential for false alarms.

Figure 6.9 shows the SPE, along with 97 percent confidence limits, for a particular test batch. During this fermentation an intermittent drift on a sensor measurement was experienced. This drift was most apparent between sample numbers 40 to 70 and 100 to 150. The SPE chart displayed in Fig. 6.9 demonstrates that the PCA model has detected this fault entering the system. Analysis of the contribution charts for this batch correctly identified the problem variable.

Further investigation of the PCA model showed that the T^2 chart failed to identify this fault. This result agrees with the previous expectations. The T^2 chart is able to detect significant changes in operation and high-impact faults. A fault such as this, a small drift on a sensor, is unlikely to be detected in the T^2 chart.

Previously, two alternative techniques for filling up the unfolded data matrix were described. The suitability of these techniques was compared with the results obtained using filling method 2. Figure 6.10 shows the SPE chart for the test batch investigated earlier, this time

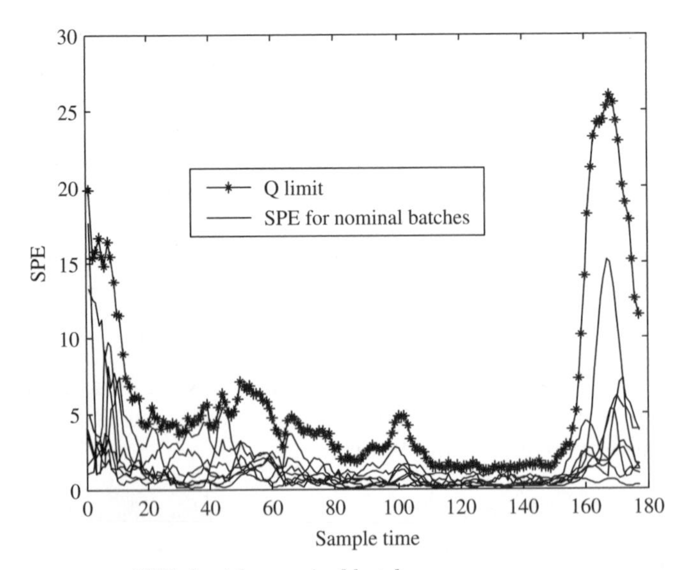

Figure 6.8 SPE chart for nominal batches.

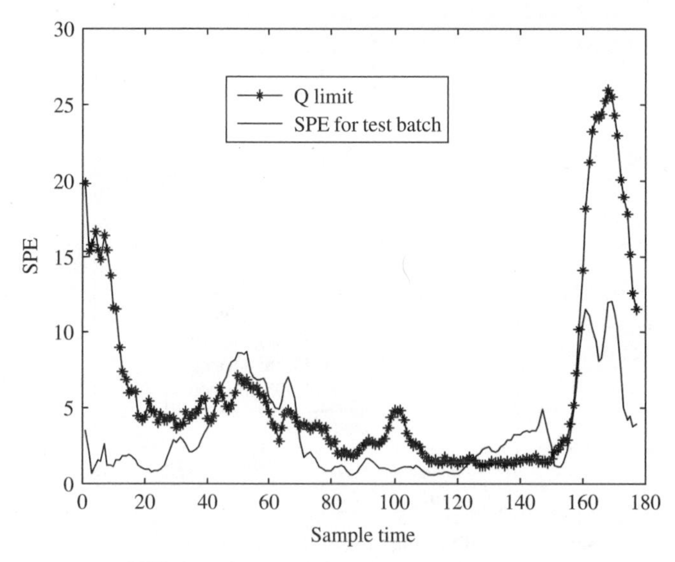

Figure 6.9 SPE chart for test batch.

using filling method 1. By using this technique the PCA model is no longer able to identify the sensor fault. A similar reduction in monitoring capabilities using this approach has been reported by Lakshminarayanan et al. (1996) and Nomikos and MacGregor (1995).

The reason for the reduction in performance using filling method 1 can be explained by considering Fig. 6.11. This figure shows a particular

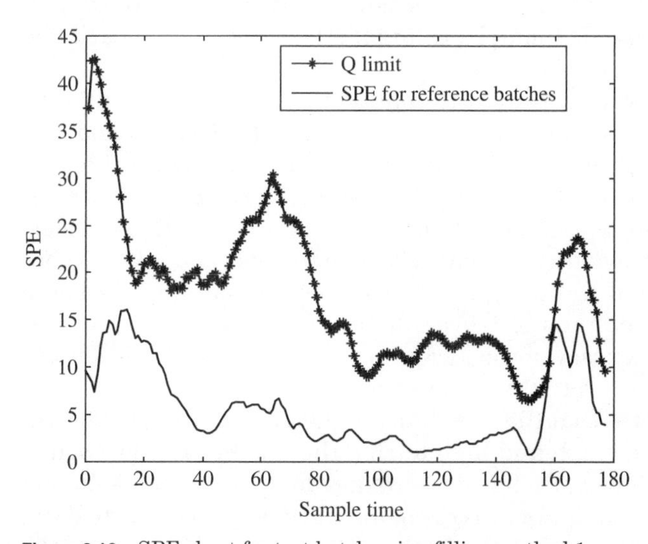

Figure 6.10 SPE chart for test batch using filling method 1.

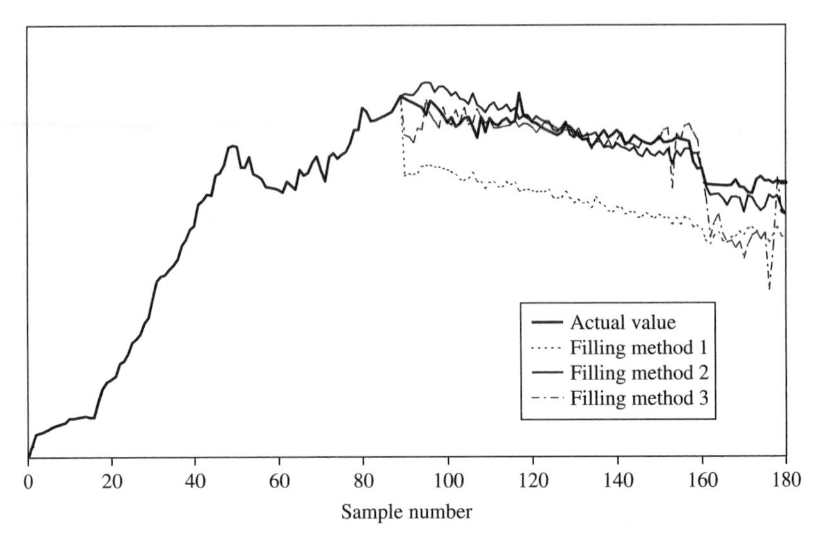

Figure 6.11 Inferred process measurement using various filling methods.

process variable measurement recorded during a complete batch cycle (bold line). The batch chosen in this example was a high-yield batch free from process disturbances. Using the data collected up to sample point 90, the value of the process variable between sample times 91 and 180 has been inferred using each of the three methods to fill up the unfolded data matrix. It is evident from this figure that filling method 2 most accurately matches the actual process measurement. Since the performance of the process monitor will be dependent on the accuracy with which it fills the unfolded matrix, filling method 2 would appear to be more suitable in this application.

Similar results were repeated in many other tests using different variables and batches, confirming the suitability of filling method 2.

The low principle components that are combined to create the SPE chart capture relationships between the process variables and as such can be considered models of the system. The more accurate these models are, the more sensitive the SPE chart will be to abnormal process conditions. The value of the Q limit, which has been specified as being greater than the maximum SPE over all the nominal batches, provides a measure of the accuracy of the low-power PCA models. The lower the Q limit, the more accurate the models are.

Figure 6.8 indicates that the sensitivity of the SPE is low at the beginning and end of the batch and high during the middle. The low sensitivity at the start is because few data are available from the batch and therefore the unfolded data matrix is being filled with inaccurate data. The sensitivity at the end of the batch is believed to be because the relationships between the variables differ during the progression of the batch. Relationships between variables that are valid at the early stages of the batch may not hold toward the end of the batch. To account for this it is possible to construct PCA submodels. Using this technique PCA models are constructed using data collected from only part of the batch. For example, one PCA model may be applied up to sample number 60, a second model between sample 60 and 150, and a third model between 150 and 180.

Figure 6.12 compares the Q limit obtained using such a submodel approach with that obtained using a single PCA model over the full duration of the batch. The figure shows that the SPE and the corresponding Q limit using submodels are significantly lower than that using the single PCA model. This indicates that the SPE chart should be more sensitive to process abnormalities, such as faults, if submodels are employed and is therefore a more suitable method for process monitoring purposes. One minor drawback with using submodels that can be seen in Fig. 6.12 is that the Q limit increases sharply following the change from one model to the next, which occurs at samples 60 and 150 in this example.

The reason for this is the same as for the high SPE experienced at the start of the single-model technique and is because so few data are available when filling up the unfolded data matrix. To reduce this problem it is possible to construct submodels that operate during particular periods but are developed using data collected both before and during the period. For example, a submodel that operates between sample numbers 80 and 120 may actually be developed using data collected between sample

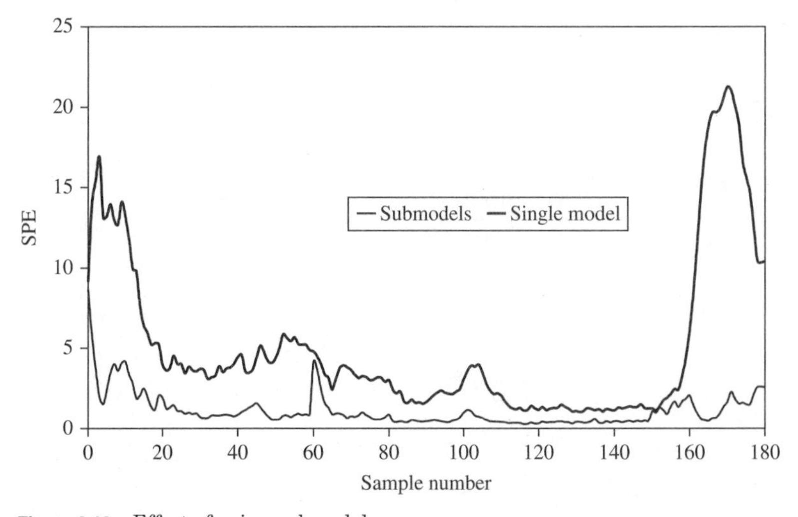

Figure 6.12 Effect of using submodels.

numbers 60 and 120. Therefore when the model is introduced at sample number 80 the data it uses to fill up the unfolded data matrix will be more reliable.

The ability of PCA to handle missing data values is demonstrated in Fig. 6.13. The solid line in this chart shows the SPE over the course of a batch.

Data collected from this batch were passed through the PCA algorithm once more; however, this time the measurement from a particular variable was set to be missing between samples 41 and 60, and 101 and 130. Using the single-component regression method to replace the missing values, the SPE chart for this batch is shown by the dotted line in Fig. 6.13. It is evident from this chart that there is little difference between the two SPE values through the duration of the batch and therefore in this particular example the routines for handling missing data values have worked very well. Similar results to that displayed in Fig. 6.13 were recorded from a number of subsequent missing data tests involving other process variables and batches.

Further tests using the moving window PCA technique showed that it produced similar results to those obtained using the projection method with submodels. It is therefore difficult to establish conclusions as to which is the most suitable technique for this application. An advantage in using the moving window approach is that there are fewer variables that require specifying, such as the number of submodels and the periods during which these submodels should be used. However, it was found that the moving window approach required accurate specification of the time delays, while the projection method was more robust with this information.

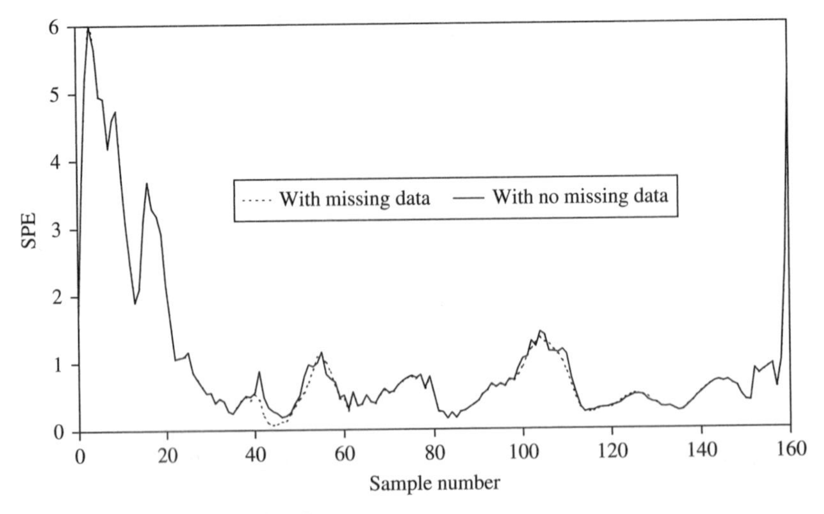

Figure 6.13 Effect of missing data.

PLS models were developed to estimate the final concentration of biomass in the fermenter. The aim of this exercise was to use this estimate to categorize the on-line fermentations into high- and low-yield production. The PLS model was developed on historical data and applied to a number of alternative batches. Figure 6.14 shows an example of a PLS monitoring chart. This chart shows the final biomass concentration (solid line) estimated at each sample time during the batch, 95 percent confidence limits (dashed line), and the actual final concentration of biomass (thin solid line). Details of the confidence limit calculations can be found in Nomikos and MacGregor (1995).

It can be seen from Fig. 6.14 that for this batch, which proceeded upset free, the PLS estimate of the final product concentration was reasonably accurate throughout. It is also noticeable that the biomass estimate during the first half of the run is relatively noisy and for the second half of the run it is reasonably consistent. This is because at the start of the batch most of the data in the unfolded data matrix have been estimated and are therefore less reliable than in the later stages of the batch.

The accuracy of the PLS model displayed in Fig. 6.14 was consistent with that produced for several other batches. This level of accuracy makes the PLS model a useful tool for monitoring the progress of the fermentation system for this particular application.

Case study conclusions. The primary conclusion from the work is that existing technologies in the field of process monitoring, such as PCA and PLS, provide suitable tools for the detection of process abnormalities. The case study took two years to complete. Many of the practical considerations

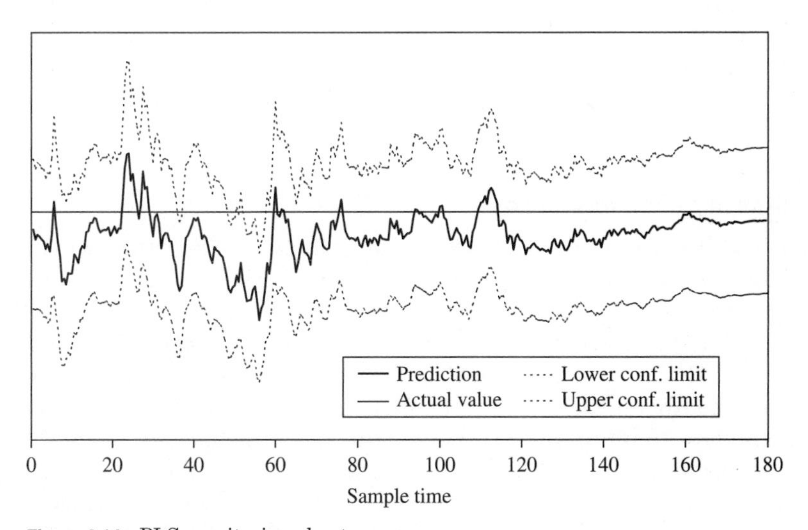

Figure 6.14 PLS monitoring chart.

that were encountered in this application took longer to accommodate than the basic MSPC methods. Such considerations have ranged from data preprocessing issues to computation problems.

Despite practical difficulties PCA has been employed successfully to detect and isolate process faults while PLS has been applied to estimate final product composition in the fermenter. The PLS estimate was shown to be sufficiently accurate to enable it to categorize on-line batches into high and low yield.

The integration of the PCA and PLS methods with information from other sources such as inferences from RBF networks serves to provide operators with a rich source of operational assistance. The monitoring procedures developed in this work are currently being applied to a research facility operated by Biochemie in Austria.

6.4 Artificial Neural Networks

Artificial neural networks (ANNs) have attracted immense interest from academia and industry in recent years due to their inherent nonlinear characteristics. A common use of neural networks is as models capturing relationships between process inputs and outputs. This is described in Sec. 6.4.2 where the use of feedforward *ANNs* (FANNs) is considered. In addition to model construction, FANNs have been extensively used for fault detection and isolation in various process industries (Hoskins and Himmelblau, 1988; Watanabe et al., 1989; Venkatasubramanian et al., 1990; Leonard et al., 1992). Besides using the FANNs, many variants of unsupervised ANNs have also been applied to process monitoring and fault detection in various process industries; Aldrich et al., 1995; Kohonen et al., 1996). An unsupervised ANN, the *self-organizing feature map* (SOM), is considered in the next section as its application is for pattern recognition and is therefore similar to MSPC described in Sec. 6.3.

6.4.1 SOM in pattern recognition

The concept of self-organization is based on competitive learning, which is implemented in an unsupervised type of network. The idea of self-organization was developed by the work of Willshaw and Von der Malsburg (1976). Several variants of self-organizing ANNs exist (Fukushima, 1975; Carpenter and Grossberg, 1987) with the most widely used being the Kohonen SOM ANN (Kohonen, 1982). The SOM ANN has only two layers, an input and SOM layer, and the dimension of the SOM has to be specified. The SOM layer is a low-dimensional array of neurons that is generally one- or two-dimensional. The mapping performs a nonlinear projection from the multidimensional space of input data onto a low-dimensional array of nodes, while preserving the topological relationships of the input space. An example of a 2-D SOM layer is presented in Fig. 6.15. The dimension of the map is 6 by 5 nodes or 30 nodes in total. Each element in the input vector **X** with dimension $[1 \times j]$, where j is the number of variables, is fully connected to each node of the SOM layer. The nodes in the map are vectors themselves with the same dimension as the input vector **X**. Technically they are called weight vectors **W**. In the figure a winning node and its first neighbors are also shown.

In contrast to the adaptation of supervised weights, the training of unsupervised ANNs is usually based on some form of competition between the nodes. The basic concept of competitive learning is that it divides a set of input patterns into clusters inherent in the input data. The nodes of the map compete for the input with their internal parameters, and the node that matches the closest becomes the winner. The selected node is updated so that in future it is even more sensitive to the previously presented input. The Kohonen SOM is a further development of competitive learning in which the best matching element also activates, to a certain degree, its neighboring elements and they are also updated during the training (Fig. 6.15). The SOM becomes ordered in such a way that clusters of similar behavior activate the nodes that are near to each other. Further details on training can be found in a number of textbooks and articles dedicated to these networks.

The Kohonen SOM possesses several desirable properties, as discussed in Dony and Haykin (1995):

1. The set of weight vectors are a good approximation to the original input space.
2. The topology is preserved, so the vectors close to each other have similar properties.
3. The density of the map corresponds to the density of the input distribution.

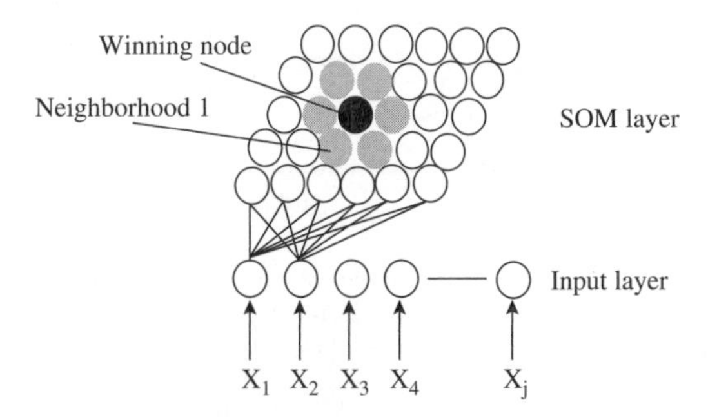

Figure 6.15 Kohonen SOM network.

However, the algorithm has some deficiencies as discussed in Kohonen (1996) and Bishop (1995). They include:

1. The absence of a cost function.
2. The absence of a theoretical basis for choosing the training parameters to ensure topographic ordering.
3. The absence of any general proofs of convergence. Nevertheless, it can be a useful tool for pattern recognition.

To demonstrate the potential of the SOM, data from the seed stage of a fermentation where the inoculum for the final stage is produced are considered. The seed fermentation operates as a batch process. All substrates necessary for growth are batched, and the pH is adjusted. After inoculation, the seed is allowed to grow and when the culture reaches a predefined state it is transferred to the main fermenters. Several on- and offline variables are measured during the seed stage. The *carbon dioxide evolution rate* (CER) and *oxygen uptake rate* (OUR) measurements are available online, while inoculum level, final pH, and initial spore count are measured in offline mode. Variables that are believed to influence the *quality* of the seed are given in Fig. 6.16. In the figure, the available on- and offline data are connected to the quality of the seed by solid lines, while the data which are not presently measured are connected by dashed lines.

CER and OUR were available throughout the batch, and a time series of their values was used as an input to the SOM. The initial spore count is available at the start of the batch while the inoculum level and the final pH are measured at the end of the batch. Data from 101 seed fermentations was collected for analysis purposes, with 41 high final performance and 60 low and mean final performance fermentations. The high final performance data are defined as nominal, while the mean and low final performance data are referred to as poor. A threshold productivity was chosen to distinguish between nominal and poor productivity

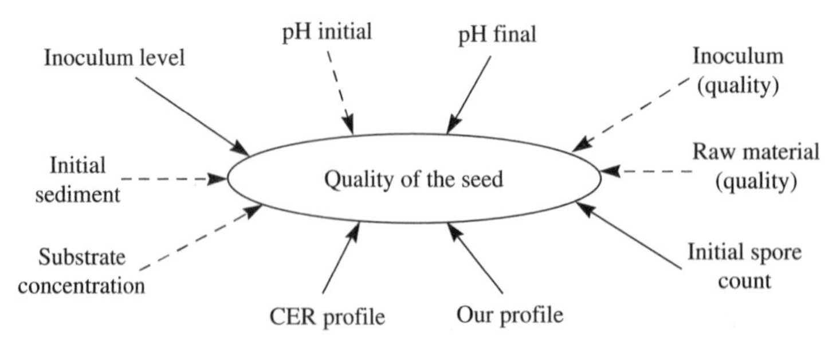

Figure 6.16 Qualitative scheme of the influence of the process variables on the seed quality.

fermentations. Note that the final productivity refers to the production batch subsequent to the seed fermentation. If patterns are observed in the seed data, it suggests that the seed fermentations influence the production fermentation.

To demonstrate the concept a 16 × 14 array was generated. The trained Kohonen SOMs is presented in Fig. 6.17. The winning nodes of the nominal data are shown as "o" on the map and the winning nodes of the poor performance data shown with the symbols "x." It can be observed in both figures that there are nodes which are winners for both nominal and poor performance data. This particular situation was expected as discussed earlier in this section. Some of the winning nodes of the poor performance data for the map lie in the second neighborhood of the nominal nodes. These nodes are assigned as poor nodes, of which there are nine in total. The poor nodes are indicated by a rectangle in Fig. 6.17*a*. The number of poor performance fermentations assigned to these nodes is 10.

The second indicator in detecting the fault in real-time is the Q value as used with MSPC. The Q values for both nominal and poor performance data are presented in Fig. 6.17*b*. Twenty-six poor performance fermentations have exceeded the 99 percent confidence limits. Using the two indicators, the winning nodes of poor performance fermentations that lie in the second neighborhood of the nominal data and those that are identified using the Q-statistic, give the total number of poor performance fermentations detected as 31.

6.4.2 Artificial neural networks in inferential estimation

The basic feedforward network is shown diagrammatically in Fig. 6.18. Process data are scaled and enter the network at the input nodes. The data are propagated forward through the network via the connections between hidden layers to the output layer. The network is fully connected (i.e., every neuron in a layer is connected to every neuron in adjacent layers). Each connection acts to modify the strength of the signal being carried.

In the basic network the signals are modified by scalar multiplication (the connection weight). At the neurons the input strengths are summed up with a bias term (Eq. 6.7) and the resulting value is passed through a sigmoidal nonlinearity (Eq. 6.8).

$$N_s = \mathrm{wt}_b + \sum_{i=1}^{N_{\mathrm{in}}} \mathrm{wt}_i \; S_i \tag{6.7}$$

$$N_{\mathrm{out}} = \frac{1}{1 + e^{-N_s}} \tag{6.8}$$

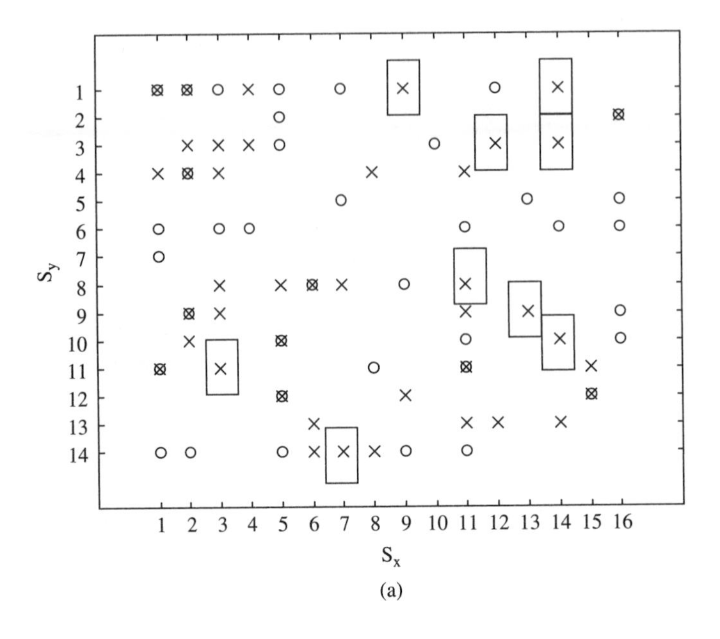

(a)

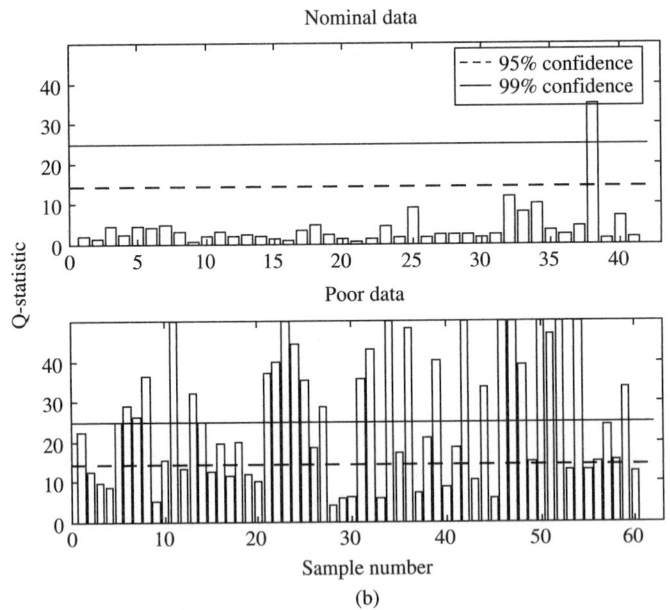

(b)

Figure 6.17 (*a*) SOM trained on nominal data and used for testing the poor performance data; (*b*) Q-statistic for both nominal and poor performance data with 95 percent (—) and 99 percent (—) confidence.

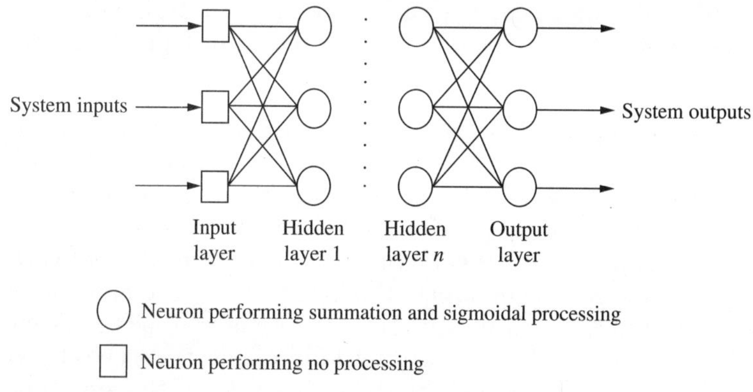

Figure 6.18 Feedforward neural network architecture.

Here N_{in} is the number of neuron inputs, wt the scalar weight associated with each connection, s the signal strength, and wtb the bias weight. Data flow forward through successive layers of the network, with the outputs of neurons in one layer (N_{out}) becoming the signal strengths (s) of connections to the following layer. Ultimately data arrive at the output layer, following which they are rescaled to engineering units.

The procedure for constructing an artificial neural network model, given process data, relies to a considerable degree on an understanding of the process problem. Given the architecture shown in Fig. 6.18, it is necessary to specify:

1. *The number of network inputs and outputs.* This choice is primarily based on an engineering appreciation of the problem. As with linear identification techniques, highly correlated inputs can degrade the quality of the resulting model. It is therefore necessary to attempt to minimize redundancy in network information.
2. *Number of hidden layers.* The literature to date provides conflicting information regarding the choice of the number of hidden layers. Cybenko (1989) postulated that two hidden layers were sufficient to model any given continuous nonlinearity, while other papers (e.g., Hornik et al., 1989) claimed that one hidden layer was adequate. Experience to date suggests that this decision is best made heuristically; if one hidden layer is insufficient then move to the use of two.
3. *Number of neurons per hidden layer.* The best policy to follow to determine the number of neurons per hidden layer is to start with a relatively small number (typically around the same as the number of

inputs), and if this does not produce “acceptable” results, increase the number of nodes in the hidden layer further searching for an improved solution.

Clearly the methods for the selection of artificial neural network topology are not refined as yet, in particular the problem of hidden unit/layer specification. Nevertheless, even using trial and error methods, topologies that lead to acceptable models can be obtained.

Once the topology of the network has been specified, the network weights and biases must be obtained. Since essentially the problem is one of nonlinear function minimization, a variety of methods exist for this task. The description of the algorithms falls outside the scope of this book but further information can be found in many of the standard texts on neural networks (e.g., Bishop et al., 1995).

A process example serves to demonstrate how FANNs can be used to build models that function as tools to infer process variables. One of the most important problems in transition of recombinant DNA technology from laboratory-scale investigation to production-scale application is the efficient monitoring and supervision of the fermentation process during large-scale production. Existing sensor technology does not allow the most significant indicators of the fermentation behavior, biomass, and recombinant protein concentrations, to be measured accurately on-line. These measurements are currently performed off-line in the laboratory and provide delayed and relatively infrequent information about the process. Thus, the early recognition of an undesirable fermentation condition is not feasible, resulting in hindered on-line control actions. This is particularly important during the pilot plant and large-scale production when the continuation of an undesirable fermentation results in a significant waste of time and resources. One solution to this problem is the development of a process model performing supervision tasks such as biomass and product estimation based on variables routinely measured on-line.

A conventional modeling approach based on fundamental understanding of the system structure is not applicable in this process due to its complexity and a frequent change in underlying process characteristics (i.e., host strain and expression vector). Thus, a neural network model for the estimation of fermentation behavior has been developed using CER and OUR together with the batch age as the input variables. These input variables were selected from a range of possible factors measured on-line based on the quality of the resulting biomass or recombinant protein concentration prediction. The number of hidden neurons was determined by increasing the number of hidden neurons until the reduction in prediction error on the test data became negligible. This resulted in an “optimum” topology of 3-3-1, that is, 3 input neurons, 1 hidden layer with 3 neurons, and 1 output neuron. Similarly, an

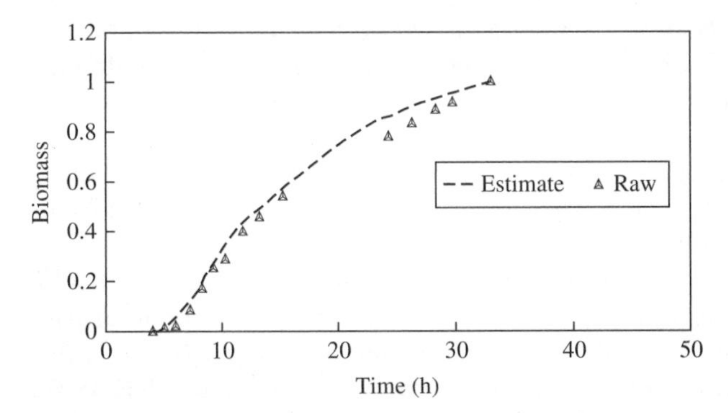

Figure 6.19 FANN biomass estimation.

"optimum" topology of FANN model for recombinant protein estimation is 3-4-4-1, that is, 3 input neurons, 2 hidden layers with 4 neurons in each layer, and 1 output neuron. Figures 6.19 and 6.20 show the neural network model fit to testing data (not used in training). In these figures only scaled biomass and recombinant protein concentration data are shown due to commercial confidentiality.

Figures 6.19 and 6.20 clearly demonstrate a satisfactory prediction of both biomass and recombinant protein, which is achieved in the majority of the fermentations tested. Following such verification, the models can be implemented on-line with real-time measurements of the model inputs to provide predictions of biomass and protein concentration.

6.5 Knowledge-Based Systems

Pattern recognition techniques described in previous sections relied almost exclusively on process data, and the wealth of human expertise was often utilized only implicitly during model development. However,

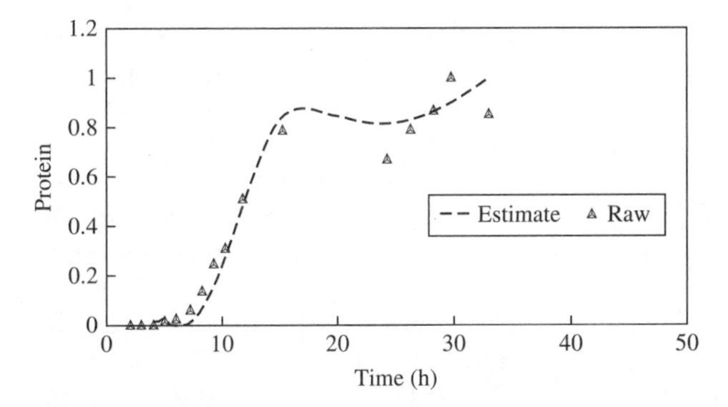

Figure 6.20 FANN recombinant protein estimation.

an *expert system* (ES) or a *knowledge-based system* (KBS) based on this often overlooked valuable source of process understanding. The driving force has been the perceived benefit of providing the operators with a *real-time knowledge-based system* (RTKBS), which uses engineering models and best operator rules to give intelligent information, higher-level decision support, and control assistance as and when required. The development of rule-based expert systems in the early 1980s was the first experience of process engineering applications of *artificial intelligence* (AI). From this starting point, the challenging new opportunities brought by the application of AI in process engineering were quickly discovered (Muratet and Bourseau, 1993). The installation of RTKBS for process manufacturing has become a practical reality over the past decade or so. The enabling technology applies human expertise, in the form of rules, procedures, and models to give real-time advice to operators and/or to affect control.

The technology of RTKBS has allowed manufacturing companies to install systems that would have been very difficult to implement using conventional software. Current industrial processes offer natural applications for RTKBSs, and they have been widely installed in the process industries, including food, steel, aluminum, paper, cement, chemicals, petroleum, and others. Among the applications are intelligent interpretation of sensor data, diagnosis of problems, analysis and validation of data, coping with process disturbances, and prediction of consequences. For example, DuPont has an extensive program of ES applications, including advisory, scheduling, and process control systems. Their process control systems include an on-line ES that receives data primarily from sensor based process monitoring and control systems and responds in real time to process problems (Rowan, 1989). More than 100 ESs for process control are deployed in DuPont, with about 50 percent being real-time systems. The total savings from all the ESs deployed in the company, according to the 1990 figures, was more than $75 million per year (Helton, 1990). Another example is 3M where system engineers simulate on a real-time basis the effect of steam energy input into a solvent recovery operation versus desired quality yield. This system allows them to get purity levels at very high standards with minimal cost (Helton, 1990). Monsanto's W. G Krummrich facility installed an ES that would assist the operators by monitoring field instrumentation and process equipment, provide real-time alarm prioritization, and provide on-line access to process troubleshooting documentation (Rehbein, 1992).

The first KBSs developed for fault diagnosis in the process industries were rule-based, using heuristic knowledge. In such KBSs, the symptoms are directly linked to their causes. However, the accuracy of the developed system depends on how rich and accurate the rules are. Also, the system is process specific and is not easily transportable to another similar type of process. Moreover, rapid development of these systems has been difficult to achieve because of the tedious nature of knowledge acquisition, inability of the system to learn or dynamically improve its performance, and unpredictability of the system outside the scope of the knowledge base it contains. Discussion concerning the unpredictability of a heuristic rule-based KBS in diagnosing faults when there is no specific knowledge about them can be found in Karim and Halme (1988). In spite of the limitations of heuristic KBSs, such as lack of flexibility, extensibility, user-computer interactions, and especially reliability, their applications continue to grow, particularly for very complex processes where other types of models are difficult to obtain (Janusz and Venkatasubramanian, 1991; Konstantinov et al., 1994; Konstantinov, 1996; Konstantinov and Yoshida, 1991, 1992a, 1992b). Chapter 4 described methods of knowledge elicitation and illustrated how they can be used to increase the speed and accuracy of the elicitation process. The first KBS case study described here will show the results of obtaining rules for KBSs through machine induction (Bakshi et al., 1994).

Potential benefits of automatic rule induction are

- Rules extracted from data could be compared to the existing rule base to determine the extent of operator adherence to standard operating procedures.
- Changes in operator behavior that could indicate a need to update the rule base.
- Rule induction might greatly assist rule-based development to remove some of the need for interviews.

Several rule induction strategies are described in literature. The results of applying a procedure based on ID3 are presented as an indication of the findings. The algorithm is incorporated in a software tool developed by *Knowledge Process Software* (KPS).

6.5.1 Case study G—RTKBS-based process parameter control

The case study selected to demonstrate the suitability of automatic rule induction and RTKBS in process control is a large-scale production of a secondary metabolite using a fermentation process. In order to maximize productivity, it is essential to tightly control the level of *dissolved oxygen* (DO_2) in the vessel. This parameter is affected by the level of carbon source supplied to the production organism in the form of sugar and oil. The dynamics of response to addition/cut of these two carbon sources are quite different and in extreme cases, where rapid response is required, water may also be added. The decisions about which variable to use (and to what extent) to control the DO_2 level are critical, yet difficult to make consistently on the basis of process data available at the time. While some operators manage to control DO_2 levels within the acceptable range, deviations often occur. This makes DO_2 level control an ideal application for RTKBS system. The rules required for DO_2 control were extracted from historical process data. To simplify the decision-tree construction each process variable was classified into several categories based on a preselected scale.

Figure 6.21 shows the tree developed from data after implementation of the Gensym (1997) G2-based control system. This example investigates the addition of sugar to the fermentation. The first split in the tree is based on the log hour (batch age). This is not surprising as a standard policy in sugar addition is adopted for the early stages of the fermentation. The other components of the tree show how sugar flow rates are set based on the condition of other process variables. Considering Fig. 6.21, one rule that can be derived from the tree is *if the log hour is 2, the water is 2, and the oil is 3, then the sugar flow is class 1.*

Here variables known to be considered by the operators were used to build the tree, but further variables could have been included if it was felt that decisions were being made regarding other measurements. Following the split based on log hours the next variable that distinguishes sugar flow rate is the water addition rate. Following the nodes progressively down the tree reveals the actions operators used to control DO_2 during fermentations from which data were collected and used to build the tree. Once the consistency of these actions is verified, they can be implemented within an RTKBS, which can then advise the operators of the appropriate action for DO_2 control of a current fermentation in real time. Automatic rule induction is also useful in postimplementation analysis of process data.

Knowledge trees can be built on the basis of the data collected after RTKBS implementation to analyze the compliance of operators with the RTKBS advice. Such trees allow management to focus on particular aspects of the control strategy to determine why decisions that are opposed to those suggested by RTKBS are taken. If there are valid reasons for such actions, this points to a deficiency in the knowledge base that will have to be rectified. If there are not valid reasons, then inappropriate control decisions may be causing a loss in product yield, and management decisions on operator performance have to be considered.

Once the validity and accuracy of the knowledge base is verified, it is important to test the ability of the RTKBS to control the selected process variable. In this case a G2 software package was used to implement the

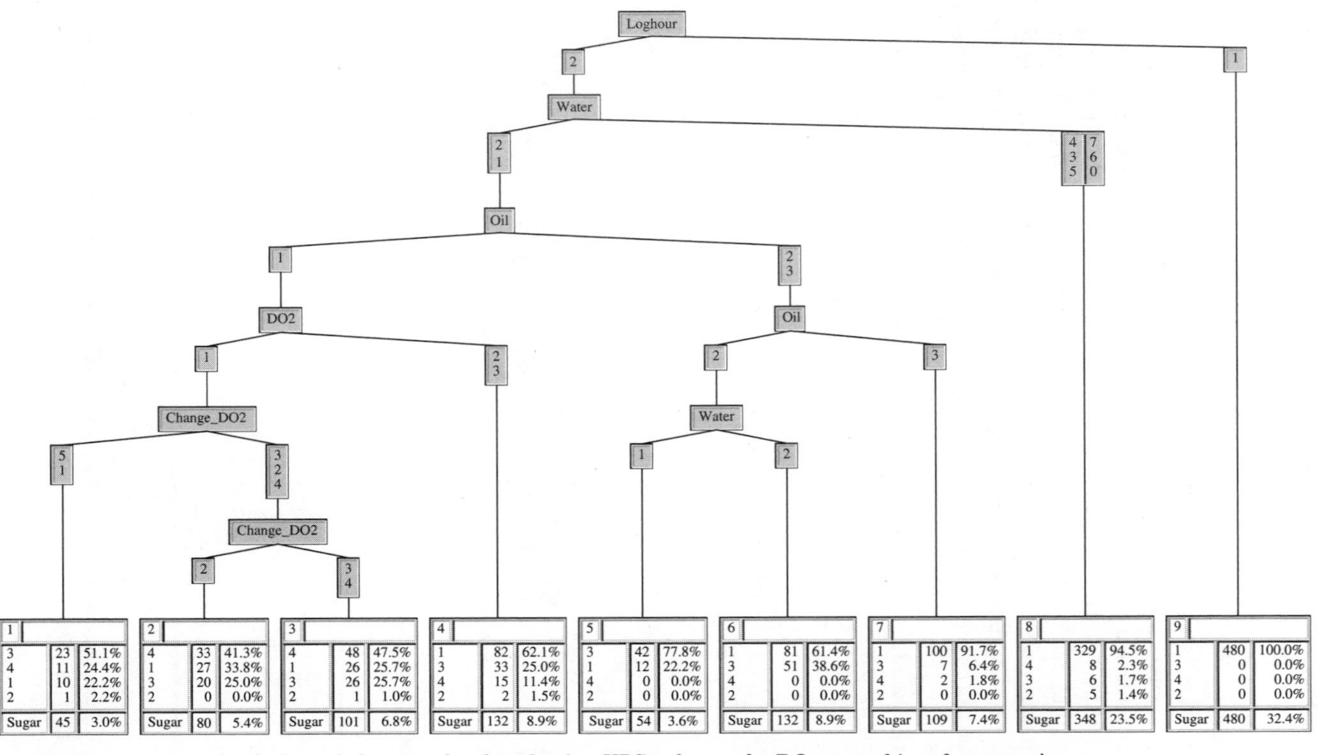

Figure 6.21 An example of a knowledge tree developed using KPS software for DO_2 control in a fermentation process.

control, since this package was already available within the company. Figure 6.22 demonstrates control improvements for a number of fermentations run with G2 implementation when compared to historical data from fermentations controlled by operators according to standard procedures.

Considering dissolved oxygen distributions in multiple batches of bioprocess data during the production period when dissolved oxygen approaches the constraint, it can be seen that improved control has been achieved. The histograms in Fig. 6.22 indicate that the variation in the data set has been reduced. The bin centers are 0.5 units apart and the histograms show much reduced variation. The standard deviation of the pre-G2 data set is 1.09 while that of the data set after G2 implementation is 0.58. This halving of the standard deviation means that it is possible to move the set point closer to the constraint yet still not violate it, possibly reducing the costs associated with the process and increasing the vessel yield.

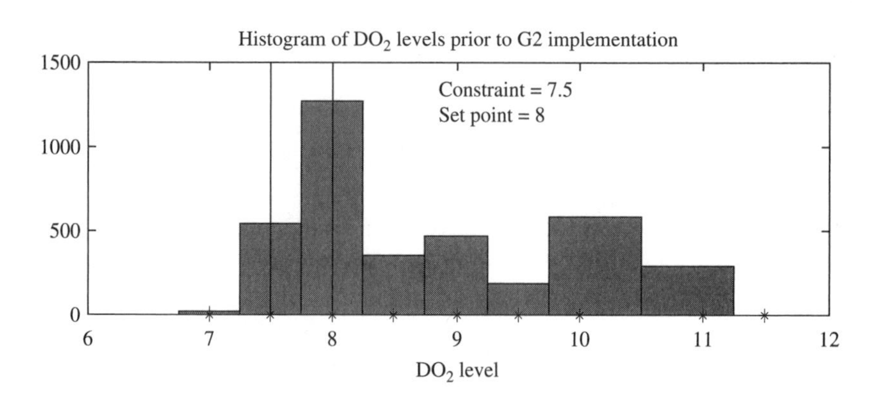

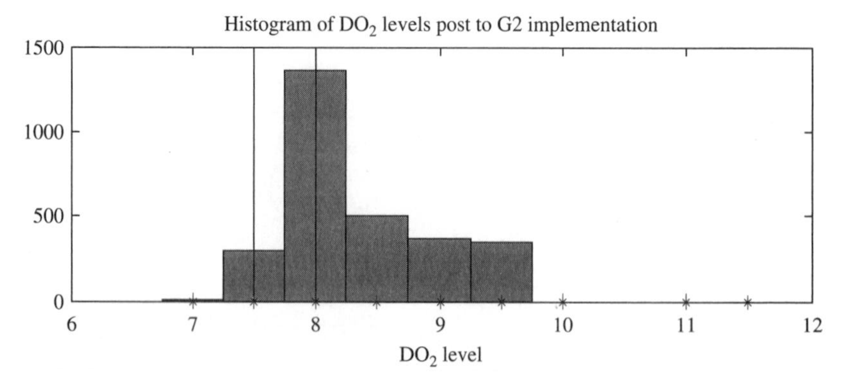

Figure 6.22 Dissolved oxygen levels in the production period prior to and postimplementation of G2.

6.5.2 Case study D—intelligent alarming using RTKBS

This section describes the development and the implementation of a RTKBS carried out in a bioprocess industry. The RTKBS was developed to provide consistent advice to the operators of a large-scale antibiotic production plant as described in case study D in Chap. 4. The objective was to develop a robust real-time advisory tool for intelligent alarming (basic alarm strategy was described in Chap. 3) that would be capable of sending messages to appropriate personnel, warning them of potential deviations during seed stage fermentation. The project was undertaken in two stages. The knowledge elicitation stage and lessons learned from this are discussed in detail in Chap. 4 and will not be further elaborated here. However, the outcome of this elicitation phase (in the form of an eGraph illustrated in Chap. 4) was converted into a rule base and implemented within G2, which then acted as an advisory/decision support system.

Gensym's G2 was installed to construct an intelligent alarming system (Glassey et al., 2000). This RTKBS application was designed to compare process variables with established univariate statistical control limits (see Chap. 5 for more detail on SPC) in a real-time environment. If a data point of a key variable was outside its control limits, G2 sent an alarm message to the screen and paged the operator. The G2 system also displayed the values of key process variables to the users and monitored the start-up and shutdown procedure for each process unit. This intelligent alarming strategy is illustrated in Fig. 6.23.

The univariate statistical control limits are useful but do not provide a total solution. For instance, this application did not incorporate the

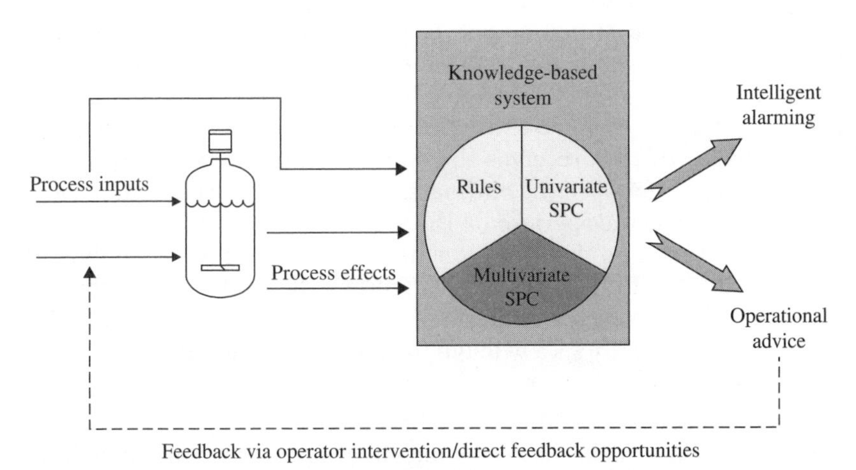

Figure 6.23 Intelligent alarming strategy.

heuristic knowledge (rules of thumb) needed by the operators to control the process. These heuristics usually reside in the minds of the experts. The experts have gathered this knowledge from their experience of the process. Typically, during a batch, the operators monitor the process by following a standard operating procedure that is valid as long as there are no deviations from the norm, which is not always the case. When the operators face an anomaly in the process, they fall back on the experts for their advice and act on it. The experts manipulate their mental model by using "rules of thumb" and advise the operators on the action to take. However, this cycle is not always effective because:

- The experts by virtue of their experience are rare and busy.
- It is difficult to get in touch with the experts out of hours.
- On a multiproduct plant, at any given time there are several processes running simultaneously and for each process there are many variables, the time course of which need to be considered to come to a decision. This puts a lot of cognitive demand on the experts.
- It is often difficult, even for a very skilled expert, to abstract a good situational assessment when he or she is under time pressure.
- If the expert is not available, the operators have to rely on their experience to come to a decision and take action. Since different operators have different levels of expertise, the decisions made by them tend to be inconsistent.

All these factors lead to inconsistent decision making, which in turn contributes to increased variability in the process. To reduce variability, it is essential to provide the operators with consistent decision-making support 24 hours a day and throughout the year. To achieve this, an on-line decision support system was developed. Capturing the decision-making knowledge or the mental rules of thumb of experts using the KAT method described in Chap. 4 and coding them in an on-line expert system achieved this. The system is interfaced with the database from which it receives the on-line data. When the process is running, the ES monitors the on-line data from the process and in case of anomaly informs the operator on the actions they should take. Ideally, the ES should come to the same conclusion as the expert by analyzing the same data. In effect this is equivalent to making the expert continually available to the process operators and allowing them to apply the expert's decision-making knowledge to give consistent advice.

Testing and validation of RTKBS. The first step in setting up the system is verifying the knowledge base on which it is based. Since this stage was described in detail in Chap. 4, we will concentrate on the subsequent steps of coding, testing, and validation of the ES. One of the advantages in using KAT for knowledge elicitation is the ease with which the resulting eGraph can be written into "if/then" rules for an ES. This can be performed manually or with the use of a KATKit software (Duke, 1998), which provides a means of automatically converting the KAT's eGraphs into rules. It was found that using KATKit for automated rule generation into G2 could almost entirely eliminate the time required for coding and testing the rule accuracy (Duke, 1999).

Regardless of the method of rule coding, it is critical to test the response of the ES in simulated process situations. The system is designed to advise the operator in case of any anomaly in the process. Since most of the time the process runs without problem, for testing it was necessary to supply values simulating anomalies for the variables and verify the output. This ensures that the system behaves in a way it should behave when there is an anomaly. However, this does not necessarily ensure that the system will behave the way it should, when there is no anomaly. To test this condition as well, extra code was added into the procedure that confirms that the "process is all right" when no deviations occur. Exhaustive testing, using manual methods was completed within three man-weeks. The resulting output was written to a file, which serves the purpose of validation.

After the coding and off-line testing was completed, the system was put on-line for testing. This system monitored eight seed tanks simultaneously and in a case of anomaly informed the operator with the relevant advice. The output was also stored in a file, which was then verified for consistency. Though care was taken in the (manual) coding of rules and procedures, during testing it was found that some of the rules did not fire the way they should have fired. In these cases, the logic of the concerned rules was rechecked and they were recoded.

Operator acceptance and training. Operator acceptance is a critical success factor in a RTKBS that aims to monitor or control the plant (Cartledge, 1992). Since operators using the system have varying levels of computer literacy, the system must be easy for the operators to use; they should feel in control of the application, not intimidated by it. These issues were identified at the beginning of the project, and therefore the operators were kept "on board" from the start of the project. The potential benefits of the application were discussed with them, and the ways it would assist them were highlighted.

Since the operators had been using G2 for several years (application discussed in Sec. 6.5.2) not much operator training was required to handle the system. Moreover, since the application developed was primarily an advisory system, wherein an advice message would "pop-up" in case of an anomaly, the handling of the output was not complex. The operator had to just acknowledge the message by pressing a button on the screen and then performing the actions suggested by the system.

Lessons learned. The aim of this study was to develop a real-time expert system, which would provide consistent on-line advice to the process operators. In this study, a relatively new knowledge elicitation technique (KAT) has been used, which has proved to be very effective and efficient in capturing knowledge. Claims that the method was particularly efficient at capturing knowledge were verified. Following the knowledge capture, the fact that KAT facilitates a computable representation of the experts' knowledge allowed the information to be easily transferred into production systems and ES shells such as G2.

The extracted knowledge was coded into G2 as rules and procedures and tested on an on-line process. The entire project took six months to complete with three knowledge engineers available for elicitation and one knowledge engineer for coding the elicited knowledge into G2 rules and procedures. This was achieved primarily because of the speed and accuracy with which the knowledge could be extracted using KAT. In total the project amounted to nine months of knowledge engineer time.

The industrial benefit is not only in operational decision making, thereby reducing variability and enforcing good and robust operations, but also in retaining (and growing) knowledge in an easy-to-maintain system. However, it is vital that the knowledge base remains up-to-date and provides the most appropriate advice. Updating will clearly need to be carried out if plant changes are made, but it is also essential to reassess plant operations periodically. Change of KBS procedures may result if, for example, operational strategy changes due to shifting market conditions. Putting in place a management strategy for control improvement and KBS maintenance is essential if investment in KBS is to result in a long-term financial gain.

Concluding Comments

This chapter has described some sophisticated approaches for delivering improvements in quality. The technologies described are established in some but not all areas of the process sector. This is not as a result of failures in the technology, and in the future it is likely that the technology will become more widespread and at the same time more straightforward to implement. Take for example the case of multivariate SPC. Since developments in the mid 1990s, the technology has developed rapidly and is now found in several supervisory control packages. In addition, there are numerous companies marketing their expertise in multivariate data analysis with off-line data analysis packages complementing on-line software. If the rise of MSPC is contrasted with other data analysis/control approaches (i.e., artificial neural networks, expert systems) then this initial push to market the technology is likely to be tempered with long-term realism when the applications that really are likely to benefit are identified as opposed to those where more traditional approaches are satisfactory. Initial indications are that there are some multivariate processes that will benefit from the use of MSPC but others where a more straightforward univariate SPC would perform acceptably. This is in line with model-based predictive control application experience where failure to recognize that a more simple approach is sufficient can be a costly mistake.

While basic MSPC and model predictive control concept are theoretically sound, further research developments in certain aspects are still required. Nonlinear variants and procedures making use of existing mechanistic structures are just two among many areas where further research will advance the scope of application. From an end-user perspective, research into easing the difficulties of problem identification and software configuration would be particularly welcome. Indeed, it could be argued that without this the methods will stay limited in application. Ease of implementation and configurability is essential to ensure that short-term gains are maintained in the face of process improvements. In essence, these control strategies can help maximize productivity of a process if applied in addition (as needed) to the critical control strategies discussed in this book.

References

Aldrige, S. (1994), "Chemical engineering meeting spotlights new technologies for bioprocess control," *Genet. Eng. News*, **14**(18)(October 15):12–18.

Aldrich, C., D. W. Moolman, J. S. J. Vandeventer, (1995), "Monitoring and control of hydrometallurgical processes with self-organizing and adaptive neural-net systems," *Comput. Chem. Eng.*, **19**(S):S803–S808.

Bakshi, B. R., G. Locher, and G. Stephanopoulos, (1994), "Analysis of operating data for evaluation, diagnosis and control of batch operations, *J. Process Control,* **4**(4):179–194.

Bishop, C. M. (1995), *Neural Networks for Pattern Recognition*, Oxford University Press, Oxford.

Carpenter, G. A., and S. Grossberg, (1987), "A massively parallel architecture for a self-organizing neural pattern-recognition machine," *Comput. Vis. Graph. Image Process.*, **37**(1):54–115.

Cartledge, J. G. (1992), "A plant operator interface for G2 process application," Air Products and Chemicals, Allentown, PA, *Gensym Technical Paper List*, Chem-005.

Cybenko, G. (1989), "Approximation by superpositions of a sigmoidal function. mathematics of control, signals and systems, **2**:303–314.

Dong, D., T. J. Mcavoy, E. Zafiriou (1996), "Batch-to-batch optimization using neural-network models,"*Ind. Eng. Chem. Res.*, **35**(7)(July):2269–2276.

Dony, R. D., and S. Haykin, (1995), "Neural-network approaches to image compression," *Proc. of the IEEE,* **83**(2)(February):288–303.

Duke, P. (1998), *KAT™ Reference Manual,* CK Design, Lancaster, UK.

Duke, P. (1999), *KAT™ White Paper*, CK Design, Lancaster, UK.

Fukushima, K. (1975), "Cognitron — self-organizing multilayered neural network," *Biol.l Cybern.*, **20**(3-4):121–136.

Gallagher, N. B., B. M. Wise, C. W. Stewart (1996), "Application of multiway principal components-analysis to nuclear waste storage tank monitoring," *Comput. Chem. Eng.*, **20**:S739–S744.
Gensym Corporation (1997), *Gensym Reference Manual*, Version 5.0, vols.1 and 2, Gensym Corporation, Cambridge, MA.
Geladi, P., B. R. Kowalski (1986), "An example of 2-block predictive partial least-squares regression with simulated data," *Anal. Chim. Acta*, **185**(July):19–32.
Gertler, J., W. H. Li, Y. B. Huang, and T. McAvoy (1999), "Isolation enhanced principal component analysis," *AICHE J.*, **45**(2)(February):323–334.
Glassey, J., G. Montague, and P. Mohan, (2000), "Issues in industrial bioprocess advisory system development," *Trends Biotechnol.*, **18**[4(195)]:136–141.
Gregersen, L., and S. B. Jorgensen (1999), "Supervision of fed-batch fermentations," *Chem. Eng. J.*, **75**(1)(August), 69–76.
Helton, T. J. (1990), *AI: Major Contributor to Profitability in the Chemical Industry,* Wiley, New York, pp. 5–7.
Hornik, K., M. Stinchcombe, and H. White H. (1989), "Multilayer feedforward networks are universal approximators," *Neural Netw.*, **2**:359–366.
Hoskins, J. C., and D. M. Himmelblau (1988), "Artificial neural network models of knowledge representation in chemical-engineering," *Comp. Chem. Eng.*, **12**(9–10):881–890.
Janusz, M. E., and V. Venkatasubramanian (1991), "Automatic generation of qualitative descriptions of process trends for fault detection and diagnosis," *Eng. Appl. in AI*, **4**(5):329–39.
Karim, M. N., and A. Halme (1988), "Reconciliation of measurement data in fermentation using on-line expert system," *Proceedings of the 4th International Congress on Computer Applications in Fermentation Technology: Modeling and Control of Biotechnical Processes*, Cambridge, UK, Ellis Horwood, Chichester, UK, pp. 37–46.
Kohonen, T. (1982), "Analysis of a simple self-organizing process," *Biol. Cybern.*, **44**(2):135–140.
Kohonen, T., E. Oja, O. Simula, A. Visa, J. Kangas (1996), "Engineering applications of the self-organizing map," *Proc. of the IEEE*, **84**(10)(October):1358–1384.
Konstantinov, K. B., and T. Yoshida (1991), "A knowledge-based pattern recognition approach for real-time diagnosis and control of fermentation processes as variable structure plants," *IEEE Trans. Sys., Man Cybern.*, **21**(4):908–14.
Konstantinov, K. B., and T. Yoshida (1992a), "A method for on-line reasoning about the time-profiles of process variables," *IFAC Symposium,* Newark, NJ, Pergamon Press, 133–138.
Konstantinov, K. B., and T. Yoshida, T. (1992b), "Real-time qualitative analysis of the temporal shapes of (Bio)process variables," *AIChE J.*, **38**(11):1703–15.
Konstantinov, K. B., W. Zhou, F. Golini, F., and W. S. Hu (1994), "Expert systems in the control of animal cell culture processes: Potentials, functions and perspectives," *Cytotechnology*, **14**:233–46.
Konstantinov, K. B. (1996), "Monitoring and control of the physiological-state of cell-cultures," *Biotech. Bioeng.*, **52**(2):271–289.
Kourti T., and MacGregor J. F. (1995), "Process Analysis, Monitoring and Diagnosis, Using Multivariate Projection Methods," *Chemometrics and Intell. Lab. Sys.*, **28**(1)(April):3–21.
Lennox, B., G. A. Montague, H. G. Hiden, and G. Kornfeld (1999), "Case study investigating multivariate statistical techniques for fermentation supervision," *Comput. Chem. Eng*, **23**[S (June 1)]:S827–S830.
Leonard, J. A., Kramer, M. A., and L. H. Ungar (1992), "Neural network architecture that computes its own reliability," *Comput. Chem. Eng.*, **16**(9)(September):819–835.
Ljung, L. (1999), *System Identification: Theory for the User*, 2d ed., Prentice Hall, Englewood Cliffs, NJ.
Luenberger, D. G. (2003), *Linear and Non-linear Programming*, Kluwer Academic, Boston.
Maciejowski, J. M. (2001), *Predictive Control: With Constraints*, New York, Prentice Hall.
Martin, E. B., and A. J. Morris (1996), "An overview of multivariate statistical process control in continuous and batch process performance monitoring," *Trans. Institute of Measurement and Control,* **18**(1):51–60.
Muratet, G., and P. Bourseau (1993), "Artificial intelligence for process engineering - state of the art," *Comput. Chem. Eng.*, **17**(S):S380–S388.

Nelson, P. R. C., P. A. Taylor, and J. F. MacGregor (1996), "Missing data methods in PCA and PLS - score calculations with incomplete observations," *Chemometrics and Intell. Lab. Sys.,* **35**(1)(November):45–65.
Nomikos, P., and J. F. MacGregor (1995), "Multiway partial least-squares in monitoring batch processes," *Chemometrics and Intell. Lab. Sys.*, **30**(1)(November):97–108.
Qin, S. J., and T. A. Badgwell (2003), "A survey of industrial model predictive control technology," *Control Eng. Pract.*, **11**(7):733–764.
Rehbein, D., S. Thorp, D. Deitz, and L. Schultz (1992), "Expert systems in process control," *ISA Trans.*, **31**(2):49–55.
Rowan, D. A. (1989), "On-line expert systems in process industries," *AI Expert*, August, 31–38.
Soeterboek, R. (1992), *Predictive Control:A Unified Approach*, Prentice Hall, Englewood Cliffs, NJ.
Venkatasubramanian, V., R. Vaidyanathan, Y. Yamamoto (1990), "Process fault-detection and diagnosis using neural networks.1. steady-state processes," *Comput. Chem. Eng.*, **14**(7):699–712.
Watanabe, K., I. Matsuura, M. Abe, M. Kubota, and D. M. Himmelblau (1989), "Incipient fault-diagnosis of chemical processes via artificial neural networks," *AICHE J.—American Institute of Chem. Eng.*, **35**(11):1803–1812.
Wold, S. (1978), "Cross-validatory estimation of number of components in factor and principal components models," *Technometrics,* **20**(4):397–405.
Weitz, R. R. and Lakshminarayanan, S., "On a heuristic for the final exam scheduling problem," *Journal of the Operational Research Society*, **47**(4)(April):599–600.
Willshaw, D. J, and C. V. D. Malsburg (1976), "How patterned neural connections can be set up by self-organization, *Proc. of the Royal Society of London Series B-Biological Sciences*, **194**(1117), 431–445.
Zhang, H. W., B. Lennox, P. R. Goulding, and A. Y. T. Leung (2000), "A float-encoded genetic algorithm technique for integrated optimization of piezoelectric actuator and sensor placement and feedback gains," *Smart Mater. Struct.*, **9**(4)(August):552–557.

Part 4

Productivity

In a competitive environment, organizations focus on productivity to generate a winning proposition for their customers. The aim of productivity improvement is to continuously or radically enhance manufacturing metrics. At the organizational level, productivity has to be envisioned at a broader level encompassing all aspects of the business. In a complex, ambiguous, and dynamic business environment, productivity is the key differentiator of a successful business. Knowledge, leadership, and synergy are the three key pillars of optimal productivity. This part focuses on the methodology and philosophy for optimal productivity.

Chapter

7

Productivity Improvement Methodologies

Productivity: The essence of corporate longevity P. MOHAN

Objective

The aim of this chapter is to present an integrated picture of productivity improvement methodologies under the overarching improvement philosophy of Six Sigma.

Learning outcomes of this chapter will include:

- Introduction to Six Sigma
- Lean Six Sigma concept
- Overview of Six Sigma tools
- Introduction to design for Six Sigma

The learning is then applied to a case study to gain practical understanding of the application of the concepts.

7.1 Introduction

Productivity improvement is an integral part of manufacturing and has its roots in the "classical and modern quality thinking movement" with Deming as the founder. This chapter focuses on consolidating tools and methodologies for productivity improvement used in this book under the banner of Six Sigma. The Deming wheel is perhaps the foundation of improvement methodology. The Deming wheel, or the *plan-do-check-act* PDCA cycle (Fig. 7.1), forms the basis of newer philosophies such as Six Sigma.

The initial phase of the cycle is the "plan" (P) phase, which implies a review of current performance to identify issues and gaps. This leads to the development of a problem statement providing a clear objective and direction. Subsequently, data are gathered on key issues/gaps to identify the root cause of the problem. Finally, an optimal solution and plan of action are devised for implementation. The plan also includes resource planning and a business case to justify pursuing the improvement process further.

Once a plan for improvement has been approved, the next step is the "do" (D) phase. This is the implementation stage during which the plan is executed in a pilot mode. This stage may involve another loop of PDCA to resolve any issues and challenges.

The next phase is "check" (C), where the deliverable from the pilot is measured and evaluated (from the pilot study), to ascertain whether the solution has the potential to achieve the intended objective.

The final phase is "act" (A), to standardize the implemented solution if found successful or to formalize the learning if found unsuccessful. It is important to document the learning from failure as it may form the basis of the plan phase for the next PDCA cycle. The continuous improvement effort is an ongoing and iterative process and a constant engagement is key to its success.

Six Sigma has its origin in the PDCA cycle and forms the basis of the modern structured analytical approach to productivity improvement. In this chapter Six Sigma is discussed as an overarching philosophy for productivity improvement with a case study example.

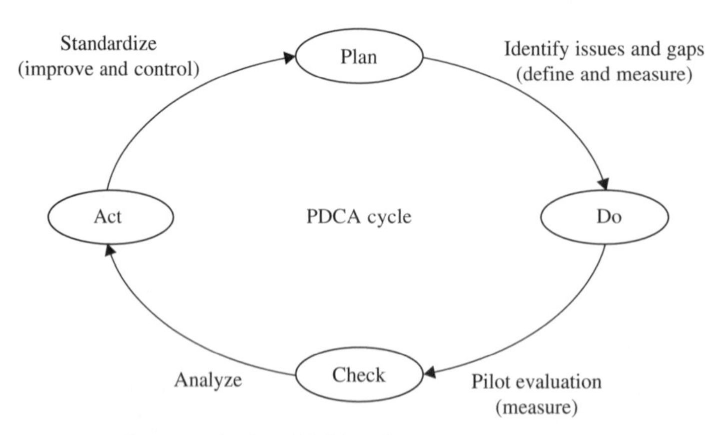

Figure 7.1 Deming wheel or PDCA cycle.

7.2 Six Sigma

Six Sigma has various meanings and definitions. Especially noticeable are the technologist and business definitions of Six Sigma, which are as follows:

- Technologists define Six Sigma as a structured technical methodology using statistical and mathematical tools, used by engineers and statisticians, to improve products and processes.
- Businesses define Six Sigma as not just a process improvement methodology but an objective management system with customer centricity, quality focus, measures, and a goal for improvement that aims for near perfection.
- Pande, Neuman, and Cavanagh (2000) describe Six Sigma as "A comprehensive and flexible system for achieving, sustaining and maximizing business success. Six Sigma is uniquely driven by close understanding of customer needs, disciplined use of facts, data, and statistical analysis, and diligent attention to managing, improving, and reinventing business processes."

Compiling these definitions, the key attributes of the Six Sigma methodology could be summarized as:

- *Customer centricity.* The voice of customers that is translated into specific, measurable *critical to quality* (CTQ) requirements. This typically manifests itself in the form of specification limits on the key quality attributes of the product.
- *Measure.* Sigma level is the measure that defines the capability of the process. This measure is based on defects, which in essence is the failure to deliver to the customer's CTQ requirements.
- *Goal.* A goal that reaches near perfection; a Sigma level of six with defects as low as 3.4 per million defect opportunities.
- *Executive engagement.* Six Sigma implementation requires executive sponsorship and engagement to engineer the cultural and paradigm shift in the organization.
- *Resource commitment and infrastructure.* Employees' participation is vital to the success of the Six Sigma approach. Six Sigma is application of the analytical methods to the design and operation of manufacturing systems. Six Sigma is not a completely new way to manage an enterprise, but it is a very different way. In essence, Six Sigma forces change to occur in a systematic way. A very powerful feature of Six Sigma is the creation of an infrastructure to assure that performance improvement activities have the necessary resources.

- *Financial results.* Financial metrics are critical to initiation of the improvement projects. Shareholder value and profits are the primary business drivers.
- Special roles such as black belt, green belt, and master black belts, who are experts in the Six Sigma process, champion and lead this process (Fig. 7.2).

The deliverables from Six Sigma take various forms, which can be categorized into process and business deliverables as illustrated in Fig. 7.3. The business deliverables includes market share growth and high profit due to high quality and low cost. The process deliverables that lead into the business deliverables include process improvement, variability reduction, cycle-time reduction, and defect reduction. Another significant benefit is that Six Sigma acts as a catalyst for culture change by creating a productivity mind-set in the organization, which eventually leads to customer retention and growth.

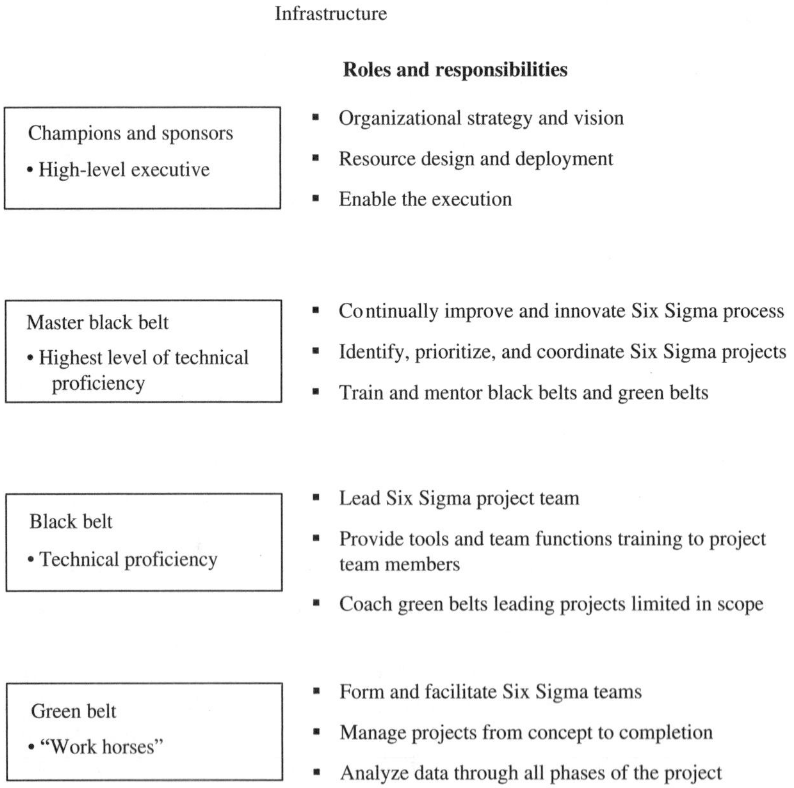

Figure 7.2 Six Sigma infrastructure.

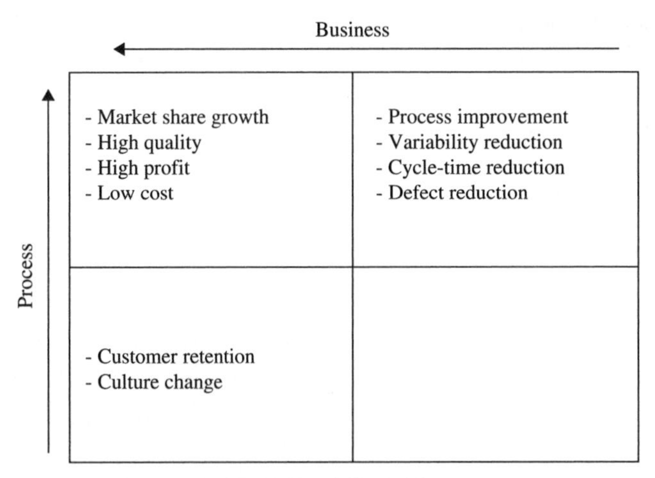

Figure 7.3 Process and business deliverables.

The primary focus in this chapter is process/productivity improvement. One of the key parameters in productivity improvement is the number of rejects or defects. Six Sigma employs statistical techniques focusing on the defects. A defect occurs when the manufactured unit is outside the specified quality limits of the product attribute. Typically, a defect is rejected, resulting in rework. On the other hand, *yield* signifies the number of units without defects as a percentage over the total number of units.

Sigma level performance is expressed in terms of *defects per million opportunities* (DPMO). DPMO has the following elements:

(a) *Defects per unit* (DPU) is the average defects over the total number of units sampled.

$$\text{DPU} = \frac{\text{Nnumber of units with defects}}{\text{Nnumber of total units}} \tag{7.1}$$

One of the advantages of Six Sigma is the capability of generating a level field for comparing complex and simple processes. The defect opportunities are the normalizing factor. Understanding the defect opportunities is also a challenge in Six Sigma methodology; the list of defect opportunities could become subjective.

For example, defect opportunities in a television (a complex system) and a light bulb (a simple system) could be listed as shown in Table 7.1

There should be a structured approach to identification of defect opportunities. The approach could include group brainstorming or use of knowledge extraction techniques. The defects should have customer centricity as the underlying theme (see Table 7.2).

TABLE 7.1 Comparison of a Simple System with a Complex System

Television (10)	Light bulb (3)
Picture tube	Element
Screen	Glass
Capacitor	Thread
Casing	
Circuit	
Remote	
Software logic	
Memory	
Magnet	
Transformer	

(b) DPMO signifies the number of defects that could arise out of one million opportunities. DPMO could also be referred to in manufacturing as *parts per million* (PPM). The formula of DPMO is:

$$\text{DPMO} = \frac{\text{number of defects}}{(\text{number of units} \times \text{number of opportunities})} \tag{7.2}$$

This example illustrates that Six Sigma provides a consistent metric that can be used across different industries to compare performance. As a process metric, sigma level has value as an indicator of how often the company fails to meet customer needs. The measure also relates to a Six Sigma improvement target, which is 3.7 DPMO—aiming for near perfection.

The financial aspect of Six Sigma improvement also becomes evident to management as defects have costs associated with them. The major benefit of defect reduction goes much beyond the savings in cost. It impacts the quality significantly (positively) leading to a sequence of events: customer satisfaction; customer retention and growth; higher profit; shareholder value growth; and credibility with regulatory agencies.

TABLE 7.2 Calculation of Sigma Level

Parameters	Television	Bulb
Number of defect opportunities	10	3
Number of units	5000	50,000
Number of defects	45	60
DPMO	$45 \times 10^6/(10 \times 5000) = 900$	$60 \times 10^6/(3 \times 50{,}000) = 400$
Sigma level	4.625	4.875

7.2.1 1.5 Sigma shift

There is a difference between standard deviation and sigma as implied in the Six Sigma methodology. For example, for normal distribution 3σ in statistical terms will mean a defect level of less than 3,000, while using Six Sigma this number would be 66,800 (see Table 7.3 or Fig. 7.4). This has been a point of debate, however, in the evolution of Six Sigma methodology at General Electric the focus was to extend the concept to include a broader improvement philosophy measuring total process performance. The broader total performance philosophy includes understanding defects over a large number of data points (million) including assessment of number of defect opportunities. Though there will always be some debate on the actual values of sigma, the important thing is to have a consistent way to measure defects.

An explanation of the differences in sigma value could be attributed to increased process variation over time. This increase in variation may be due to small variations with process inputs, the way the process is monitored, changing conditions, and so forth. The increase in process variation is often assumed for the sake of simplicity to be similar to temporary shifts in the underlying process mean. The increase in process variation has

TABLE 7.3 Six Sigma Conversion Table

Yield (%)	DPMO	1.5-σ shift in mean
6.68	933200	0
10.56	894400	0.25
15.87	841300	0.5
22.66	773400	0.75
30.85	**691500**	**1.00**
40.13	598700	1.25
50	500000	1.50
59.87	401300	1.75
69.15	**308500**	**2.00**
77.34	226600	2.25
84.13	158700	2.50
89.44	105600	2.75
93.32	**66800**	**3.00**
95.99	40100	3.25
97.73	22700	3.50
98.78	12200	3.75
99.38	**6200**	**4.00**
99.7	3000	4.25
99.87	1300	4.50
99.94	600	4.75
99.977	230	5.00
99.987	130	5.25
99.997	30	5.50
99.99833	16.7	5.75
99.99966	3.4	6.00

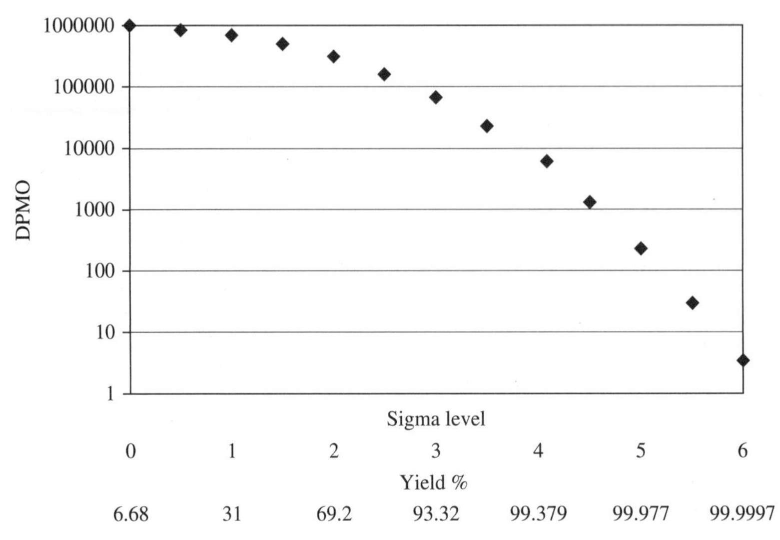

Figure 7.4 Graphical representation of Six Sigma conversion.

been shown in practice to be equivalent to an average shift of 1.5 standard deviations in the mean of the originally designed and monitored process (Gitlow and Levine, 2004). If a process is originally designed to be twice as good as customer requirements (i.e., the specifications representing the customer requirements are six standard deviations from the process target), then even with a shift of 1.5σ, the customer demands are likely to be met. If the process shifted off target by 1.5 standard deviations, there are 4.5 standard deviations between the process mean ($\mu + 1.5\sigma$), which results in at worst 3.4 ppm at the time the process has shifted or the variation has increased to have a similar impact as a 1.5 standard deviation shift. In the 1900s, Motorola demonstrated that a 1.5 standard deviation shift was in practice, which was observed as the equivalent increase in process variation for many processes that were benchmarked.

A statistical purist would argue that the genesis of the sigma metric is flawed because it is based on a shift factor. The engineers viewed the metric as a worst-case ppm for a process because it was assumed that any shift factor significantly larger than 1.5 would be caught by the common usage of *statistical process control* (SPC), a point beyond a three sigma control limit. If the shift is less than 1.5σ, that is good, because the ppm is less.

7.3 Six Sigma Improvement Process

The intention for the comprehensive measurement of sigma level is to identify the current state of operation. This helps set the company target for improvement. For example, for a production unit the current operation

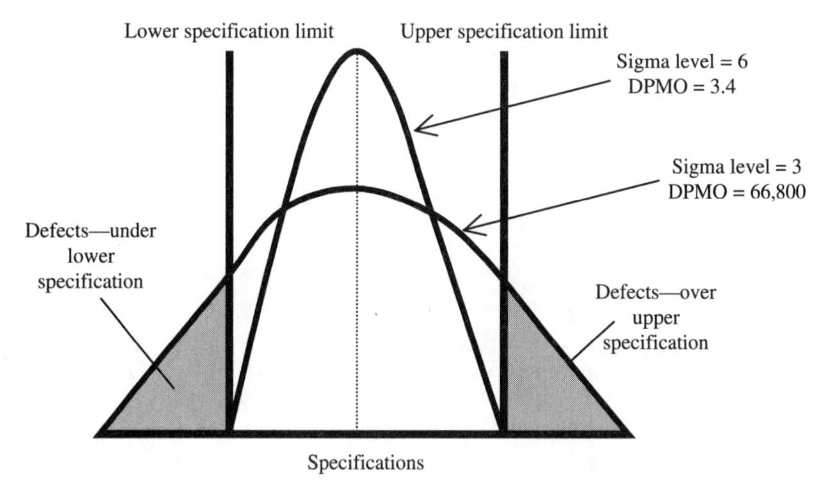

Figure 7.5 Six Sigma improvement.

reflects a 3.4σ leading to about 66,800 DPMO (see Fig. 7.5). The target for the company should be to move toward a Six Sigma level leading to only 3.4 DPMO. This improvement in the performance has significant impact on the viability of the business. The first impact of course is the dollar savings on rework for the rejects; this is called *cost of poor quality* (COPQ). Besides COPQ, the major aspects are customer satisfaction, retention, and growth, which by far are more significant benefits

The preceding discussion indicates that Six Sigma is a structured approach to achieve productivity improvement. It has five phases: define, measure, analyze, improve, and control (Table 7.4). This follows the PDCA model and is cyclic in nature. Each phase has tools and methodologies to carry out the overall Six Sigma process. Table 7.5 summarizes the Six Sigma process with the tools and technology.

7.3.1 Define

The purpose of the define phase is to "set the stage." The output of this phase is a comprehensive "problem statement" outlining clear goals and objectives. Einstein once said that a "problem well defined is problem half solved." The problem statement provides several key aspects for success of the improvement efforts, which include clarity of purpose, hence eliminating confusion; identification of resources; and an alignment objective. The tools applied in the define phase serve two purposes: (1) documenting key information about the project (project definition form) and (2) providing a high-level overview. Some of the tools and methodologies of this phase include project definition, SIPOC, process mapping, and flowcharting.

TABLE 7.4 Six Sigma Process

Six Sigma is a five-phase improvement cycle based on PDCA. The five phases are **DMAIC** which stands for:

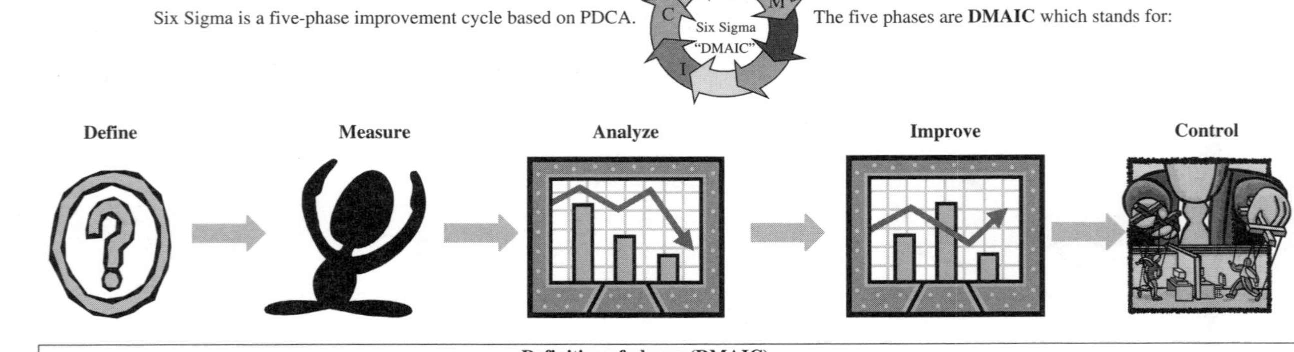

Definition of phases (DMAIC)				
Problem statement	Identification of key parameters	Identify root causes	Generate solution/s	Set expectations for maintenance
Identify resources and requirements	Measurement	Causal hypothesis	Test solution/s	Establish control mechanism
Setting objective	Data integrity	Hypothesis verification from data	Implement and standardize	Monitor control performance
Tools and methodology				
Project definition	Cause and effect diagram	DOE	Mistake proofing	SPC
SIPOC	Pareto analysis	Statistical analysis and hypothesis testing	New control strategy	Alarm strategy
Process mapping	Control chart	SPC/MSPC	Project management	Process audit
Flowchart	Risk analysis			
Pareto charting		Process capability analysis		
Financial justification	Root cause analysis			

SIPOC stands for:

Supplier. A pool of resources to the process, which includes people, material, and equipment.

Input. The actual resources used in the process.

Process. The transformational step that converts input into desired output adding value.

Output. The final desired product resulting from the process.

Customer. The receiver of the output.

SIPOC provides a high-level overview of the core process displaying a cross-functional set of activities in a single, simple diagram. Systems thinking could be used as an integral part of SIPOC, illustrating the problem comprehensively. In case study H (Sec. 7.6), SIPOC is used to describe the problem.

Process mapping. A process map is a graphic representation of a process, showing the sequence by using a modified version of standard flowcharting symbols.

Flow charts. A process flowchart is simply a tool that graphically shows the inputs, actions, and outputs of a given system.

7.3.2 Measure

Once the project is initiated, the next step is to collect data on the problem. The main aspects of the measure phase include the quantification of the current sigma level of the process, that is, the quantitative characterization of the baseline (current operation) and the narrowing of the root causes to be analyzed for the improvement.

There are various measurement methodologies including root cause analysis, Pareto chart, risk analysis, control chart, and so forth used in the measure phase. Root cause analysis is a structured approach that could cover both the measure and the analyze phases of Six Sigma. This methodology is discussed in detail in Chap. 5 along with control charting. Risk analysis is discussed in detail in Sec. 7.5.

7.3.3 Analyze

The analyze phase focuses on identifying the root cause of defects, generating a causal hypothesis for the root cause, and hypothesis generation using data. In the analyze phase, data contributing to the root cause are studied in detail to generate clues for improvement. Analysis tools such as hypothesis testing, *analysis of variance* (ANOVA), scatter plot, and regression methodologies are used on the data to understand the causal relationships. Statistical process control as detailed in Chap. 5 (as well as MSPC—Chap. 6) could be used to analyze the data using control charts. *Design of experiment* (DOE) is a powerful analysis tool to study the critical parameters identified in the measure phase. DOE is discussed in detail in Sec. 7.5.2.

7.3.4 Improve

The implementation of the output from the analyze phase leads to improvement. The improve phase may require modification of the process based on the scientific justification resulting from the analyze phase. The process control strategy may change as a result of this implementation. An important aspect of the improve phase is to implement "mistake proofing" to eliminate mistakes relevant to the root cause and to enhance the robustness of the process. Mistake proofing is discussed in Sec. 7.5.4. The improvement must be quantified in terms of the new sigma level.

7.3.5 Control

The control phase signifies that the improvement should be standardized and become part of the day-to-day process. The necessary actions are taken to sustain this improvement for the long term. These actions may include setting expectations for maintenance, establishing the new control mechanism, and monitoring control performance. The tools that can help facilitate the control stage include control charting (SPC) of the new process to monitor the performance and setting the alarm strategy to alert the operators of any unplanned deviation in the process.

The Six Sigma methodology discussed in the previous sections is illustrated in case study H. The next section introduces "lean Six Sigma" that helps drive lean manufacturing, a key aspect of productivity.

7.4 Lean Six Sigma

Lean Six Sigma uses all the traditional Six Sigma concepts (DMAIC) and infrastructure (master black belts, black belts, and so forth) with the refined focus in the following areas (see Fig. 7.6):

1. Customer focus
2. Elimination of waste
3. Improvement
4. Transformational leadership

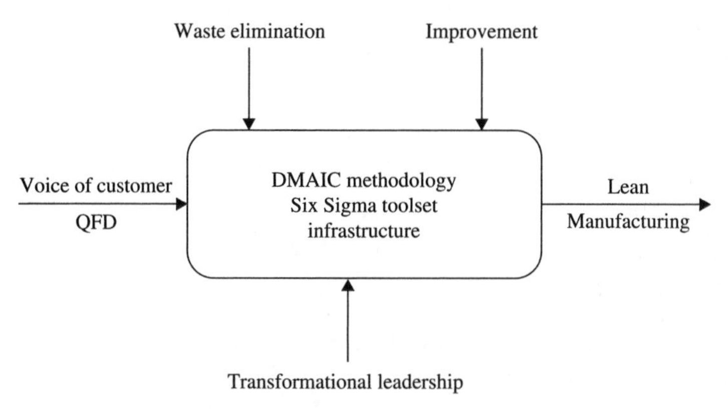

Figure 7.6 Lean Six Sigma model.

Customer focus for lean Six Sigma stands for delivering higher quality service in less time. The voice of customer is critically important in decisions about products or services. *Quality function deployment* (QFD) is an important tool for translating customers' needs to product design parameters. Chapter 2 discusses QFD in detail.

Transformational leadership is critical to the success of lean manufacturing. This leadership will transform the voice of customer into lean manufacturing by optimal engagement of methods, toolset, and infrastructure. The details of transformational leadership are discussed in Chap. 8. Chapter 8 also discusses the improvement aspect including both incremental and breakthrough improvement.

The focus on elimination of waste is unique to lean Six Sigma. Lean Six Sigma focuses beyond defects to include waste-elimination strategies and factory physics concepts (Hopp and Spearman, 2000). The waste-elimination strategies for lean manufacturing include the following elements:

1. *Inventory management.* A *just-in-time* (JIT) inventory management with zero inventory is an ideal state. However, some controlled excess inventory may be required to adjust for uncertainties and variability.
2. *Overproduction.* A production system should be modeled as a "pull production system" (Kanban) minimizing overproduction (Hopp and Spearman, 2000).
3. *Cycle time and queue management.* Cycle time and queue should be minimized; this is discussed in Chap. 8.
4. *Conveyance.* Any unnecessary transport must be identified and eliminated.

5. *Avoid system decay.* Equipment maintenance must be a priority, engaging reliability-centered maintenance (see Chap. 5).
6. *Optimize flows.* Flows of people, process, and material must be optimized, minimizing waiting time and cross flows.
7. *Overburden.* Machine overburden must be avoided and 85 percent of machine capacity must not be exceeded.
8. *People.* People must be respected, engaged in value-added activities, and burnouts (overburdening) must be avoided.

Further suggested reading for lean Six Sigma includes George et al. (2004, 2005) and Lareau (2000).

Elimination of waste is an important aspect of productivity improvement. The next section describes the Six Sigma toolset that is required to drive a systematic approach to productivity improvement.

7.5 Six Sigma Toolset

The evolution of Six Sigma over the years has led to incorporating and integrating numerous tools into the overarching philosophy of improvement. Some of these tools are shown in Table 7.4 and the ones covered in this book are shown in Table 7.5.

Risk analysis, design of experiment, root cause analysis, and hypothesis testing are discussed in this chapter, followed by a case study illustrating the systematic application of Six Sigma.

7.5.1 Risk analysis

Profiling of risk is instrumental to the success of the productivity improvement efforts. There are four types of risk discussed in this section, namely quality, financial, technology, and organizational (see Fig. 7.7). The overriding aspect of risk assessment is quality. Experiments should be

TABLE 7.5 Six Sigma Tools

Phase	Tools	Chapter number
Measure	Control chart	5
Measure	Risk analysis	7
Analyze	Statistical process control	5
Analyze	Multivariate statistical process control	6
Analyze	Design of experiment	7
Improve	Mistake proofing	7
Improve	Control strategy	3
Control	Alarm strategy	6
Analyze	Hypothesis testing	7

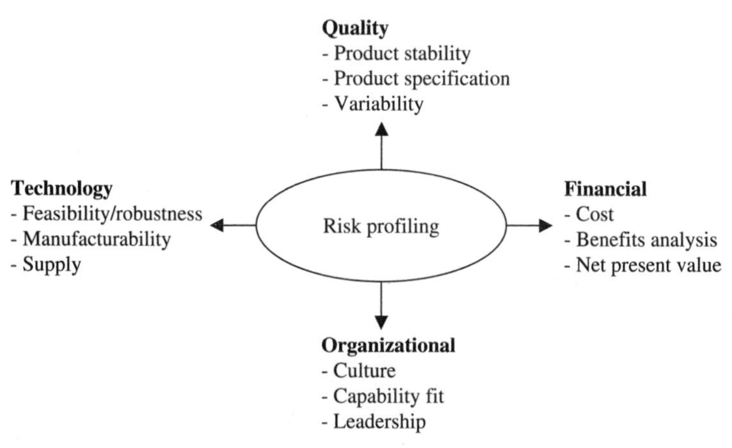

Figure 7.7 Risk profiling.

performed to assess the impact of quality in relation to relevant quality parameters including product stability, specification, and variability. The detailed aspects of quality risk management are given in Chap. 2. The capability to deliver a quality product is the primary requirement of the productivity improvement process.

Technology risk relates to the feasibility/robustness, manufacturability, and supply of the technology being assessed. Assessment of emerging technology is vital for continued success of the manufacturing operation especially in relation to competitive edge. Lack of technology assessment programs can lead to missed opportunities, for example, the *Encyclopedia Britannica* suffered a loss in sales when it failed to adapt to the paperless approach (new technology) to encyclopedias (CD for computers) in a timely manner. Day et al. (2000) have proposed a technology assessment process that includes four stages: scoping, searching, evaluating, and committing. Scoping involves assessing a company's core capabilities, defining a company's target market, and predicting a company's ability to change and redeploy resources. Searching involves surveying, discovering, searching for emergent technology, and predicting and assessing market need. Evaluating involves planning the development of new technology in light of market needs and competitive goals. Finally committing involves rigorous risk profiling from the standpoint of all other aspects: quality, financial, and organizational. When evaluating technology risks it is important that a structured risk analysis methodology be used, such as FMEA (Chap. 2). Financial risk analysis is discussed in detail in the following section.

Financial risk analysis. Organizational risk relates to the culture, appetite for change, and quality of the transformational leadership driving the change agenda. In Chap. 8, leadership is discussed with emphasis on transformation. Change is typically resisted in any organization, as it is well known that "only a wet baby likes change." The success of driving productivity improvement relies on how well the change management process is implemented. Change management requires an agent (doer), a champion (visionary), and a sponsor (top management support). The change management activities include defining change, assessing the resistance to change, generating sponsorship, communicating clearly about the change, implementing the change, reinforcing the change, and communicating on the observed improvement.

In a business environment the financial yardstick is the best prioritization tool. Project prioritization in a business environment requires a rigorous financial analysis. Investment in productivity improvement projects should undergo the same financial scrutiny as an investment appraisal of the stock market. There are various tools for financial decision making, some of which are indicated in Fig. 7.8. Risk and return is important in the financial aspect of risk profiling.

Payback is a simple financial tool that focuses on the time needed to pay off the investment. A shorter time to pay off an investment is deemed better in prioritizing the projects. However, the payback method has some serious drawbacks that include time value of money, risk, and consideration of cash flow beyond the payback period.

The financial tool used for investment appraisal is the *discounted cash flow* (DCF). DCF methods illustrate the present value of the investment, and there are various investment appraisal methodologies for DCF, which include *internal rate of return* (IRR), *accounting rate of return* (ARR), payback, *economic value added* (EVA), and *net present value* (NPV). The details on these methodologies can be found in standard financial management literature (Brealey and Myers, 1991; Stern et al., 1991).

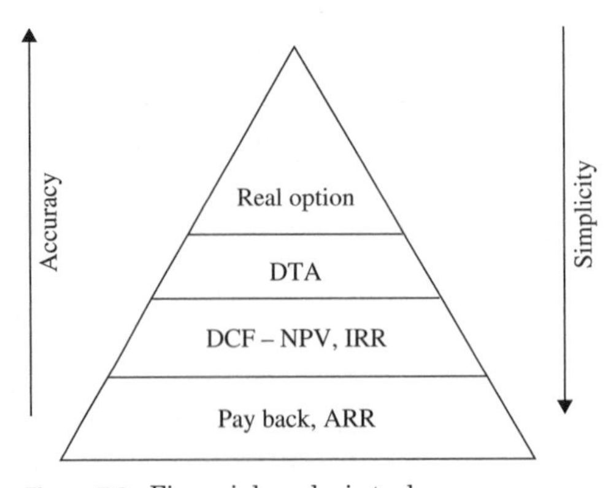

Figure 7.8 Financial analysis tools.

NPV is a widely used DCF technique, which is used in this section as a project prioritization tool. *Present value* (PV) is calculated as:

$$\mathrm{PV} = \sum_{t=0}^{n} \frac{A_t}{(1 + r)^t} \tag{7.3}$$

$$\mathrm{NPV} = \text{benefit of the investment} - \mathrm{PV}$$

where A = net cash flow in one year
r = discount rate
t = number of years used for the NPV calculation

Discount cash flow accounts for time value of money, inflation, and risk. Time value of money can be computed from the risk-free rate of return for that geographic area. However, discount cash flow has shortcomings such as the assessment of the risk premium. It considers the whole project in its entirety and commits the management to the entire investment.

Decision tree analysis (DTA) divides a project into various milestones with decision gates providing flexibility to the management to respond to uncertainties, thus adapting the decision-making process. While DTA is a major improvement over DCF it does not adjust for changes in a project's risk over the project's life. A real option is an extension of DTA that uses NPV as the base discount methodology with additional flexibility to make staged investments, delay further investment until more information is available, expand operations if initial results are successful, and switch from one mode of operation to another. Unlike DTA, real option is designed to value projects where the risk changes as new information becomes available. Details on real options are covered in various financial texts including Copeland and Vladimir 2003 and Amran and Kulatilaka (1999).

The simple version of real option is discussed in this section with the help of an example as a powerful tool in understanding the financial risk involved in the productivity improvement projects. For assessment of a new technology, the improvement project could be divided into various stages such (see Fig. 7.9) as:

- Test the new technology in a small-scale setup for proof of concept, an investment of $150K.
- If the proof of concept fails (probability of failure is 60 percent) the technology when implemented still delivers a perpetual cash flow of $500K. The implementation of the suboptimal version of new technology would need an investment of $300K.
- If the proof of concept for optimal engagement of new technology succeeds (probability of success is 40 percent) then the investment in implementing the technology at the manufacturing scale would be $1000K. The perpetual return would be $900K.

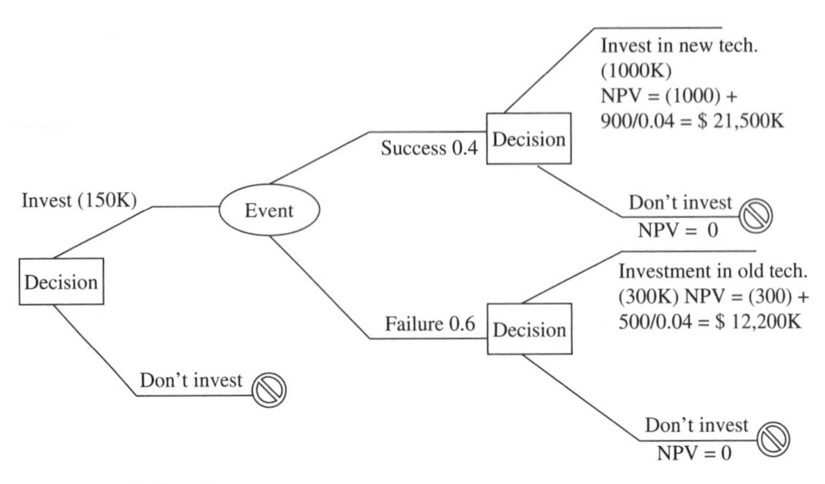

Figure 7.9 Real options.

Options for management

1. Do not invest in the emerging technology which does not carry any risk; however it may be a lost opportunity.
2. Invest in the optimal version of the new technology given the proof of concept has been established. The inflation rate was assumed to be 2 percent and the risk-free premium was assumed to be 2 percent. Therefore, the discount rate will be equal to 4 percent. The risk of success is reflected in the probability of 0.4 obtained by brainstorming.

   $$\text{Expected value for option 2} = (150) + 0.4 \times 21{,}500 = \$\ 8450\ \text{K} \quad (7.4)$$

3. Invest in the suboptimal version of technology given that the new technology proof of concept fails.

   $$\text{Expected value for option 3} = (150) + 0.6 \times 12{,}200 = \$\ 7170\ \text{K} \quad (7.5)$$

This analysis leads to a decision tool that can accommodate uncertainties. It is important to ascertain the financial risk and uncertainties prior to embarking upon the next phase of Six Sigma. Once the project is financially approved, design of experiment is a toolset in the analysis mode of Six Sigma methodology, which will be discussed in the following section.

7.5.2 Design of experiments

Design of experiment (DOE) is a very valuable Six Sigma tool as it generates knowledge to optimize the process. DOE is a collection of statistical methods for studying the relationships between independent variables or factors, the Xs (also called input variables or process variables), and their interactions on a dependent variable, the critical to quality (CTQ) or Y. Design of experiment is a statistical technique aimed at developing and implementing experimental strategies to study variables (factors) optimizing experimental space and the consequent knowledge for the process under investigation. The importance of rational design and analysis of experiments is demonstrated by the extensive literature available on this research area (e.g., Montgomery, 1997; Roy, 2001; Wu and Hamada, 2000). While the intention of this book is not to reiterate the detailed theoretical principles of each of the methods described in the literature, it is important to outline the most frequently used methods and discuss their benefits and limitations with regard to industrial application.

One of the major advantages of DOE is that multiple factors (variables) can be investigated in a comprehensive design. Such multifactor investigation also leads to an understanding of interactions between factors. The purpose of an experimental design depends on the level of knowledge available concerning the process, product, or service being studied. The type of DOE is selected based on the process knowledge and the intended objective as indicated in Fig. 7.10.

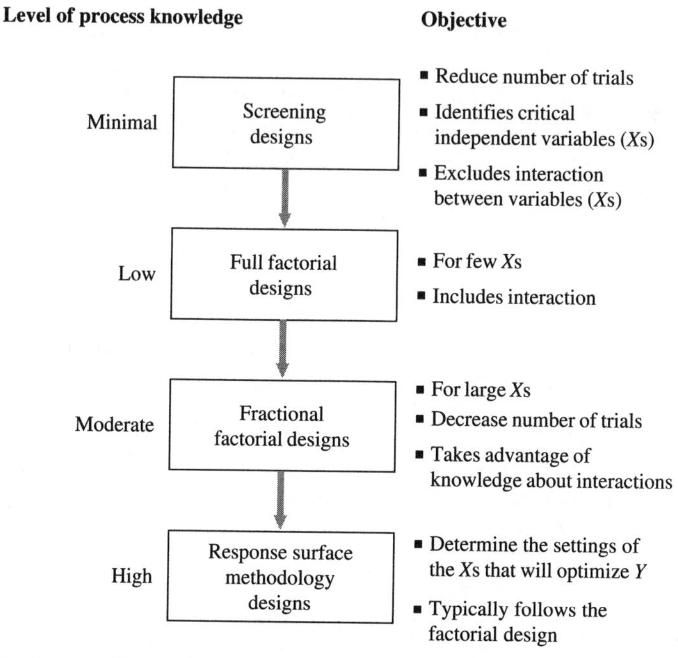

Figure 7.10 Process knowledge and objective of DOE.

Identify factors and levels. The first step in DOE is to identify the factors (variables) that need to be investigated. Identifying factors is the most important aspect of DOE. Identification of wrong or inappropriate factors can lead to unnecessary experiments consuming time and resources with no return. Factors must be identified in a structured manner, using first principles and brainstorming.

Factors are changed systematically during experimentation, and whether continuous or discrete, their different settings are referred to commonly as levels. A factor may have several levels. For example, an experimental evaluation of impact of temperature as a factor may have three levels for experimental evaluation: 25, 30, and 40°C. To understand the concept of factorial design, let us take an example from the metal industry. The tensile strength of iron is dependent on temperature, time, and pressure. Here the response is tensile strength, and the factors are temperature, time, and pressure. Further, for understanding the optimum condition for a desirable response (tensile strength) an experimental strategy should involve the three factors and two levels per factor. The lower level is denoted as "–" and the upper level is denoted as "+." See Table 7.6.

$$\text{Total experimental run} = (\text{levels of factor a}) \times (\text{levels of factor b}) \times (\text{levels of factor c})....$$

For the given example, total number of experiments $= 2 \times 2 \times 2 = 8$.

A full factorial would mean running eight experiments randomly and studying the response. Full factorial design allows detailed analysis of effects and interactions. However, in an industrial environment time, speed, and resource constraints drive the experimental designs to explore statistical ways of minimizing the number of experiments by resorting to fractional factorial design (see "Multivariate test" in Sec. 7.5.3).

Block designs—screening design. In many industrial projects, where certain sources of variability are known, it is necessary to design experiments in a way that allows systematic control of these sources. This improves the accuracy of the comparisons between experimental factors for each value of known source of variability. Returning to case study A

TABLE 7.6 Factors and Level

Factors	Level 1, –	Level 2, +
Temperature (°C)	400	500
Pressure (Pa)	10	20
Time (min)	100	200

of antibiotic production, described in Chap. 1, let us consider the applicability of block designs to optimize the yield of the antibiotic.

Randomized complete block design. In this design each block contains all the tested treatments. In our example of the antibiotic production suppose we wish to determine whether yield improvements can be achieved by introducing modified production microorganisms. To assess this, four different strains of microorganisms may be tested in a series of small-scale experiments. However, the researchers realize that each time they perform the experiments (or indeed the production runs) there may be a change in the supply of the raw materials used to sustain the growth and production. Just as different diets in human society can manifest themselves very clearly, changes in media composition can influence the growth and production of any microorganism. Thus, it is important to perform each of the experiments with different strains for each batch of raw materials. Thus, a set of experiments as indicated in Table 7.7 may be performed and the yield of antibiotic recorded for each experiment.

There is one observation per treatment in each block, and the order in which the treatments are run within each block is determined randomly. Usually, we are interested in testing the differences in the means of each of the treatments. Standard statistical tests can be used to determine whether there is a significant difference between the performances within each block of raw materials. We may also be interested in comparing the block means, since if they do not differ greatly, blocking may not be necessary in the future. An important decision when setting out a randomized block design is the choice of sample size and the number of blocks. Increasing the number of blocks increases the number of replicates and the number of error degrees of freedom, which leads to a more sensitive design. Indeed, blocking may lead to a significantly reduced number of replicates needed to achieve the same sensitivity as with a completely randomized design only. In general, if the importance of the block effect is not obviously clear, it is still advisable to block since the detrimental effects of blocking would be negligible.

TABLE 7.7 Randomized Complete Block Design for Antibiotic Production

	Raw material batch			
Strain	1	2	3	4
1	13.9	14.1	14.4	15.0
2	14.1	13.9	14.7	14.8
3	13.8	14.1	14.2	14.6
4	14.6	14.4	15.0	15.3

Latin square design. The randomized complete block design addresses the issue of a single source of variability being removed by blocking. Let us assume that except for the raw material batch, the operator performing the experiment has a significant influence on the yield of the process. In such a case, the blocking is performed in two directions and the rows and columns of Table 7.8 represent two restrictions on randomization with operators being labeled as A, B, C, and D.

Numerous statistical texts (e.g., Montgomery, 1997) offer standard Latin square design layouts of various sizes. Usually one design is selected and then the order of the rows, columns, and letters is arranged at random to achieve a random order of experimentation. Once the observations are obtained, it is possible to analyze the results using standard statistical techniques. If there are additional known variables contributing to the variability, it is possible to use a different block design, for example, the Graeco-Latin square design, where an additional nuisance factor is coded using Greek letters to provide a third direction of blocking. In the example this may be the feed rate of a main source of nutrient fed into the process. Once again standard statistical procedures are used to assess the influence of these factors on the process yield.

Factorial designs. Many experiments involve investigating the effects of two or more factors on performance. In such cases, factorial designs are often the most effective designs. In these designs all possible combinations of factor levels are investigated in each complete trial or replication of the experiments. In such full factorial designs the number of experiments to be performed increases rapidly. For example, if the researchers in our case study of antibiotic production wish to investigate the effect of temperature, feed rate, and feed initiation time of a critical nutrient on the yield, they may use high and low levels of each of these factors. If they carry out a full factorial design, 2^3, that is, eight experiments would have to be carried out and all factor interaction effects on the response variable can be assessed. However, with the increasing number of factors and levels, the number of experiments may become prohibitive, given time and resource pressures in an industrial

TABLE 7.8 Latin Square Design for Antibiotic Production

	Raw material batch			
Strain	1	2	3	4
1	A = 13.9	B = 14.1	C = 14.4	D = 15.0
2	B = 14.1	C = 13.9	D = 14.7	A = 14.8
3	C = 13.8	D = 14.1	A = 14.2	B = 14.6
4	D = 14.6	A = 14.4	B = 15.0	C = 15.3

setting. If six or seven factors are to be investigated, the number of experiments increases significantly; for example, 2^7 results in 128 experiments. It is possible to reduce the number of experiments by ignoring interactions between factors. For example, the design known as the Plackett-Burman design (Plackett and Burman, 1956) uses a Hadamard matrix to define the minimal number of experiments to run. While the main effects of each variable can be evaluated using a small number of runs, this design will produce a biased estimate of effects where interactions among the variables exist.

Response surface methodology. The designs of the experiments discussed earlier are most often used to screen for important factors and their direct additive effects (and sometimes interaction effects) on the studied output. However, in the later stages of the experiments it becomes important to optimize the levels of the factors identified in screening experiments. In practice, "response surface designs" are used for this type of investigation and the methodology is often referred to as *response surface methodology* [RSM, formally developed by Box and Wilson (1951)]. The methodology encompasses:

- Design of experiments that will yield adequate and reliable measurements of the response of interest
- Determination of a mathematical model that best fits the data collected from the chosen design by conducting appropriate tests of hypotheses concerning the model parameters
- Determination of the optimal settings of experimental factors that produce the maximum (or minimum) value of the response

The basic assumption of RSM is that the response variable is a function of the levels $x_1, x_2, \ldots, x_k$ of k experimental factors

$$\eta = \phi(x_1, x_2, \ldots, x_k)$$

The true response function ϕ is assumed to be a continuous, smooth function of the x_i and it is approximated with a Taylor series expansion that can be reduced to a polynomial of degree of d. In recent research it has been shown that other modeling approaches [such as artificial neural networks, Glassey et al. (1994)] can be used to approximate this relationship with more accuracy and ease of application.

Response surface designs yield data for this relationship range from full factorial designs described in Sec. 7.6.4, although the number of experiments carried out is excessive in order to fit the model. Two most commonly used designs are central composite designs and Box-Behnken designs.

Central composite designs (CCD). These designs can fit a full quadratic model and contain embedded factorial or fractional factorial designs with center points that are augmented with a group of "star points" that allow the estimation of curvature. There are three basic types of CCDs: circumscribed, inscribed, and face centered. Let us assume that two factors are investigated at levels ranging from –1 to +1. The circumscribed design adds star points at some distance α from the center and these points establish new extremes for the low and high settings for all factors (see Fig. 7.11). On the other hand, the inscribed CCD does not violate the original extremes of the factors—a property that may be important where process factor levels were set at the extremes of operability and their violation may lead to an inoperable process (see Fig. 7.11). Finally, the face-centered CCD uses star points that are at the center of each face of the factorial space and thus requires only three levels of each factor (compared to the five levels required by the circumscribed and inscribed CCDs). These designs are illustrated in Fig. 7.11.

Box-Behnken designs. This design is a quadratic design that does not contain an embedded factorial or fractional factorial design. In this design the treatment combinations are at the midpoints of the edges of the process space and at the center. Figure 7.12 demonstrates this for the case of three factors ranging from –1 to +1.

The benefit of this design for the case of three factors is the reduced number of experiments to be carried out compared to CCDs (see Table 7.9); however, this benefit diminishes with the increasing number of factors. An additional benefit compared to circumscribed and inscribed CCDs is that this design uses only three levels of each factor, just like the faced CCD. However, compared to the faced CCD this design is rotatable, which is desirable since it does not bias an investigation in any direction. In a rotatable design the variance of the predicted values is a function of the distance of a point from the center of the design, but not of the direction in which the point lies.

It is important to understand the basic statistics to maximize the value of DOE. The following section focuses on this aspect.

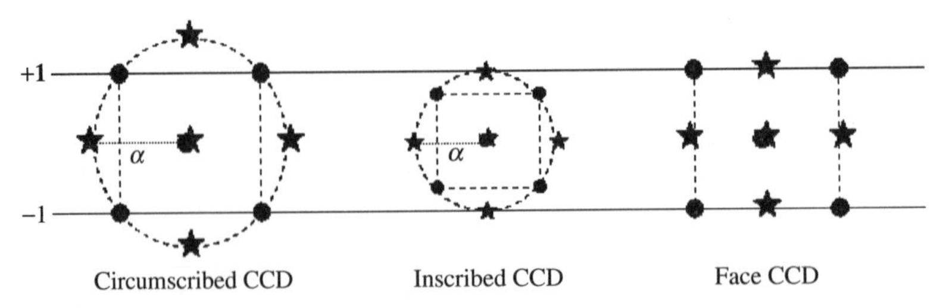

Figure 7.11 Diagrammatic representation of central composite designs.

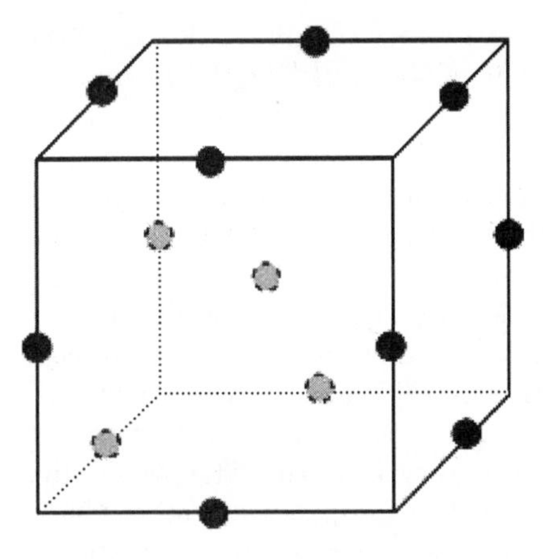

Figure 7.12 Diagrammatic representation of a Box-Behnken design for three factors.

7.5.3 Interpreting the statistics

It is important to understand the underlying statistics. Key aspects of statistics are hypothesis testing, regression methods, and analysis of variance.

Hypothesis testing. Hypothesis testing is used to check if two samples of p measurements (or variables) belong to the same population or not. This technique determines whether a change in the population is real

TABLE 7.9 Number of Experiments

Circumscribed (Inscribed)				Face CCD				Box-Behnken			
Repeat	X_1	X_2	X_3	Repeat	X_1	X_2	X_3	Repeat	X_1	X_2	X_3
1	−1	−1	−1	1	−1	−1	−1	1	−1	−1	0
1	+1	−1	−1	1	+1	−1	−1	1	+1	−1	0
1	−1	+1	−1	1	−1	+1	−1	1	−1	+1	0
1	+1	+1	−1	1	+1	+1	−1	1	+1	+1	0
1	−1	−1	+1	1	−1	−1	+1	1	−1	0	−1
1	+1	−1	+1	1	+1	−1	+1	1	+1	0	−1
1	+1	+1	+1	1	+1	+1	+1	1	−1	0	+1
1	$-\alpha$	0	0	1	−1	0	0	1	+1	0	+1
1	$+\alpha$	0	0	1	+1	0	0	1	0	−1	−1
1	0	$-\alpha$	0	1	0	−1	0	1	0	+1	−1
1	0	$+\alpha$	0	1	0	+1	0	1	0	−1	+1
1	0	0	$-\alpha$	1	0	0	−1	1	0	+1	+1
1	0	0	$+\alpha$	1	0	0	+1	3	0	0	0
6	0	0	0	6	0	0	0				
Total runs = 19				Total runs = 19				Total runs = 15			

or if it is an event that happens only by chance (Cunha-Bakeev, 2001). This can be done by testing the validity of the *null hypothesis*, usually indicating that the two means are equal. If the null hypothesis is rejected, the two classes tested are statistically different from each other.

The mean hypothesis test can be performed using one variable at a time (in a univariate way), some variables at a time, or all variables at once (a multivariate approach). The importance of a multivariate test is highlighted by some authors (Manly, 1994; Moore, 1995) as the ability to control *type I* error rates (discussed in Sec. 5.2 in relation to statistical process control). A type I error occurs if a significant difference between classes is found when, in reality, they belong to the same population. In other words, the null hypothesis is rejected when, in reality, it is true. The probability of a type I error occurring is measured by the significance level α. Using a numerical example, Manly (1994) demonstrates that the probability of a type I error occurring using a multivariate test such as the Hotelling's T^2 statistic is much smaller than that provided by a series of univariate tests, particularly when the number of variables is large. Another advantage of the multivariate test is that it takes into account the correlation between the variables and deals with it properly.

Univariate test. Traditionally, the univariate test is performed using the Student's t statistic. The null hypothesis to be tested here is given by Eq. 7.6

$$H_0 : \mu_1 = \mu_2 \tag{7.6}$$

and the alternative hypothesis is:

$$H_1 : \mu_1 \neq \mu_2 \tag{7.7}$$

where μ_1 and μ_2 are the population means for classes 1 and 2, respectively.

Given two sample observation vectors $\mathbf{x}_1$ and $\mathbf{x}_2$, the t statistic is calculated according to Manly (1994) using Eq. 7.8:

$$t = \frac{(\overline{\mathrm{x}}_1 - \overline{\mathrm{x}}_2)}{s\sqrt{\dfrac{1}{n_1} + \dfrac{1}{n_2}}} \tag{7.8}$$

where $\overline{\mathrm{x}}_1, \overline{\mathrm{x}}_2$ = sample means for each class
n_1, n_2 = number of samples for each class
s = pooled estimate of variance, given by Manly (1994) in Eq. 7.9

$$s^2 = \frac{(n_1 - 1)s_1^2 + (n_2 - 1)s_2^2}{n_1 + n_2 - 2} \tag{7.9}$$

The value of the t statistic obtained by this equation is compared with the t distribution with $n_1 + n_2 - 2$ degrees of freedom to determine whether the result is significantly different from zero.

According to Manly (1994), this test is particularly robust if the two sample sizes are equal or nearly so.

Multivariate test. The Hotelling's T^2 statistic is usually applied for the multivariate test. According to Morrison (1990), the reason for using the Hotelling's T^2 statistic is the fact that this test is invariant under coordinate changes and has null distributions free of the unknown population covariance matrix. The null hypothesis in this case is given by the Eq. 7.10:

$$H_0 : \begin{bmatrix} \mu_{11} \\ \vdots \\ \mu_{1p} \end{bmatrix} = \begin{bmatrix} \mu_{21} \\ \vdots \\ \mu_{2p} \end{bmatrix} \text{or } \boldsymbol{\mu}_1 = \boldsymbol{\mu}_2 \tag{7.10}$$

And the alternative hypothesis is:

$$H_1 : \begin{bmatrix} \mu_{11} \\ \vdots \\ \mu_{1p} \end{bmatrix} \neq \begin{bmatrix} \mu_{21} \\ \vdots \\ \mu_{2p} \end{bmatrix} \text{or } \boldsymbol{\mu}_1 \neq \boldsymbol{\mu}_2 \tag{7.11}$$

where $\boldsymbol{\mu}_1$ and $\boldsymbol{\mu}_2$ are the vectors of means for classes 1 and 2, respectively. The number of variables is represented by p.

If there are two independent sample matrices of observations, $\mathbf{X}_1$ and $\mathbf{X}_2$, where the rows represent the samples collected for each class and the columns contain the p variables, then the Hotelling's T^2 statistic for the two-sample case is given by Eq. 7.12 (Morrison, 1990):

$$T^2 = \frac{n_1 n_2}{n_1 + n_2}(\overline{\mathbf{x}}_1 - \overline{\mathbf{x}}_2)'\mathbf{C}^{-1}(\overline{x}_1 - \overline{\mathbf{x}}_2) \tag{7.12}$$

where $\overline{\mathbf{x}}_1$ and $\overline{\mathbf{x}}_2$ are the vectors of means for each class and $\mathbf{C}$ is the pooled covariance matrix given by the Eq. 7.13 (Manly, 1994):

$$\mathbf{C} = \frac{(n_1 - 1)\mathbf{C}_1 + (n_2 - 1)\mathbf{C}_2}{n_1 + n_2 - 2} \tag{7.13}$$

where $\mathbf{C}_1$ and $\mathbf{C}_2$ are the covariance matrices of the classes 1 and 2, respectively.

The quantity calculated from Eq. 7.14 has the variance ratio F distribution with degrees of freedom p and $n_1 + n_2 - p - 1$.

$$F = \frac{n_1 + n_2 - p - 1}{(n_1 + n_2 - 2)p} T^2 \tag{7.14}$$

The decision rule for a test at the significance level (has the following form:

Do not reject $H_0 : \mu_1 = \mu_2$ if:

$$T^2 \leq \frac{(n_1 + n_2 - 2)p}{n_1 + n_2 - p - 1} F_{\alpha;p,n_1+n_2-p-1} \tag{7.15}$$

and reject otherwise.

Regression model. Regression models fit a curve through the points. The curve is an equation (a linear model) that is estimated by least squares. The least squares method minimizes the sum of squared differences from each point to the line (or curve). The fit assumes that the Y variable is distributed as a random scatter above and below a line of fit.

The summary of fit table. The summary of fit (see Fig. 7.19) shows the numeric summaries of the response for the linear fit and polynomial fit of 2° for the same data. You can compare the summary of fit tables to see the improvement of one model over another as indicated by a larger RSquare value and smaller root mean square error.

RSquare. RSquare measures the proportion of the variation around the mean explained by the linear or polynomial model. The remaining variation is attributed to random error. RSquare is 1 if the model fits perfectly. An RSquare of 0 indicates that the fit is no better than the simple mean model.

RSquare Adj. RSquare Adj adjusts the RSquare value to make it more comparable than models with different numbers of parameters by using

Human	**Equipment**
- Action not followed - Wrong action taken - Information misinterpreted	- Damaged - Worn out - Uncalibrated
Procedure	**Material**
- Inadequate - Wrong - Ambiguous	- Wrong material - Damaged - Wrong quantity

Figure 7.13 Some examples of mistakes.

the degrees of freedom in its computation. It is a ratio of mean squares instead of sums of squares.

Root mean square error. Root mean square error estimates the standard deviation of the random error. It is the square root of the mean square for error.

Mean of response. Mean of response is the sample mean (arithmetic average) of the response variable. This is the predicted response when no model effects are specified.

Observations. The number of observations used to estimate the fit. If there is a weight variable, this is the sum of the weights.

Analysis of variance (ANOVA, Fig. 7.19). The ANOVA table displays the following quantities:

Source. Lists the three sources of variation, called model, error, and C total.

DF. Records the associated *degrees of freedom* (DF) for each source of variation. A degree of freedom is subtracted from the total number of nonmissing values (N) for each parameter estimate used in the computation. The error DF is the difference between the total DF and the model DF.

Sum of squares. Records an associated *sum of squares* (SS) for each source of variation. The total (C total) sum of squared distances of each response from the sample mean, that is, the sum of squares for the base model (or simple mean model) used for comparison with all other models. The total SS less the error SS gives the sum of squares attributed to the model.

Mean square. A sum of squares divided by its associated degrees of freedom.

F ratio. The model mean square divided by the error mean square. The underlying hypothesis of the fit is that all the regression parameters (except the intercept) are zero. If this hypothesis is true, then both the mean square for error and the mean square for model estimate the error variance, and their ratio has an F distribution. If a parameter is a significant model effect, the F ratio is usually higher than expected by chance alone.

Prob > F. The observed significance probability (p-value) of obtaining a greater F value by chance alone if the specified model fits no better than the overall response mean. Observed significance probabilities of 0.05 or less are often considered evidence of a regression effect.

The parameter estimates table. The parameter estimates table displays the following:

Term. Lists the name of each parameter in the requested model. The intercept is a constant term in all models.

Estimate. Lists the parameter estimates of the linear model. The prediction formula is the linear combination of these estimates with the values of their corresponding variables.

Std error. Lists the estimates of the standard errors of the parameter estimates. They are used in constructing tests and confidence intervals.

***t* ratio.** Lists the test statistics for the hypothesis that each parameter is zero. It is the ratio of the parameter estimate to its standard error. If the hypothesis is true, then this statistic has a Student's t distribution. Looking for a t ratio greater than 2 in absolute value is a common rule of thumb for judging the significance because it approximates the 0.05 significance level.

Prob>|*t*|. Lists the observed significance probability calculated from each t ratio. It is the probability of getting, by chance alone, a t ratio greater (in absolute value) than the computed value, given a true hypothesis. Often, a value below 0.05 (or sometimes 0.01) is interpreted as evidence that the parameter is significantly different from zero.

Besides DOE another important Six Sigma tool is mistake proofing. The following section discusses mistake proofing in detail.

7.5.4 Mistake proofing

The focus of Six Sigma is to minimize defects and mistakes that contribute to defects. Webster's dictionary defines *mistake* as, "fail to identify correctly." In a manufacturing operation the system can fail to correctly identify human action, state of equipment, robustness of procedure, and selection of material (see Fig. 7.13).

The consequences of mistakes could be numerous: high defects, low sigma, loss of customer confidence, human suffering, and the like. It is important to control defects due to mistakes by using mistake proofing. Poka-yoke (i.e., mistake proofing) developed by Shingo (1986) is the most effective method to minimize mistakes. Poka-yoke is based on 100 percent inspection, since mistakes are so rare that they cannot be detected and corrected by any other means.

Mistake-proofing methods have three dimensions: warning, control, and shutdown. Warning indicates that a mistake has been detected and prompts the correction. Control prevents mistakes, defects, or the flow of defective items to the next process. Shutdown stops the normal function when mistakes or defects are detected. A simple example is using counters to count the number of units processed for the next stage of operation, eliminating the mistakes caused by manual counting.

It has been demonstrated that product defect rates are strongly related to complexity. One way of quantifying complexity is the number of defect per opportunities (DPO). Mistake proofing simplifies the process and reduces the number of defect opportunities prior to engaging mistake-proofing methodologies. Once the process has been simplified, various mistake-proofing devices can help eliminate the mistakes (Hinckley, 2001).

Warning devices attract the attention of the operator and alert them to the mistake. These warning devices include visual (e.g., flashing light, flashing display), aural (e.g., sirens, whistles, chimes), tactile (e.g., vibrators), olfactory (e.g., odor), and in special cases a human barrier could be used (e.g., door, lockout/tagout).

Control devices include sensors for detection and the follow-up process control mechanism. Sensors play a critical role in mistake proofing by identifying the mistake. There are numerous sensors that measure the health of key variables of the process, that is, temperature, pressure, motion, strain, force, level, and so forth. As described in Chap. 3 (Critical Control Strategy) the outputs from the sensors are analyzed for errors (mistakes) and a control sequence is initiated based on the algorithm (feedback control). In the control sequence, the alarming strategy plays an important role in generating warnings using warning devices.

Shutdown means stopping the operation to address the mistake and avoid any further mistakes. There are various devices used for shutdown, such as valves and switches. For an integrated mistake-proofing approach the elements of the mistake-proofing methodology must be harmonized, that is, warning, control, and shutdown.

7.6 Case Study H—Application of Six-Sigma: Parenteral Filling Operations

For parenteral products, solutions or suspensions are filled in bags, syringes, cartridges, or vials. For suspension filling in cartridges the content homogeneity is a critical quality parameter. Suspensions have fine solid particles dispersed in the liquid phase and it is important to have the appropriate concentration of solids filled in the final product. A mass balance is generally established throughout the filling line to ensure homogeneity of the filling operation. In a manufacturing line a part of the batch of filled cartridges were rejected, as the amount of solids was not in control and varied out of specification. A process improvement was required to make the process more robust, reproducible, and reduce the rejects improving yield and reducing DPMO. Six Sigma was used as the improvement methodology.

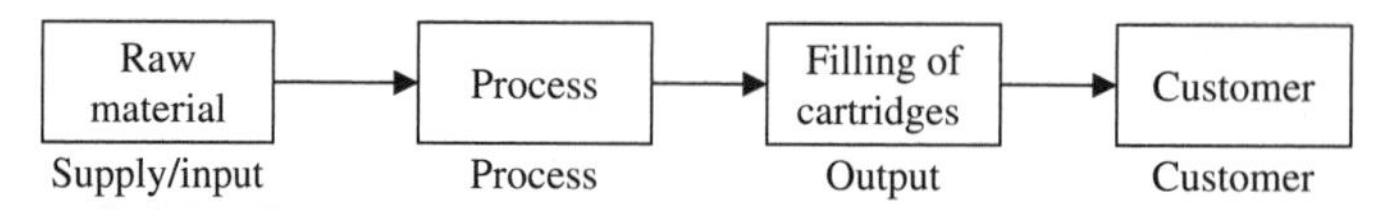

Figure 7.14 Overview.

7.6.1 Define problem identification

Problem identification involves a good understanding of the process, and use of SIPOC is typically the preferred methodology to provide an overview of the problem, which for case study H is shown in Fig. 7.14.

The problem statement was "Rejection of 5,000 cartridges per year from a total production of 100,000." The rejection was based on final quality control tests for percentage of solids. The flowchart of the process is shown in Fig. 7.15.

The next step in the definition phase is to identify the sigma level of the current operation, for which the defect opportunities need to be identified. This is typically done in a brainstorming session involving a cross-functional team with representatives from operations, engineering, quality, maintenance, and so forth. The defect opportunities identified totaled four, namely volume, recirculation rate, mixing intensity, and pumping rate. The sigma level is then calculated as shown in Table 7.10. The sigma level of the current operation was calculated as 3.0.

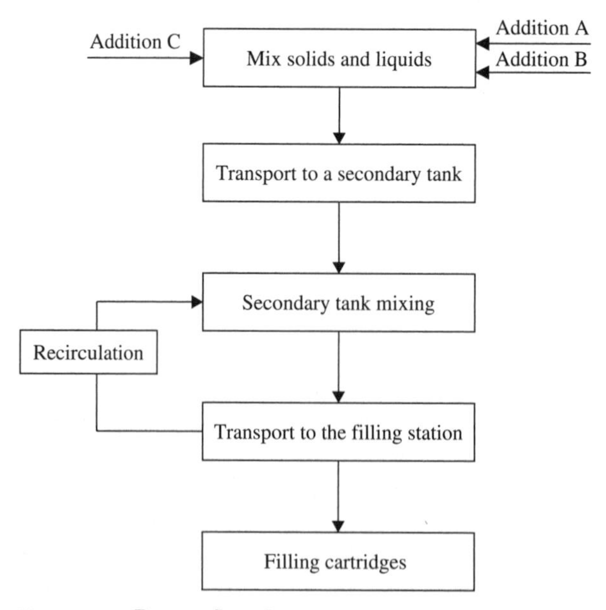

Figure 7.15 Process flow sheet.

TABLE 7.10 Sigma Level

Parameters	Case study
Number of defect opportunities	4
Number of units	100,000
Number of defects	26,720
DPMO	$26720 \times 10^6/(4 \times 100000) = 66{,}800$
Sigma level	3.0

7.6.2 Measure

Control charting is a very valuable method to make variability visible. Figure 7.16 shows the control chart for the percentage of solids in the paint cans.

The control chart indicates that the process is out of control and that it is a barely capable process. There are indications of special causes that must be analyzed further.

7.6.3 Analyze

Root cause analysis was performed for the special causes. The analysis indicated that the root cause was attributed to three defect opportunities and their manual control and settings (see Fig. 7.17).

Typically these parameters were manually adjusted within a specified range and a design of experiment was performed to assess the impact of their levels on the process response—concentration of solids.

7.6.4 Design of experiment

The design of experiment was carried out with three factors and two levels as shown in Table 7.11.

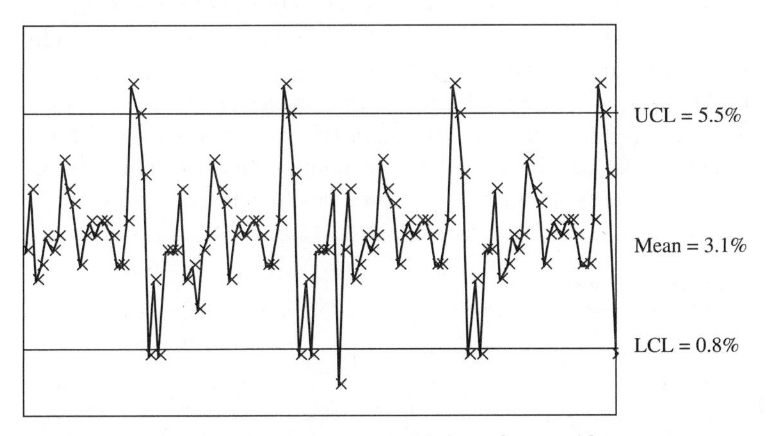

Figure 7.16 Control chart of percentage solids in the cartridges.

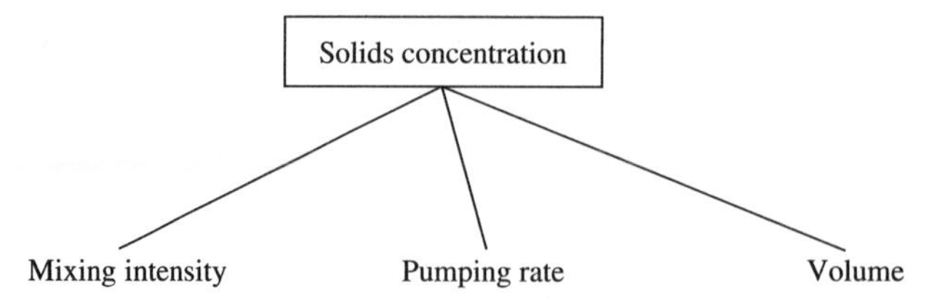

Figure 7.17 Analysis of root cause.

The total number of experiments (full factorial) $= 2^k = 2^3 = 8$.

A full factorial (Table 7.12) was executed to understand the impact of major factors and their interactions.

7.6.5 DOE output

Figure 7.18 shows the output from the full factorial design. This is followed by analysis of the statistics to understand the output. Some of the important aspects of statistical analysis are:

- The RSquare value of 0.86 indicates a good fit for the model.
- The root mean square error is 2.4.
- The parameter and the interactions are estimated based on which predictor profile is established (Fig. 7.19).

7.6.6 The prediction profiler

The prediction profiler displays *prediction traces* for each *X* variable. A prediction trace is the predicted response as one variable is changed while the others are held constant at the current values. The prediction profiler recomputes the traces as one varies the value of an *X* variable. The low and high values are shown on the *X*-axis for each factor. The vertical dotted line for each *X* variable shows its *current value* or *current setting*. The horizontal dotted line shows the *current predicted value* of each *Y* variable for the current values of the *X* variables. Prediction profiles are especially useful in multiple-response models to help judge which factor values can optimize a complex set of criteria.

TABLE 7.11 Factors and Level

Factors	Level 1, –	Level 2, +
Mixing intensity (rpm)	50	100
Volume (L)	50	70
Pumping rate (L/min)	20	28

TABLE 7.12 Experimental Design

Number	Mixing intensity	Volume	Pumping rate	Response
1	−1	−1	−1	6.5
2	−1	−1	+1	7.1
3	−1	+1	−1	3.3
4	−1	+1	+1	2.7
5	+1	−1	−1	1.5
6	+1	−1	+1	1.7
7	+1	+1	−1	1.3
8	+1	+1	+1	0.7

The required levels for the factors to achieve a required mean response of 3.1 are shown in Table 7.13.

7.6.7 Improve and control

Once the required level of the factors is identified with the help of the design of experiment the next step then is to implement. Mistake-proofing concepts were used, which included selection of sensors, and a new control strategy was implemented to control those parameters at the required set point. The control strategy used was feedback as discussed in Chap. 3. To further mistake proof the process an alarm strategy was used to alert the operators for any corrective intervention. Another aspect of mistake proofing is 100 percent inspection, which was also built in the system. An optical density meter was incorporated on-line to give indication of the solids concentration equipped with a comprehensive alarm strategy.

Such improvements also impact the number of defect opportunities, help to simplify points of intervention, and hence reduce mistakes.

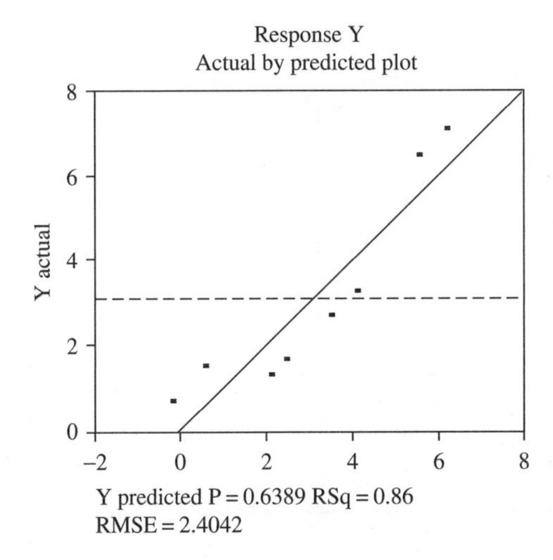

Figure 7.18 DOE output from full factorial design.

Summary of fit	
RSquare	0.859981
RSquare adj	0.019864
Root mean square error	2.404163
Mean of response	3.1
Observations (or sum Wgts)	8

Analysis of variance

Source	DF	Sum of squares	Mean square	F ratio
Model	6	35.500000	5.91667	1.0236
Error	1	5.780000	5.78000	Prob > F
C. total	7	41.280000		0.6389

Parameter estimates

Term	Estimate	Std error	*t* ratio	Prob>|t|
Intercept	3.1	0.85	3.65	0.1704
X1(50,100)	−1.45	0.85	−1.71	0.3375
X2(5070)	−0.4	0.85	−0.47	0.7200
X3(2028)	0.3	0.85	0.35	0.7840
X1(50,100)* × 2(5070)	−0.05	0.85	−0.06	0.9626
X1(50,100)* × 3(2028)	−0.35	0.85	−0.41	0.7513
X2(5070)* × 3(2028)	1.4	0.85	1.65	0.3474

Effect tests

Source	Nparm	DF	Sum of squares	F ratio	Prob > F
X1(50,100)	1	1	16.820000	2.9100	0.3375
X2(5070)	1	1	1.280000	0.2215	0.7200
X3(2028)	1	1	0.720000	0.1246	0.7840
X1(50,100)* × 2(5070)	1	1	0.020000	0.0035	0.9626
X1(50,100)* × 3(2028)	1	1	0.980000	0.1696	0.7513
X2(5070)* × 3(2028)	1	1	15.680000	2.7128	0.3474

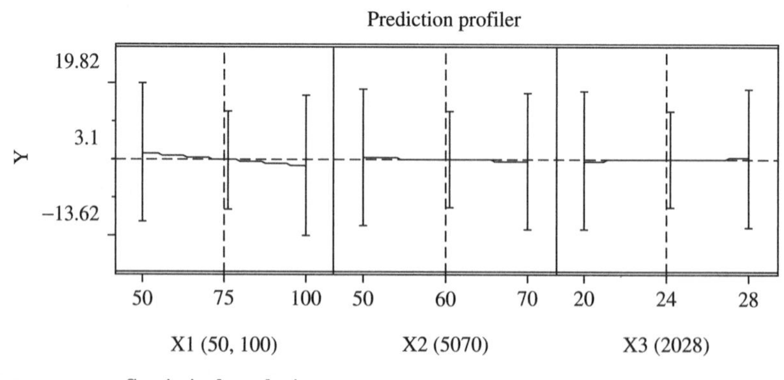

Figure 7.19 Statistical analysis.

TABLE 7.13 Optimal Levels

Levels	Required level
Mixing intensity (rpm)	75
Volume (L)	60
Pumping rate (L/min)	24

As a result of this improvement the improved control chart looks like the chart shown in Fig. 7.20.

The business summary from the Six Sigma effort can be summarized as shown in Table 7.14.

The Six Sigma effort not only reduced the production cost by over $1 million (COP) in the short term, but also more importantly enhanced quality by variability reduction, improving customer satisfaction, and enhancing sales.

7.7 Design for Six Sigma

Six Sigma is an improvement methodology and is instrumental in ensuring success to a manufacturing operation, as discussed in the previous sections. A complementary approach is also needed to focus on creating a new and better design. This approach is called *design for Six Sigma* (DFSS). DFSS consists of engineering and statistical methods used during product or process design. DFSS engages a process called IDDOV (Chowdhury, 2002), which stands for: identify the opportunity for improvement; define the requirements; develop the concept; optimize the design; and verify it.

It starts with identifying the design project, followed by a process to rigorously define the customer requirement and capture the voice of customers. The identify and define phases include generation of business case, project plan, defining and prioritizing customer needs, defining product requirement, and identifying CTQ measures. The next phase

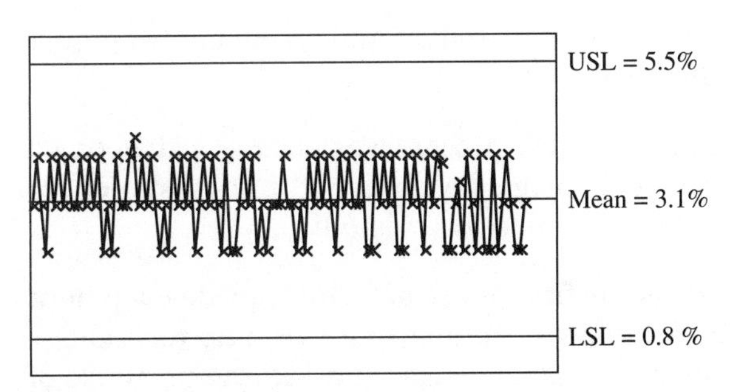

Figure 7.20 Final control chart.

TABLE 7.14 Business Summary

Parameters	Before Six Sigma application	After Six Sigma application
Number of defects	26720	390
Number of defect opportunities	4	3
DPMO	66,800	1300
Sigma level	3.0	4.5
Cost of rework at $40/unit	$1.07 M	15,600

is to develop the concept. This conceptual design involves brainstorming and other techniques for generating concepts. The conceptual design is optimized followed by verification. The verification could be carried out as a pilot plant study.

Creveling et al. (2003) extended the concept of DFSS in technology and product development. They applied different approaches for technology development and product development.

For product development (commercialization) DFSS methodology was built upon a balanced portfolio of tools and best practices that enable a product development team to develop the right data to achieve the following goals (called CDOV):

1. Conceive new product requirements and system architectures based on a balance between customer needs and the state of technology that can be effectively and economically commercialized.
2. Design baseline functional performance that is stable and capable of fulfilling the product requirements under nominal conditions.
3. Optimize the design performance so that the measured performance is robust and tunable in the presence of realistic sources of variation that the product will experience in the delivery, use, and service environments.
4. Verify systemwide capability of the product and its elements against all the product requirements.

For technology development, I^2DOV process was used as a structured and disciplined approach to technology development. I^2DOV stands for invent and innovate, develop, optimize, and visualize:

Invention and *innovation* is the first phase of technology development and involves defining technology roadmap and trends, technology benchmarking, generating new knowledge, and leveraging reuse of existing knowledge.

The *develop* phase of the process is designed to produce three distinct deliverables: superior technology and measurement system concepts, underlying performance model, and physical model or prototype.

Optimization of a new technology focuses on two distinct activities: robust optimization plan with regard to variability in the system and capability to meet critical functional requirements.

The *verify* phase focuses on the confirmation that the technology design intent is met.

Concluding Comments

Productivity/process improvement is vital for survival of a manufacturing operation. A consistent effort to enhance productivity leads to higher quality, reduced cost, and increased throughput. Productivity/process improvement has its roots in the "classical and modern quality philosophy" with Deming as the founder. The Deming wheel or the PDCA cycle forms the basis of Six Sigma. The key aspects of Six Sigma are customer centricity, measure, goal, executive engagement, resource commitment and infrastructure, elimination of waste, and improvement in financial results.

It is critically important to embrace a productivity mindset throughout the organization. Six Sigma, especially the lean aspect, has evolved as an overarching methodology for productivity improvement.

References

Amran, M., and N. Kulatilaka (1999), *Real Options*, Harvard Business School Press, Boston, MA.

Box, G. E. P., and K. B. Wilson (1951), *On the Experimental Attainment of Optimum Conditions*, *J. Roy. Statist. Soc., Ser. B.,* **13**:1–45.

Brealey, R. A., and S. C. Myers (1991), *Principles of Corporate Finance*, McGraw-Hill , New York.

Copeland, T., and A. Vladimir (2003), *Real Options*, TEXERA, New York.

Chowdhury, S. (2002), *Design for Six Sigma*, Dearborn Trade, Dearborn, MI.

Creveling, C. M., J. L. Slutsky, and D. Antis, Jr. (2003), *Design for Six Sigma*, Prentice Hall PTR, Upper Saddle River, NJ.

Cunha-Bakeev, C. (2001), Multivariate Analysis of Bioprocesses, PhD Thesis, University of Newcastle, Newcastle, UK.

Day, G. S., P. J. H. Schoemaker, and R. E. Gunther (2000), *Wharton on Managing Emerging Technologies*, Wiley, New York.

George, M., D. Rowlands, and B. Kastle (2004), *What Is Lean Six Sigma*, McGraw-Hill, New York.

George, M., D. Rowlands, M. Price, and J. Maxey (2005), *The Lean Six Sigma Pocket Toolbook*, McGraw-Hill, New York.

Gitlow, H. S., and D. M. Levine (2004), *Six Sigma for Green Belts and Champions*, FT Prentice Hall, Upper Saddle River, NJ.

Glassey, J., G. A. Montague, A. C. Ward, and B. V. Kara (1994), Artificial Neural-Network-Based Experimental-Design Procedures for Enhancing Fermentation Development, *Biotechnol. Bioeng.,* **44**(4)(Aug 5):397–405.

Hinckley, C. M. (2001), *Make No Mistake*! Productivity Press, Portland, OR.
Hopp, W. J., and M. L. Spearman (2000), *Factory Physics,* 2d ed., McGraw-Hill, New York.
Lareau, W. (2000), *Lean Leadership: From Chaos to Carrots to Commitment*, Tower II Press, Carmel, IN.
Manly, B. F. J. (1994), *Multivariate Statistical Methods: A Primer*. Chapman and Hall, London.
Montgomery, D. C. (1997), *Design and Analysis of Experiments*, 4th ed., Wiley, New York.
Moore, D. S. (1995), *The Basic Practice of Statistics,* W. H. Freeman, New York.
Morrison, D. F. (1990), *Multivariate Statistical Methods*, McGraw-Hill, New York.
Pande, P. S., R. P. Neuman, and R. P. Cavanagh (2000), *The Six-Sigma Way*, McGraw-Hill, New York.
Plackett, R., and J. Burman (1956), *The Design of Optimum Multifactorial Experiments, Biometrika*, **33**(4):305–325.
Roy, R. K (2001), *Design of Experiments Using the Taguchi Approach:16 Steps to Product and Process Improvements*, Wiley, New York.
Shingo, S. (1986), *Zero Quality Control:Source Inspection and the Poka-yoke System*, Productivity Press, Portland, OR.
Stern, J. M., J. S. Shiely, and I. Ross (1991), *The EVA Challenge-Implementing Value Added Change in an Organization*, Wiley, New York.
Wu, C. F. J., and M. Hamada (2000), *Experiments: Planning, Analysis, and Parameter Design Optimization*, Wiley, New York.

Chapter

8

Productivity Improvement Philosophy

Knowledge + Leadership + Siloless Synergy = Optimal Productivity P. MOHAN

Objective

Productivity is the underlying message in the book applied systematically to "process life cycle management." This final chapter focuses on pulling together the concepts discussed in the previous chapters in tune with an overarching philosophy for productivity improvement. The overarching philosophy focuses on key elements such as knowledge, leadership, and siloless synergy.

Learning outcomes of this chapter will include:

- Understanding productivity improvement in the larger organizational context. While Six Sigma methodology is critically important for productivity, it is a subset of a larger design that includes a well-balanced cocktail of knowledge, leadership, and siloless synergy.
- This book so far has focused on the analytical intelligence aspect of leadership. In this chapter other equally critical aspects of leadership are introduced.
- Logistical strategy to engage the organization in the productivity drive.

The learning is then applied to a comprehensive case study to gain practical understanding of the application of the concepts.

8.1 Introduction

In a competitive environment, organizations focus on productivity to generate a winning proposition for their customers. The aim of productivity improvement is to continuously or radically enhance manufacturing metrics. At the organizational level productivity has to be envisioned at a broader level encompassing all aspects of the business. In a complex, ambiguous, and dynamic business environment productivity is the key differentiator of a successful business. Productivity improvement methodologies have been discussed and applied in various stages of the manufacturing life cycle. Manufacturing life cycle management includes process design (Chap. 1), manufacturability (Chap. 2), critical control strategy (Chap. 3), knowledge management (Chap. 4), variability reduction (Chap. 5), and emerging monitoring and control strategies (Chap. 6). The productivity improvement methodologies (Chap. 7) and productivity improvement philosophy (this chapter) are key to successful process life cycle management.

The entire book could be summarized in Fig. 8.1. This simple systems diagram of the process life cycle indicates the outcome from each phase of the manufacturing life cycle. The commercialization phase (Chaps. 1 and 2) includes process design, economies of scale, scale-up/scale-down, risk management, and validation and manufacturability. The objective for the commercialization phase is to focus on quality by design. The next phase of the manufacturing life cycle is process capability (Chaps. 3 and 4). It includes critical control strategy, critical process parameters, and knowledge and data management. The output from the process capability phase is a capable manufacturing process.

The next phase of the life cycle is variability reduction (Chaps. 5 and 6). Variability reduction includes fundamental strategies for variability reduction and advanced control strategies and focuses on executing a robust manufacturing process. While productivity enhancement techniques for a learning organization are applied in all aspects of the process life cycle, they are equally significant for improving the process on an ongoing basis leading to optimal productivity.

In this chapter the productivity improvement philosophy is discussed in the broadest sense, and the key productivity enablers have been categorized into knowledge, leadership, and synergy.

$$\text{Knowledge} + \text{leadership} + \text{siloless synergy} = \text{optimal productivity} \tag{8.1}$$

Innovation ensures longevity of a business. Manifestation of organizational knowledge into innovative (breakthrough) products and services is a very important factor in the growth of a business. This knowledge base is the foundation for the organization's *intellectual property* (IP).

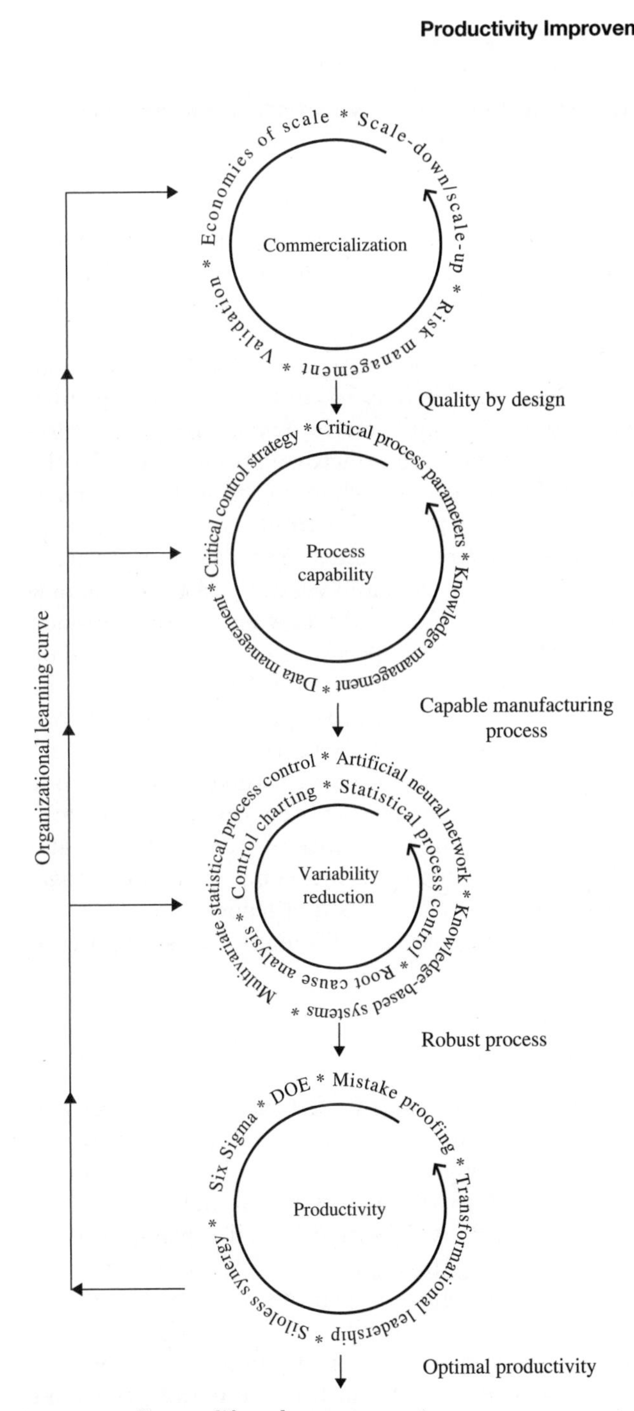

Figure 8.1 Process life cycle management.

Knowledge is the fundamental productivity metric. Knowledge is gathered throughout the evolution of the manufacturing process. Building the knowledge base and the learning curve takes years of investment, and preservation of the knowledge for posterity is critical for the survival of the business. A good knowledge management strategy is becoming a prerequisite for world-class corporations; this also includes training as a planned activity for knowledge enhancement. Chapter 4 discusses knowledge management and its critical value to an organization.

Besides knowledge, the other key aspect responsible for driving optimal productivity is leadership (discussed in Sec. 8.2) and siloless synergy (discussed in Sec. 8.3). Delivering successfully on the customer-focused scorecard (Fig. 8.4) is a key objective of a manufacturing organization in the competitive landscape. The factors that are shaping the evolving competitive landscape include global dynamics, the Internet revolution, and deregulation. The dynamic business environment in this postindustrial era is characterized by ambiguity, complexity, and volatility. Some of the new realities include: the Internet has unleashed a new era of transparency, connectivity, and customer power; the importance of branding based on credibility and trust; the breakup of old structures and categories; the value of cultural dynamics in globalization; the rapidly changing geopolitical situations; the development of new core competencies; and the complexity of internal governance. The old paradigm, orthodoxies and frame of reference (DNA code) of an organization, is being continually challenged. A successful organization is in a continual state of transformation and renewal so as to transform the new dynamics into competitive advantage. Such an organization is run by a special breed of leadership—*transformational leadership*. The following section discusses transformational leadership in detail.

8.2 Transformational Leadership

Jim Collins (*Good to Great,* 2001) suggests a five-level hierarchy of leadership. Level 1 is a highly capable individual contributing through talent, knowledge, skills, and good work habits. Level 2 is a team member contributing individual capabilities to the achievement of group objectives and working effectively with others in a group setting. Level 3 is a competent manager organizing people and resources toward the effective and efficient pursuit of predetermined objectives. Level 4 is an effective leader who catalyzes commitment to and vigorous pursuit of a clear and compelling vision, stimulating higher performance standards. Level 5 is an executive who builds enduring greatness through a paradoxical blend of personal humility and professional will.

This section introduces a higher level of leadership called transformational leadership. To produce transformation in an organization, leadership of a particular nature is required. Transformational leadership is defined as leadership that goes beyond ordinary expectations by transmitting a sense of mission, stimulating learning experiences, and inspiring new ways of thinking. The approach to transformational leadership is one that synthesizes the current body of knowledge using a single integrated model, which is represented in Fig. 8.2. This model captures the need for transformational leaders to create personal and collective consciousness, possess an internal and external focus, understand that a continuum of change exists, grasp the perspectives from which to view change, and recognize the polarity of skills required to lead transformational change.

Transformational leaders are change agents and must possess an internal and external focus on the effects of change. The internal view points to an understanding and an appreciation of change as experienced by the people and systems within the organization. The external view requires transformational leaders to understand how change impacts the people and systems outside the organization. The vertical axis of the transformational leadership model speaks to the continuum of change in organizations. At the bottom of the axis is a focus on standardization, at the top, a focus on change. Transformational leaders are able to hold this continuum in mind and make decisions about the kinds of change

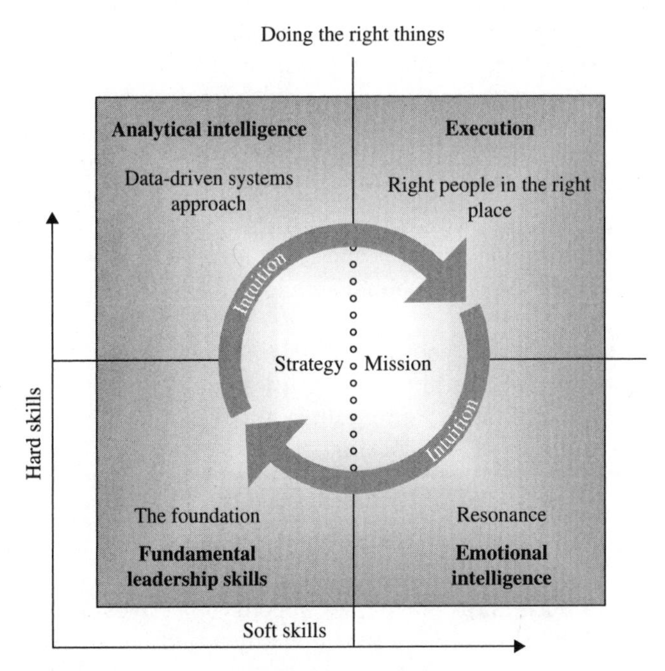

Figure 8.2 Transformational leadership model.

needed to achieve the vision and goals of the organization. Simply put, they cut through the confusion about the kinds of change needed.

The transformational leadership competencies required to lead postindustrial era organizations are illustrated as a model in Fig. 8.2. At the heart of transformational leadership lie the mission, vision, and strategy. Together they form the essence of the business, and are the prime responsibility of leadership. Intuition is required to guide leaders through the realm of uncertainty, ambiguity, and complexity.

> Intuition is when you know something without knowing how you know it. Intuition is an internal guidance system that is part association and memory, part experience, and part unknown—Nancy Rosenoff.

Transformational leaders require both hard and soft skills to deliver the essence of the business as enshrined in its mission, vision, and strategy. The soft skills include fundamental management skills and emotional intelligence. The hard skills include analytical intelligence and execution.

Analytical intelligence is an important aspect of transformational leadership. Analytical intelligence is the ability to engage appropriate objective analyses in reaching critical business decisions. Analytical intelligence is typically acquired through education, training, and experience. As manufacturing systems become more complex and the environment (regulatory, competition, and so forth) becomes increasingly challenging, leadership success hinges on knowledge of systems and analytical methods. Analytical intelligence has three elements: the understanding of basic analytical tools and methodology, systems thinking capability, and the capability of deep dives.

This book's focus has been on analytical tools and methodologies under the overarching Six Sigma approach. Chapters 1 through 7 of this book have focused on the importance of a data-driven analytical approach. It is critically important that the transformational leadership has a good understanding of these methodologies and their potential benefits.

Systems thinking is another key competency of transformational leadership. It is a scientific way of thinking, communicating, learning, and acting more effectively in a complex, ambiguous, and dynamic business climate. A system is a group of interacting, interrelated, or interdependent components that form a complex and unified whole. A system can be tangible (e.g., a car) or it can be intangible (e.g., processes, information flow, and so forth). Systems have several characteristics, which include boundaries (structure) with input and output, subsystems, patterns and alignment for optimal performance, specific objectives within the larger system, stability through fluctuations and adjustments, and feedback.

There are various interpretations of systems thinking. In this book, systems thinking denotes a data-driven scientific framework for approaching a complex situation to uncover and model generic patterns for learning and optimization. Flowcharting has been extensively used in this book to indicate systems and subsystems representing a larger complex manufacturing system with a key theme of productivity. Leaders with systems thinking competencies can break a complex challenge into simple systems with boundaries and interdependencies, overlaying performance metrics and its alignment. Such a system is easy to communicate as it has illustrations in terms of flowcharts and a stream of interrelationships. The complex pattern of engineering manufacturing productivity detailed in this book is uncovered using a systems thinking approach. One fundamental principle exists that is at the heart of all systems approaches. That is, a system should be structured and created to achieve the required emergent properties. The emergent properties are those of the system as a whole, which cannot be replicated by the simple addition of its parts. It is the interaction between the various systems and subsystems that yields the emergent properties. The concept of systems thinking and systems engineering is discussed in detail in various compilations including the works of Kossiakoff and Sweet (2003) and Blanchard and Fabrycky (1998).

Transformational leaders engage analytical intelligence in analyzing complex manufacturing operations on a day-to-day basis. They design the performance metrics in a hierarchical form: day-to-day operational data; plant operational data, for example, composite control charts; multiple plant metrics; and global business unit metrics (see Fig. 8.3). The ability to gauge the health of the business by interpreting the data

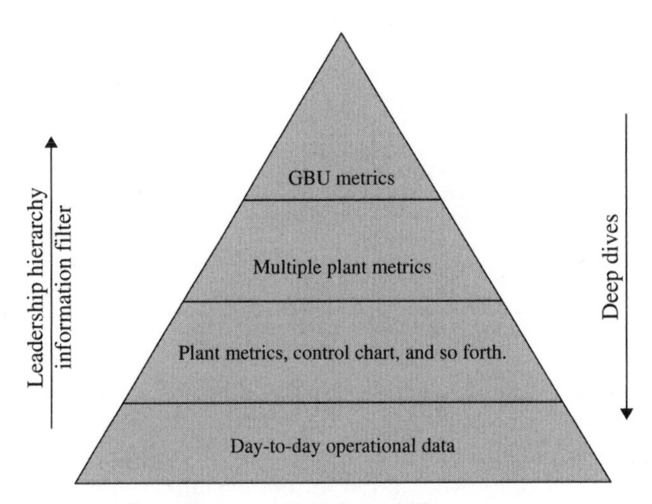

Figure 8.3 Deep dives—analytical capability.

and using the knowledge to further enhance the success of the business is a key competitive advantage. Knowledge does not mean that the leaders should become technical experts in every aspect of the process; instead it means a well-balanced understanding of the manufacturing system (as detailed in this book). This knowledge will help leadership to interface at various levels in the organization and unravel innovative approaches to the transformational needs of the organization. The ability to clearly understand and appreciate the technical challenges and the ability to package them into business-aligned objectives is key to success. Such leaders command respect (rather than demanding it) and provide the much needed inspirational leadership crucial for motivation. Equally important is the ability to dive deep to ascertain the root cause of poor performance.

Though analytical intelligence is sufficiently discussed throughout the book, it is equally important to understand other key aspects of transformational leadership, which are introduced in the following sections.

8.2.1 The essence of transformational leadership—mission, vision, and strategy

Mission and vision. Mission, vision, and strategy form the essence of the business. Mission defines the identity of the business—why do we exist? Vision defines the ultimate goal—what do we want to be? Whereas strategy sets the direction to achieve the vision. Mission, vision, and strategy are key differentiators of a business. Visionary leaders master the guts of the vision by creating a mission worth striving for, a vision that is achievable and a strategy designed to optimize the return from financial and human capital. Visionary companies are created and run by visionary leaders. Visionary companies prosper over long periods of time, through multiple product life cycles and multiple generations of active leaders. Visionary companies display a remarkable resiliency, an ability to bounce back from adversity.

Vision and mission state the core purpose for which a position, team, or organization is created. It is summarized in a clear, short, inspiring statement that focuses attention in one clear direction by stating the purpose of the individual, business, or group. It is a compass and a stretch. Gast's law (O'Halloron and O'Halloron, 1999) summarizes the key essence of a business. Gast's law implies that in order for a business to be successful in the long term, it must not only provide a just return on capital, but it must also do things such as produce a useful commodity or service, increase the wealth of society, provide productive employment opportunities, help employees find meaningful and satisfying work, and pay fair wages. In return, employees have an obligation to make the most productive use of their labor and the organization's capital.

Gast's law could be stated as:

Law 1. A business must produce a want-satisfying commodity or service, and continually improve its ability to meet needs.

Law 2. A business must increase the wealth or quality of life of society through the economic use of labor and capital.

Law 3. A business must provide opportunities for productive employment of people.

Law 4. A business must provide opportunities for the satisfaction of normal occupational desires.

Law 5. A business must provide just wages for labor.

Law 6. A business must provide a just return on capital.

Law 2 indicates that value creation in society is an important objective of a successful organization. Organizations that have done it well have created a brand name that has become a foundation for their success. J&J, Lilly, Merck, Pfizer, Nike, and so forth, are some of the companies that have successfully engaged in social focus by following Gast's law. Strong leadership is key in generating shareholder, social, learning, and employee focus.

Core ideology. The heart of the mission and vision is a core ideology that transcends short-term financial objectives and projects a timeless horizon. Core ideology is the nontangible part of mission and vision that captures the fundamental belief system of the organization that describes its genetic code. The notion of a timeless horizon is manifested in its core ideology—the north star of the business. A company must have a core ideology to become a visionary company. It must also have an unrelenting drive for progress with all the key pieces working in alignment.

A key step in building a visionary company is to articulate a core ideology. Core ideology has two elements: core values and *big hairy audacious goals* (BHAGs). Core values are the essence of the business, the foundation of the culture, and the genetic code of the organization.

$$\text{Core ideology} = \text{core values} + \text{BHAGs}$$

BHAGs—the organization's fundamental reasons for existence beyond just making money. A BHAG engages people, it reaches out and grabs them in the gut. It is tangible, energizing, and highly focused.

Core ideology is inherent in the mission and vision statements. Following are the mission and vision statements of some of the leading companies.

Lilly

Eli Lilly and Company is a leading, innovation-driven corporation committed to developing a growing portfolio of best-in-class and first-in-class pharmaceutical products that help people live longer, healthier and more active lives. We are committed to providing answers that matter—through medicines and information—for some of the world's most urgent medical needs. (www.lilly.com)

Merck

"To provide society with superior products and services by developing innovations and solutions that improve the quality of life and satisfy customers needs, and to provide employees with meaningful work and advancement opportunities, and investors with a superior rate of return."

Merck values include preserving and improving human life; ethics and integrity; scientific excellence; profits only from work that satisfies customers' needs and benefits humanity; integrity, knowledge, imagination, skill, diversity, and teamwork of our employees. (www.merck.com)

Dow

Dow's mission statement can be broken down into three components:

- Constantly improve – This concept is bedrock to Dow's culture and has been since H. H. Dow first said, "If you can't do better, why do it?" It underscores our drive to become an ever better and bigger company.
- Essential to Human Progress – The products we make find their way into products that provide people the world over with improved lifestyles. All of us at Dow must understand and take pride in this. We must also use this concept to further connect Dow with the external markets we serve. When we think in terms of the markets we serve, we become more outside-in focused and we can better seek growth opportunities.
- Mastering Science and Technology – We must put our science and technology to work to create solutions for our customers and for society. (www.dow.com)

Dow's values include integrity, respect for people, unity, outside-in focus, agility, and innovation. (www.dow.com)

Dupont

We, the people of DuPont, dedicate ourselves daily to the work of improving life on our planet.

We have the curiosity to go farther ... the imagination to think bigger ... the determination to try harder ... and the conscience to care more.

Our solutions will be bold. We will answer the fundamental needs of the people we live with to ensure harmony, health, and prosperity in the world.

Our methods will be our obsession. Our singular focus will be to serve humanity with the power of all the sciences available to us.

Our tools are our minds. We will encourage unconventional ideas, be daring in our thinking, and courageous in our actions. By sharing our knowledge and learning from each other and the markets we serve, we will solve problems in surprising and magnificent ways.

Our success will be ensured. We will be demanding of ourselves and work relentlessly to complete our tasks. Our achievements will create superior profit for our shareholders and ourselves.

Our principles are sacred. We will respect nature and living things, work safely, be gracious to one another and our partners, and each day we will leave for home with consciences clear and spirits soaring. (www.dupont.com)

These mission and vision statements though from different companies consist of generic key elements of BHAGs and core values. Maximizing profit or shareholder wealth was not the dominant driving force or primary objective through the history of most of these visionary companies. They have tended to pursue a cluster of objectives, of which making money is only one, and not necessarily the primary objective (following Gast's law). Values are the principles, standards, and actions that people in an organization take, which they consider inherently worthwhile and of utmost importance. They include how people treat each other; how people, groups, and organizations conduct their business; and what is more important to the organization.

Strategy. Strategy is derived from the Greek word "strategos " meaning "art of the general." Strategy was used by generals in ancient times as a "game plan" for winning wars. Today, strategy denotes the game plan required to achieve vision and mission focused on winning in a competitive business landscape. Developing a good strategy is key to the success of an organization as it optimally engages its scarce resources—labor, capital, and time—generating a winning proposition.

Hambrick and Fredrickson (2001) define strategy as an integrated, overarching concept of how the business will achieve its objectives with external focus. Strategy recognizes a business opportunity and a plan for seizing it. Hambrick and Fredrickson (2001) present a framework for strategy design consisting of five elements with an example of a personal computer:

1. *Economic logic.* Economic logic is the central element focusing on generating returns above the cost of capital. If return on asset is the key economic leverage, then the strategy would be focused around pricing, cost, and asset leverage. This is the essence of maximizing shareholder wealth and the fulcrum for profit generation.

   The economic logic for a PC manufacturer could include scale, JIT, direct to customers, and low cost.

2. *Arenas.* Arenas identify the playground where the business will be active. This will include product categories, market segments, geographic areas, core technologies, and value creation stages.

The arenas for a PC manufacturer could be global with special business focus on United States, India, and China. The product class could be customized personal computers.

3. *Vehicles.* Vehicles focus on the question: *how will we get there?* The approach may include organic growth, joint ventures, partnerships, acquisitions, and licensing/franchising.

   Vehicles for a PC manufacturer could be partnership for sales and marketing outside the United States and organic growth within the United States. Another aspect could be the use of the Internet for direct sales to customers.

4. *Differentiators.* Differentiators focus on the question: *how will we win?* In other words what are the key core competencies that differentiate a firm from another giving it competitive advantage?

   Differentiators for a PC manufacturer could include build to order, direct to customers, and world-class services.

5. *Staging.* Staging focuses on the question: *what will be the speed and sequence of moves?* In essence this entails project management to prioritize resources and capital by strategic initiatives.

   The staging approach for a PC manufacturer could include focus on establishing in the United States first and then focus on India and China. Focus first on the home market followed by small businesses. Not to engage in large corporate sectors.

Executives replace formal reporting structures with strategic themes and priorities that enable a consistent message and set of priorities to be used across diverse and dispersed organizational units. In the 1990s, companies extended the financial framework to embrace financial metrics that correlated better with shareholder value, leading to *economic value added* (EVA) and value-based management metrics. The strategy maps and customer-centric balanced scorecards constitute the measurement technology for managing in a knowledge-based economy.

Manufacturing productivity metrics—customer-centric scorecard. The concept of a balanced scorecard (Kaplan and Norton, 2001) is a key strategic enabler of productivity. The Kaplan and Norton version of the balanced scorecard includes four perspectives, namely, learning and growth, internal, customer, and financial. Vision and strategy are also integrated with the measurement. A critical enabler to achieve desired performance goals is the ability to measure performance. "What gets measured gets done."

In this section a customer-focused scorecard, which forms the central aspect of productivity, will be discussed. With customers at the center of the scorecard (see Fig. 8.4), the scorecard has two dimensions, namely, internal (within the organization) and external (outside the organization).

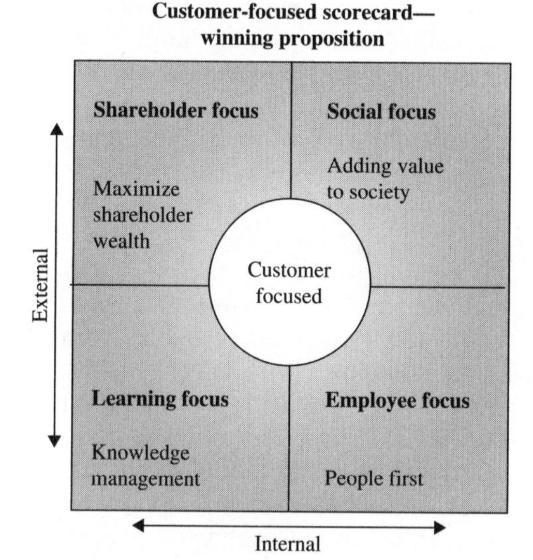

Figure 8.4 Customer-focused scorecard.

The internal dimension includes employee and learning focus whereas the external dimension includes shareholder and social focus.

Customer focus drives higher product quality and lower cost. As described in this book the manufacturing organization's response to customer focus lies in implementing a robust process design (with quality built into the design), reduced variability, and a productivity improvement program.

Good organizations focus on their employees. People are the cornerstone as well as the foundation of a business.

> We build great people, who then build great products and services.—Jack Welch

The challenge for leadership is to develop a win-win partnership with employees. The business should provide a fertile ground for maximization of an individual's potential and create a sense of ownership amongst employees maximizing their commitment to the company's success. Crisis of commitment is real when people are not working at their full potential. Competitive advantage comes from the effort workers put in above and beyond "just doing their job." If an experienced operator identifies ways to improve robustness in the operation but does not share it with management, technically he is still doing his job. Sharing the information could be framed as "walking the extra mile" that cannot be forced by leadership. Employees provide it at their discretion, and if they choose to give it often enough, the power it gives an enterprise is truly awesome.

Maximization of shareholder wealth drives a rigorous financial discipline. Until a business returns a profit that is greater than its cost of capital, it does not create wealth, it destroys it. There are various financial indicators, which are extensively discussed in various textbooks (Brealey and Myers, 1991; Stern et al., 1991). However, financial measures are lag indicators: they report on outcomes, the consequences of past actions. Exclusive reliance on financial indicators promotes short-term behavior that sacrifices long-term value creation.

In an economy dominated by tangible assets, financial measurements were adequate to record investments in inventory, property, plant, and equipment on a company's balance sheets. Income statements could also capture expenses associated with the use of these tangible assets to produce revenues and profits. But a postindustrial economy, where intangible assets have become the major sources of competitive advantage, calls for tools that describe knowledge-based assets and the value-creating strategies that these assets make possible. Opportunities for creating value are shifting from managing tangible assets to managing knowledge-based strategies that deploy an organization's intangible assets: customer relationships, innovative products and services, high quality and responsive operational processes, information technology and databases, and employee capabilities, skills, and motivation. The knowledge management aspects are discussed in Chap. 4.

Social focus requires organizations to focus on adding value to society. An important source of adding value to society is the product itself. Equally important, social aspects include employment generation, enhancement of local economy, and charitable donations.

Critical alignment. The mission, vision, BHAGs, strategy, scorecard, and objectives as discussed in the preceding sections must be aligned to maximize the success of the business. Creating alignment is a key part of the effort to enable companies to transform themselves into visionary companies. Critical alignment requires two key processes: (1) developing new alignments to preserve the core and stimulate progress and (2) eliminating misalignments—those that drive the company away from the core ideology and those that impede progress toward the envisioned future. Figure 8.5 depicts a critical alignment at a strategic and logistical level summarizing this discussion. Such an alignment between the mission/vision all the way to the employees' personal objectives is critical to the success of the business.

Strategy and mission form the essence of the business and are central to the transformational leadership model, as depicted in Fig. 8.2 and discussed in the preceding sections. The first quadrant of this leadership model is "fundamental leadership skills." This quadrant is discussed in detail in the following section.

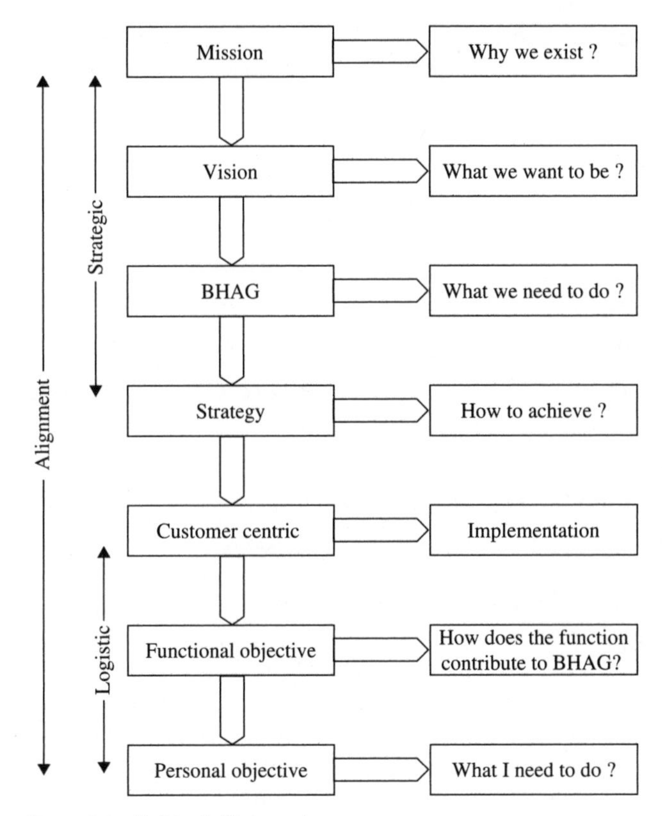

Figure 8.5 Critical alignment.

8.2.2 Fundamental leadership skills

Fundamental leadership skills (FLS) are the bedrock of transformational leadership. The prominent skill set is character. Character breeds trust, respect, and integrity. Character is the foundation on which the other leadership attributes are built.

Fundamental leadership quality has been studied by leading thinkers like W. Edwards Deming and Stephan R. Covey. Deming in his classic compilation *Out of Crisis* (2000) suggests "14 points for management" to lead the transformation of American industry. These are:

1. Create constancy of purpose toward improvement of product and service, with the aim to become competitive, stay in business, and provide jobs
2. Adopt the new philosophy (understand the need for change)
3. Cease dependence on inspection (build quality in design)
4. Minimize total cost

5. Improve constantly and forever the system of production and service, to improve quality and productivity, and thus decrease cost (improvement strategy to include both innovation and continuous improvement)
6. Institute training on the job
7. Institute leadership
8. Drive out fear
9. Break down barriers between departments
10. Eliminate slogans
11. Eliminate work standard and management by objective on the factory floor by substituting it with leadership
12. Remove barriers
13. Training education and self improvement
14. Participative transformation program

Covey's (1992) concept of *principle centered leadership* has the primary focus on people and the interrelation with natural laws. Covey describes principle-centered leadership as an inside-out approach with personal trustworthiness at the core of this concept. Trustworthiness creates the basis for the managerial style of empowering others to unleash more of their potential and become independent and self-managing. The three levels of personal, interpersonal, and managerial relationships form the necessary conditions for working on issues of alignment—harmonizing the organization's shared mission and values, structures, and systems as it responds to its customers, suppliers, competitors, and other stakeholders.

Individuals with fundamental leadership skills tend to develop the softer aspect of leadership that is called "emotional intelligence." Emotional intelligence forms the second quadrant of the transformational leadership model (Fig. 8.2).

8.2.3 Emotional intelligence—connect and resonate with people

Emotional intelligence is a combination of two competencies—personal and social. Conscious individuals forming conscious groups working toward a shared purpose and vision are vital ingredients in creating a conscious enterprise. Consciousness of enterprise is seen when the entire organization has the ability to reflect and learn. For leaders of transformational change, there are three primary perspectives in the ring of mastery to be considered: *self-mastery*, *people mastery*, and *enterprise mastery*. Self-mastery includes clarity of purpose, vision, planning, reflection, and feedback. Transformational leaders are engaged in the lives of the people in their organization, encouraging personal growth, feedback, continuous learning, and mentoring. The most successful transformation initiatives are fostered by leaders who are committed to their own self-awareness or personal mastery and to assisting others in the same process. People mastery requires both individual consciousness and an understanding of the attitudes, behaviors, beliefs, and assumptions of the collective in addition to developing skills in group dynamics. Relationship management, trust, management of agreements, and effective communication are vital to the development of consciousness, mastery, and effectiveness at the group level. Understanding the business is a key component of enterprise mastery. Enterprise mastery includes the skills and capacity of the organization's leaders as chief architects of business, benchmarking, organizational culture, process improvement, motivation systems, evaluation and measurement of performance, and the importance of understanding the organization's value and how it can change over time.

Perhaps the emotional intelligence aspect of leadership is best represented by great leaders (typically in the social arena—Mahatma Gandhi, Martin Luther King, Mother Teresa) who move us. They ignite our passion and inspire the best in us. Great leadership works through emotions. To formulate a vision that will resonate with others, leaders start by tuning in to their own feelings and the feelings of others. The key aspects of great emotional leadership are resonance, self-awareness, self-management, self-regulation, motivation, empathy, development of others, and organizational savvy.

Resonance. Being in synch with people's emotional centers in a positive way. One of the most powerful and direct ways to make that resonant brain-to-brain connection is through laughter. Social awareness, particularly empathy, is crucial in driving resonance.

Self-awareness. Reading one's own emotions and recognizing their impact. It also relates to knowing one's strengths and limits and having a sound sense of one's self-worth and capabilities. People with strong self-awareness are realistic, neither overly self-critical nor overly optimistic. Rather they are honest with themselves about themselves. And they are honest about themselves with others, even to the point of being able to laugh at their weaknesses.

Self-management. Allowing the mental clarity and concentrated energy that leadership demands, and keeping disruptive emotions from throwing them off track. Leaders with such self-mastery embody an upbeat, optimistic enthusiasm that tunes resources to the positive range. Self-discipline is key to self-management.

Self-regulation. Managing one's internal status, impulses, and resources.

Practicing self-control. Keeping disruptive emotions and impulses in check.

Trustworthiness. Maintaining standards of honesty and integrity.

Conscientiousness. Taking responsibility for personal performance.

Adaptable and innovative. Handling changes flexibly and being comfortable with novel ideas, approaches, and new information.

Motivation. Having an internal engine of emotional tendencies that guide or facilitate reaching goals. This includes drive, commitment, initiative, and optimism.

Empathy. Being aware of others' feelings, needs, and concerns. This is a key social competence for understanding others, sensing others' feelings and perspectives, and taking an active interest in their concerns. Great leaders are wired for empathy.

Developing others. Sensing their development needs and bolstering their abilities. Services orientation anticipating, recognizing, and meeting customer's needs. Leveraging diversity—cultivating opportunities through different kinds of people.

Organizational savvy. Being in tune with an organization's emotional state, current priorities, and power relationships. Power relationships include networking with people in positions of authority and appropriately leveraging them for business and personal advancement.

The success of an organization hinges on strong transformational leadership; the leadership that masters the essence of the business (mission and strategy) and has the leadership attributes as depicted in Fig. 8.2, namely, fundamental leadership skill, emotional intelligence, analytical intelligence, and execution. Execution is the "doing" part of leadership, which will be discussed in the following section.

8.2.4 Execution

Bossidy and Charan (2002) define execution as the discipline of getting things done. It is the most challenging aspect of leadership and only a few organizations excel in this area. The inability to execute strategy is a major weakness. Strategy execution can be more important than the strategy itself. Bossidy and Charan (2002) have identified three building blocks and three core processes for effective execution. The three building blocks are the leader's seven essential behaviors, creating the framework for cultural change, and having the right people in the right place. The three core processes focused on linking the strategy to operations include the people process, the strategy process, and the operation process.

Mastery of the self is a key aspect of execution that can be manifested in the leader's seven essential behaviors (Bossidy and Charan, 2002), namely, know your people and your business, insist on realism, set clear goals and priorities, follow through, reward the doers, expand people's capabilities, and know yourself. The second building block of execution focuses on the cultural system or the social software. Linking rewards to performance is an important element of the social software and in enabling employees to succeed. To achieve high performance and develop a result-oriented culture is an opportunity for management. The third building block for effective execution is having the right people in the right place (Bossidy and Charan, 2002). The right people energize others, are decisive on tough issues, effectively delegate, and follow through.

The people process emphasizes the importance of people first in linking strategy and operations. The people process should include linking people to strategy and operations; developing the leadership pipeline through continuous improvement of succession planning, and reducing retention risk; dealing with nonperformers; and linking *human resources* (HR) to business results (Bossidy and Charan, 2002). The strategy process focuses on continual evaluation of strategy with respect to external environment, customers and markets, obstacles of growth, competition, execution potential, balance between short term and long term, critical issues facing the business, and sustainable basis of profit generation. The operation process focuses on the logistics of building operation plans, developing budgets, promoting siloless synergy, setting realistic goals, contingency planning, and quarterly reviews. The competencies of execution could be broadly categorized into personal effectiveness and organizational design. Personal effectiveness includes mastering the essence of the business, laser sharp priorities, organizational savvy, consistent execution, and managing the social system.

Transformational leadership is a key enabler of any successful organization focused at optimizing productivity. Another key element of optimizing productivity (Eq. 8.1) is siloless synergy. This will be discussed in the following section.

8.3 Siloless Synergy

Synergy is the overarching goal of organization design. Organizations consist of numerous sectors, business units, and specialized departments, each with its own strategy. For organizational performance to become more than the sum of its parts, individual strategies must be linked and integrated. The leaders define the linkages expected to create synergy and ensure that those linkages actually occur—a task that is easier said than done. Organizations are traditionally designed around functional specialties such as finance, manufacturing, marketing, sales, engineering, and purchasing. Each function has its own body of knowledge, language, and culture. Functional silos arise and become a major barrier to strategy implementation as most organizations have great difficulty communicating and coordinating across these specialty functions.

> The boundary-less company would remove all the barriers among the functions: engineering, manufacturing, marketing, and the rest. It would recognize no distinctions between "domestic" and "foreign" operations—Jack Welch
>
> A boundary-less company would knock down external walls, making suppliers and customers part of a single process. It would eliminate the less visible walls of race and gender. It would put team ahead of individual ego—Jack Welch
>
> Boundary-less would also open us up to the best ideas and practices from other companies—killing NIH (Not-Invented-Here)—Jack Welch

Figure 8.6 models four states of siloless synergy with vertical and horizontal integration as two enablers. On one extreme a high level of vertical integration signifies that the function has achieved a high level of functional excellence though its ability to integrate horizontally is severely limited, creating a functional silo. Such functions or global business units have difficulty generating synergistic values. On the other hand a function with low vertical and a high horizontal integration lacks functional identity and functional excellence. Low vertical and low horizontal integration lead to a dysfunctional organization.

The ideal state is a good balance between vertical and horizontal integration generating synergistic value leading to cooperation, collaboration, convergence, and competence. The opportunity for leadership is to identify the present state and devise and execute strategies to move toward an ideal state of generating value to the customers. As an example under

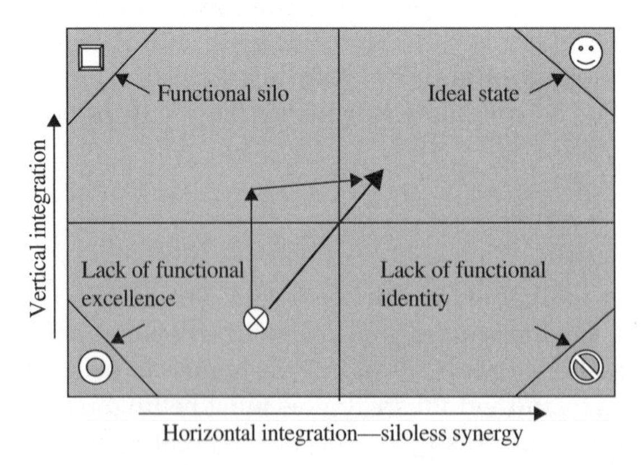

Figure 8.6 Siloless synergy—four states.

Jack Welch's leadership, the global business units within GE were asked to become either number 1 or 2 in their respective market segment—promoting vertical integration. The global business units who could not get to number 1 or 2 were either closed or sold out. Thereafter the functional silos between various global business units were synergized by Jack Welch's initiative of "boundary-less collaboration" maximizing and leveraging value creation for customers (Welch, 2003).

The previous sections introduced productivity improvement philosophy, and the following sections will introduce logistical strategies to execute improvement, illustrated by a case study.

8.4 Logistical Aspect of Productivity Improvement

Productivity improvement strategies are traditionally split into two types, which represent different and, to some extent, opposing philosophies (Imai, 1986). These two philosophies are continuous, incremental improvement and radical, step-change improvement (as shown in Fig. 8.7).

Continuous improvement entails a mindset geared toward problem solving—determining root causes of inferior performance, converging on a solution, and systematically implementing that solution. Continuous improvement involves modest but continual changes to an existing process. Continuous improvement is an evolutionary approach to improvement and is synonymous with the concept of total quality management (Imai, 1986; Oakland, 1989). It is a philosophy of business that aims to root ongoing improvement into the basic strategies, culture, and management systems (Deming, 1986).

Continuous improvement is often referred to as *kaizen*. "Kaizen means improvement. When applied to the workplace, kaizen means continuous improvement involving everyone—managers and workers alike" (Imai, 1986).

Radical change, in contrast, is a revolutionary approach. It is not about improving existing processes, but reinventing them. Radical change

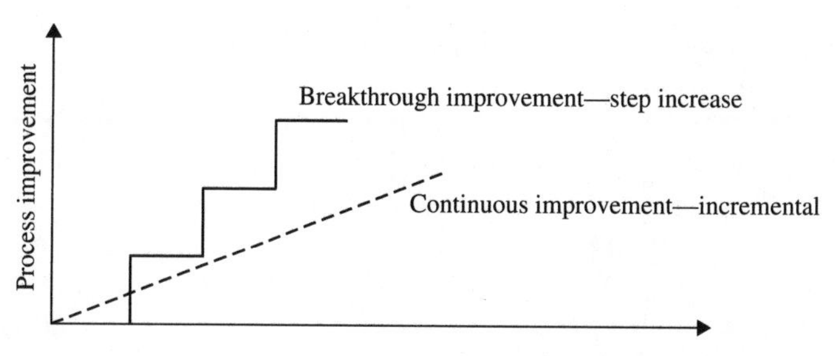

Figure 8.7 Continuous versus breakthrough improvement.

involves streamlining, reorganizing, and integrating activities to create new ways of working to support a process management orientation. This is essentially a fundamental rethinking of processes with the basic aim of getting dramatic improvements (Hammer, 1996). Processes for radical change are usually classified in broad terms so that they cut across functional boundaries. Those with narrow scope and falling within one functional area provide little opportunity for radical change and radical improvement (Hammer and Champy, 2001). Radical change would also typically require redefining information flows, their points of use, and work roles (Hammer, 1996).

Several key differences between continuous and step-change strategies have been suggested. Step change seeks radical changes, indeed the total redesign of existing processes coupled with a significant improvement in performance (Hammer, 1996). The benefits from small, successive, continuous improvements are expected to be attained over a long period of time unlike radical change, which aims to create major improvements in the short to medium term (Imai, 1986). Continuous incremental improvement involves everyone in an organization, and the changes are driven by them, thus requiring little senior management time and effort. Radical change is usually driven by a senior management champion and requires substantial senior management time and effort (Imai, 1986). Both these approaches involve processes as the primary unit of analysis, and rigorous measurement of process performance is necessary for either approach to succeed. Both process innovation and improvement also require significant organizational and behavioral change to be successful. At the most basic level, all process management approaches flourish only in an environment intent on implementing operational change—improving the way work is done—rather than making quick fixes in financial results or organizational structure.

Finally, both continuous process improvement and process innovation programs require a substantial investment of time, often as much as one or two years, before significant results can be seen. Continuous improvement requires time-consuming training and cultural change, while process innovation typically requires time for construction of new information systems and organizational structures.

Surprisingly, the differences between the two approaches are greater than the similarities. Process innovation or reengineering programs strive for radical, sometimes tenfold levels of improvement in the cost, time, or quality of a process. Improvement programs are considered successful if they achieve a 10 percent improvement in any given year. Improvement programs start from the current state of the process and chip away at it. Innovation programs urge participants to imagine they are starting with a clean sheet of paper. Improvement programs are highly participative. Innovation programs tend to be addressed from the top-down in terms of how the new work design is created. Improvement programs stress the rigor of statistical process control to minimize unexplained variation in a process. On the other hand, process innovation programs attempt to identify the technological or organizational process factors that will maximize variation and create fruitful changes.

8.4.1 Integrated approach

When there is little understanding of the differences between improvement and innovation, the wrong techniques can be applied. In certain cases expectations of innovation could be created with access to the tools of improvement. However, employing innovation-oriented tools to achieve continuous improvement is a recipe for failure. This can also lead to miscommunication with parties outside the inner circle of operational change. Here, it is essential that concepts and terms be crystal clear. In some firms, for example, all initiatives to improve processes are called *reengineering* regardless of their change in goals or methods. One solution is to combine continuous improvement and breakthrough improvement as an integrated strategy in the organization's improvement plan.

It is critically important to generate a corporatewide "integrated improvement vision" using Six Sigma as an overarching methodology to integrate continuous and breakthrough improvement (see Fig. 8.8). Unless process improvement and innovation approaches are integrated in organizations, employees become confused about the differences between change programs. Unless the subtle, but important distinctions between innovation and improvement are made apparent to them, they are likely to view the different programs as passing fads. Failing to integrate these approaches can also be quite demoralizing for those who participate in process change teams. Employees can spend time improving processes that may later be eliminated through innovation.

The first step of the integrated strategy is hypothesis generation (Fig. 8.8) followed by risk profiling and prioritization. Risk profiling is discussed in detail in Chap. 7, and prioritization is typically driven by financial consideration, which is also discussed in Chap. 7. The following section introduces hypothesis generation.

8.4.2 Hypothesis generation

Hypothesis generation aims at developing a list of ideas for improvement and innovation. However, hypothesis generation processes for innovation and improvement follow different approaches.

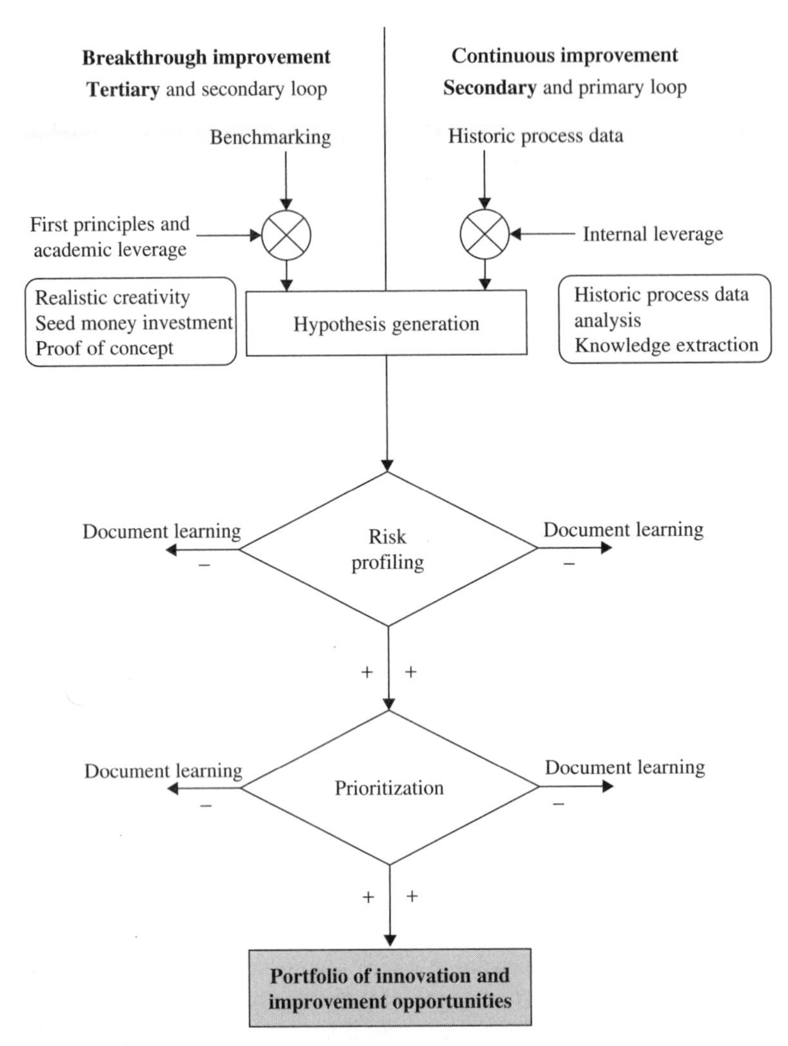

Figure 8.8 Define phase—innovation and improvement opportunities.

For innovation (breakthrough improvement) a "think tank" should be assembled in partnership with academics and consultants. The think tank also sponsors external benchmarking and literature search. The task for the think tank is to develop a list of ideas containing hypothesis derived from mechanistic approach (a "deep knowledge-based approach"). Deep process knowledge includes underpinning technical knowledge and experience with running the process. The cultural base for breakthrough improvements is generally different from that of day-to-day manufacturing troubleshooting. This is a medium- to long-term investment in enhancing process performance. The culture entails risk taking, rigorous analytical environments, process analysis and discussions, out-of-the-box thinking, sharing thoughts, collaborating with external centers of excellence in various aspects of the process, and so forth. It is more akin to a research and development environment. The staffing requirements are also similar. Highly qualified Ph.D.-level individuals, experienced in the subject area relevant to the process, are required as technical leaders. As an example, for a bioprocess the list of ideas could include (in no particular order) strain improvement, feeding strategy, continuous sterilization, robotics, and lights-out operation.

For continuous or incremental improvement the targets for process performance measures can be set by comparison with a variety of benchmarks, which, although not necessarily mutually exclusive, essentially fall into two types—internally based and externally based. Internal benchmarking uses historical standards based on the past performance of internal processes. The internal process might be the one undergoing improvement or it could be another similar internal process. The key disadvantage of using the process itself as the base for comparison, while undoubtedly encouraging improvements in performance, is that it only provides information as to whether the operation is getting better over time rather than whether performance is satisfactory. Comparison with other internal processes has the additional advantage that it provides a relative position for each process within the organization.

External benchmarking involves comparison with other organizations, using either competitor-based targets and/or best-in-field benchmarks. Competitor-based targets are those with similar operations in other, similar organizations. Best-in-field benchmarks are set based on the performance achieved by organizations, which may or may not be in the same field, but where the performance is considered to be outstanding. Organizations undertaking continuous improvement of operational processes will primarily employ internally based targets, whereas organizations undertaking radical change will primarily employ externally based targets. This is because organizations using these strategies wish to see incremental improvement relative to their historical achievements. Furthermore, as organizations using these strategies tend to be both successful and competitive, they may have already outperformed competitors or be the best-practice leader focused on building on their existing strengths.

It is important to define the organizational concept that can effectively execute the integrated approach. Process engineering organization is key to management of productivity improvement opportunities. It is an opportunity for transformational leadership to create an effective process engineering infrastructure that can lead productivity improvement in the organization.

8.4.3 Create process engineering infrastructure—holistic approach

One way of establishing the infrastructure is to consider all aspects of the manufacturing life cycle as entailed in this book, namely, process development, capability, variability reduction, and productivity improvement. A compelling reason for this holistic approach is to ensure a synergistic organization with complementary parts cohesively moving with clear direction and alignment. Since it is typical in larger organizations to have disjointed efforts competing with each other, creating wasted efforts adding less value (business and quality) than expected, the process engineering functionality could be subdivided into primary loop, secondary loop, and tertiary loop (see Fig. 8.9).

Primary loop (plant-based engineering) signifies a vital process engineering function responsible for "keeping the fire burning." The role of the primary loop, which is closest to the manufacturing floor (plant technology), is to maintain equipment in a qualified state, implement and sustain process improvements, and continually reduce variability and the required troubleshooting. This loop should have a mix of engineering talent and experience, typically 0 to 10 years, and should be the entry point for new engineers. Organizationally this loop should be an integral part of the local site manufacturing management well integrated with the local quality, operations, and science functions. In the pharmaceutical industry, typically this function gets buried in *current good manufacturing practices* (cGMP) documentation and only occasionally gets to fulfill its core functionality. A possible remedy could be to align this organization into the integrated system of the secondary and tertiary loops so that a clear functional vision could be achieved, without losing sight of the need for local control. Additionally, other resources and functions (quality) can contribute to sharing the much needed cGMP documentation function.

The focus of the secondary loop (product-based engineering) should be on the product, which should include ownership of continuous improvement (including capacity), partnering for breakthrough improvement, new process implementation (commercialization), and also closely mentoring the primary loop. This is a vital link that takes a step back from day-to-day operation and focuses on short- to medium-term improvements focused on individual product. Ownership of the continuous improvement is a major function, which includes enhancement of product capacity. The secondary loop should also be involved in the commercialization of new processes. Working together with the tertiary loop, they should deliver a robust new process in the plant (commercialization).

The tertiary loop would do the process development at pilot scale. The integration with the secondary loop is critically important in scaling-up the engineering aspects from pilot to production scale. Educating the local

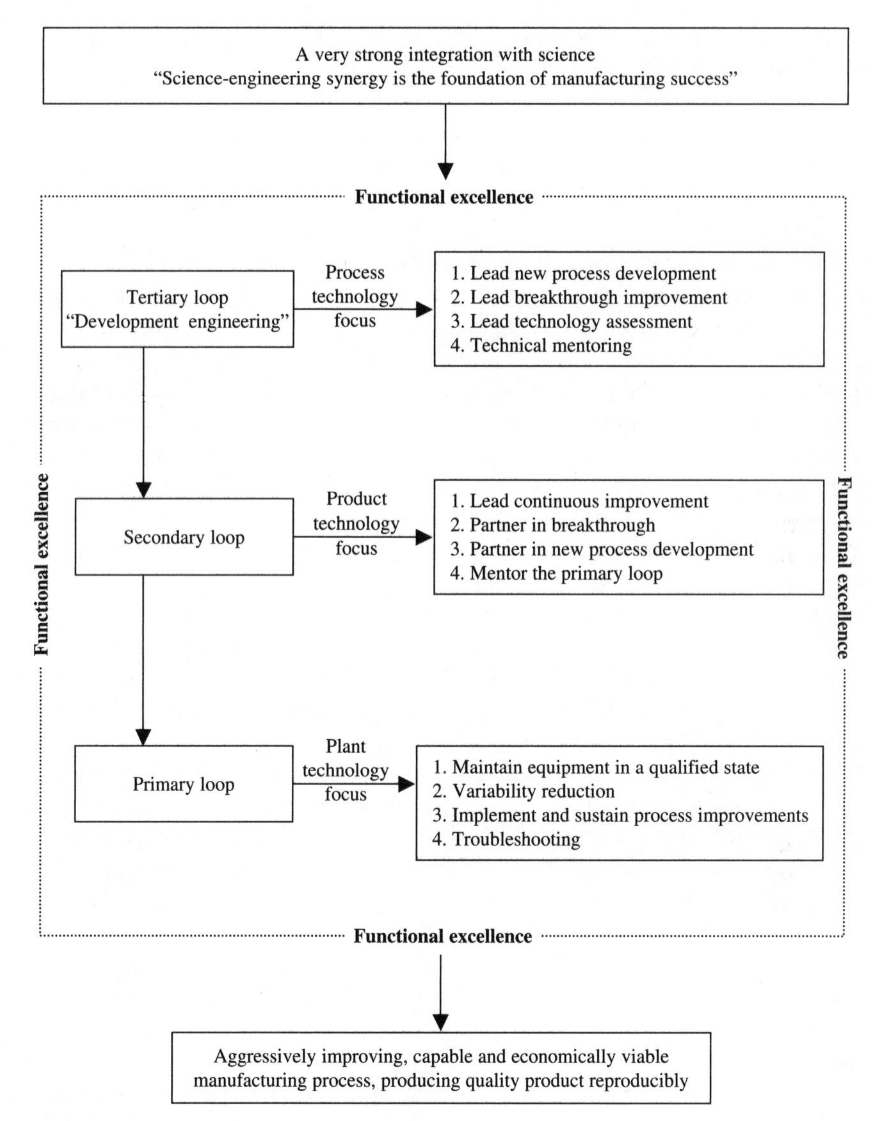

Figure 8.9 Process engineering infrastructure.

site concerning the technology challenges and opportunity along with maintaining a close connection with the tertiary loop is critical to success of this loop. Mentoring of the primary loop is an important activity, which involves regular exchange of technical information and coaching for technical professional growth. It has been a subject of debate in industry as to where this organization should fit; while a close integration with local site management has the advantage of integration by organizational design;

however, a major disadvantage is that there is always a danger that this function could become a primary loop extension. A compromise could be that a matrix management be used to ensure that this critical function is linked to the overall functional design with accountability toward the local site.

The tertiary loop (development engineering) should have a technology focus, the fundamentals of which should be applied to any process development or breakthrough improvement. The tertiary loop should have a strong external focus with links to academia and industry. The tertiary loop should be a well-balanced team with highly experienced generalists and *subject matter experts* (SME). The SME should be industry leaders in their subject of expertise and many of them may have higher qualification, Ph.D.s in the relevant subject area. They should be leading "centers of excellence" in their subject of specialization with access to engineering labs. Another major objective of the tertiary loop is to evaluate emerging technologies and assess the risk of the improvement processes in pilot plant or engineering lab facilities. Technical mentoring is another important role of the tertiary loop, which involves technical training in the areas of specialization to the primary and secondary loops. Though often not given enough priority training and knowledge, sharing is a proactive approach of building a learning organization.

The tertiary loop should be the engineering department responsible for late stage development of new processes closely integrated with the scientific community. It is generally debated whether they should be a part of the scientific organization or a part of the engineering organization. Being a part of the scientific organization ensures that close integration with scientists is built into the organizational design. However, the lack of integration with the engineering function may result in suboptimal functional excellence, which may hinder application of rigorous engineering. There is no right answer and different companies have different philosophies. A compromise for late stage development could be to ensure that the functional integration is maintained with the integration with scientific community driven by a "process-centric" approach with clear accountability. Dual reporting structure and matrix management may also aid in integration with scientific community and with the functional excellence.

The infrastructure of process engineering and the portfolio of innovation and improvement opportunities define the productivity landscape. The next step is to understand the productivity target.

8.5 Productivity Target

Productivity target is based on the needs of the plant. The define phase of Six Sigma can be used as an overarching methodology to set the productivity target (see previous sections). In this section productivity avenues in relation to cost and capacity are discussed.

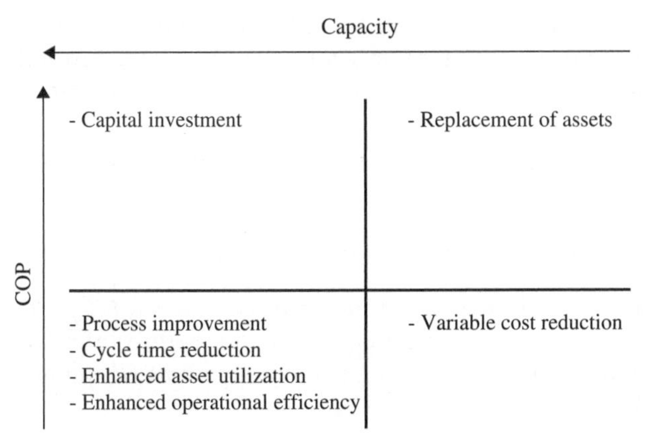

Figure 8.10 Capacity versus cost of product.

8.5.1 Cost and capacity

The interaction between cost and capacity creates interesting dynamics and there are various levels of cost and capacity. Enhancement of capacity with reduction in *cost of product* (COP) is a preferred option. Capacity versus COP could be represented as shown in Fig. 8.10.

1. *Variable cost reduction.* Unit cost decreases as throughput increases and variable unit cost decreases. The variable cost includes (as described in Chap. 1) raw material, utilities, consumables, and maintenance labor cost. There is always pressure to find a cheaper or better substrate (medium). Fermentation industries may have an advantage over some other manufacturing industries in that their raw materials can sometimes be altered, within limits, and some buffering against increasing world prices may be possible. Another major advantage as compared to the conventional chemical industry is that the fermentation raw materials are generally cheap agricultural products (waste, i.e., molasses, corn steep liquor, and the like).

$$\begin{aligned}\text{Unit cost} &= (\text{fixed cost} + \text{variable cost})/\text{throughput}\\ &= (\text{fixed cost}/\text{throughput}) + \text{variable unit cost}\end{aligned}$$

2. *Process improvement/optimizing throughput/productivity.* Optimizing the process to maximize its full potential is a key focus for manufacturing. For example, strain improvement using a mutation/selection program for an organism being used in an established process or a potential process can be very cost-effective. The process improvement could be continuous or breakthrough; these aspects have been covered in detail in the previous sections of this chapter.
3. *Operational efficiency.* This is primarily related to scheduling of both man and machine. The shifts are planned to maximize the

manufacturing potential. Along with that, the campaign strategy for a multiproduct manufacturing facility is also closely examined. (For example, the number of units of a product that are produced before the facility switches to another product must be carefully planned.) Given the changeover and cycle time of a particular product, unit cost can be expressed as:

Unit cost = (changeover cost/units per run) + running cost per unit

An aspect of efficiency is waiting time in queue called CT_q. For manufacturing equipment in series (with no redundancies), Kingman's equation defines the waiting time in queue as (Hopp and Spearman, 2000):

$$CT_q = V\,U\,T$$

where V = variability term
U = utilization term
T = mean effective process time

The objective should be to minimize the cycle time in queue. Therefore, the variability should be minimized, the utilization must be enhanced [$U = u/(1 - u)$, where u is related to the utilization of the equipment, i.e., probability that the equipment is busy] and the processing time T should be minimized.

4. *Reducing the cycle time for production.* Cycle time is best defined by Little's law (Hopp and Spearman, 2000):

$$CT = \frac{WIP}{TH}$$

where CT = cycle time, which is expected time spent at processing stage [= sum of queue time (CT_q) plus process time]
WIP = work in progress
TH = throughput

Reduction in WIP or increase in throughput or both could cause reduction in cycle time. Cycle time can be affected by setting up less frequently (facilitated through setup reduction), by dedicating equipment (so that some product families can be continually run without changing over), and by using specialized equipment (e.g., flexible manufacturing systems). Of course, some of these options can result in larger inventories.

5. *Asset utilization.* Long-term economies of scale are functions of plant equipment itself. Economists have long noted that the cost of the equipment tends to be proportional to its surface area, while capacity is

more closely proportional to volume. To illustrate the implication of this, suppose the equipment is a cube with side length l. Then cost could be expressed as

$$K = a_1 l^2$$

And the capacity as

$$C = a_2 l^3$$

where a_1 and a_2 are proportionality constants. To express cost as a function of capacity, solve for l in terms of C, to get

$$l = a_2 C^{1/3} \quad K = a_3 C^{2/3}$$

with a_3 representing another constant. Thus, in general, an increase in capacity leads to a reduction in cost of products.

The capital assets should be fully utilized to enhance asset productivity. For example, in typical antibiotic production, increasing the volume of the batch leads to enhancement of the asset utilization.

6. *Capital investment.* For a capacity-limited product there is always the dilemma of whether to continue to squeeze more capitalless capacity or to invest capital to build another manufacturing facility. The best way to address this is to develop models to help understand the ideal capacity of the manufacturing facility and then to conduct a financial analysis of building a new facility. The modeling aspect is covered in detail in Chap. 2.
7. *Replacement of assets.* Capital assets have a limited lifetime, and replacement of assets, generally with improved capability (may help enhance capacity), has cost implications.

This chapter summarizes the productivity philosophies along with logistical aspects. The following section applies these concepts to an industrial productivity improvement case study. The systematic approach to productivity improvement using Six Sigma methodology is illustrated in this case study—using the *define, measure, analyze, improve, and control* (DMAIC) approach.

8.6 Productivity Improvement Case Study

The case study from Chap. 3 is used to illustrate the application of Six Sigma to productivity improvement. The operations involved (as shown in Fig. 8.11) in the production process are raw material storage, raw material treatment, three unit operations, packing, and shipment.

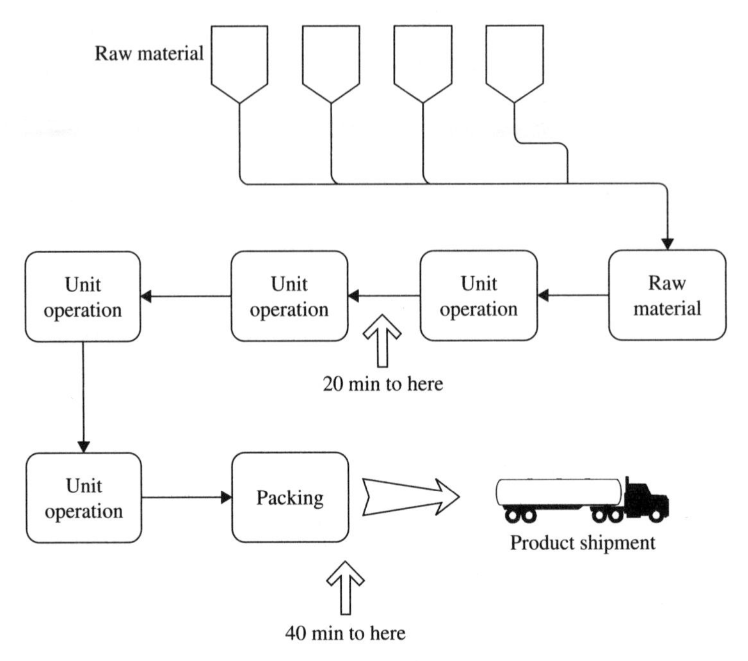

Figure 8.11 Overview of the production line.

8.6.1 Define

An improvement opportunity was identified based on external benchmarking with input from external consultants. The quality, organizational, technological, and financial risks were minimal. A well-thought-out cost justification was prepared for this case study. Inevitably, some assumptions about performance enhancement will have to be made to arrive at an estimate of savings. One possibility is if the objective is to improve the performance of a large-scale manufacturing plant, an estimate of the ultimate performance could be obtained by applying the strategy to a pilot-scale plant. Assuming the availability of a pilot plant, this does not overcome the necessity for engineering work and hardware and software costs. Thus, while it would provide confidence in the likely applicability of the method to large scale, it would not satisfy the fundamental objective of assessing likely savings before purchasing and implementing the system. The major benefit gained by adopting this approach would be minimizing the risk of lost production, but the question of scale on system performance is a major issue. Under these circumstances the more presumptive approach of cost-benefit analysis must be applied.

The strategy for cost-benefit analysis can be applied to process improvements achieved through changes in operating procedures or plant equipment modifications. The emphasis in this section is on operating policy improvement through improved process control. Cost benefit analysis is possibly one of the most difficult areas to be considered in the development of a control system for process improvement. Here the term *control system* is used in its widest sense as referring to any technology that delivers improvement in operation.

8.6.2 Measure

When considering the implementation of control scheme modifications, it is essential to establish a base case against which improved operation can be assessed. This involves building a historical record of operation prior to improvements and comparing this against resulting performance. In doing so it is vital to agree on the method of quantifying performance at the outset to avoid conflict in interpretation of the future results. Historical process operating records usually provide the necessary information. The technical requirements for a particular control system improvement can generally be ascertained prior to purchase by using a combination of past experience and process tests. However, what is much more difficult is to determine how much improvement in control is possible and what the financial consequences of this improvement are.

An approach to cost-benefit analysis that has been widely applied in the chemical industry to justify process control system investment is presented by Anderson (1996) and Anderson and Brisk (1992). The basic assumption of the procedure is that improvements in control will at least halve the existing variance of the output. This is a tried and tested statistic, and the extent to which it can be exceeded obviously depends on the existing quality of control; where little attention has been paid to control, it is likely to be an underestimate. On well-controlled plants, this level of reduction of variance may require the implementation of some complex control schemes as described in Chaps. 3 and 6. In the chemical process industries, the average improvements from implementing various levels of control have been estimated. Table 8.1 uses information from Anderson (1996) and gives an indication of savings achieved through control improvements.

Table 8.1 provides a useful insight into the typical costs associated with control system developments and the likely return on investment. The base case is a process primarily under operator control. If the costs and benefits of implementing a full optimization-based control scheme are considered to be 100 percent gain levels, then it can be seen that the basic monitoring and control system constitutes a major investment for little benefit (i.e., 70 percent of the overall cost only achieves 20 percent of the potential savings). Implementing some of the more advanced control procedures discussed in Chaps. 3 and 6 provides significant gain for

TABLE 8.1 Common Control System Benefits

Control system improvement	Savings gained (%)	Overall cost of control system (%)
Implementation of regulatory control systems and basic hardware	20	70
The use of advanced control procedures such as feed forward and model based	75	80
The application of optimization methods to the process	100	100

little extra investment. To achieve the 100 percent level requires a reasonable amount of investment. The key questions are

- What does the 100 percent level refer to in terms of financial gain?
- Where do these benefits arise from?
- What technology is required to achieve the benefits and what will it cost?
- Do the benefits outweigh the costs sufficiently to invest in change?

Figure 8.12 shows a typical and by no means exhaustive list of areas where benefits can be found by making improvements in the control system. Identifying areas of potential benefits for a particular process

Figure 8.12 Sources of potential of benefits.

requires careful consideration and is usually best achieved by establishing a team with process knowledge and process systems expertise.

The team identifying improvement opportunities should have the following skills:

- *Business and economics understanding*, as without this it would be impossible to assess the implications of improvement.
- *Process knowledge* (e.g., process engineer) to provide knowledge of the engineering limitations and implications of any process modifications.
- *Plant operation* (e.g., senior operator), as they possess the greatest insight into the current plant operating policy and are in the best position to advise on the practicality of modifying existing systems.
- *Control expertise* probably from outside the plant as control systems engineers with the depth of experience necessary tend not to be plant based.
- *The "outsider"* provides the opportunity to question the conceived wisdom.
- *The "champion"* is instrumental in establishing the team and with the need to ensure long-term usage of any systems change.

To achieve savings, it is necessary to make a prior financial commitment. Costs can arise from a number of sources such as those shown in Fig. 8.13.

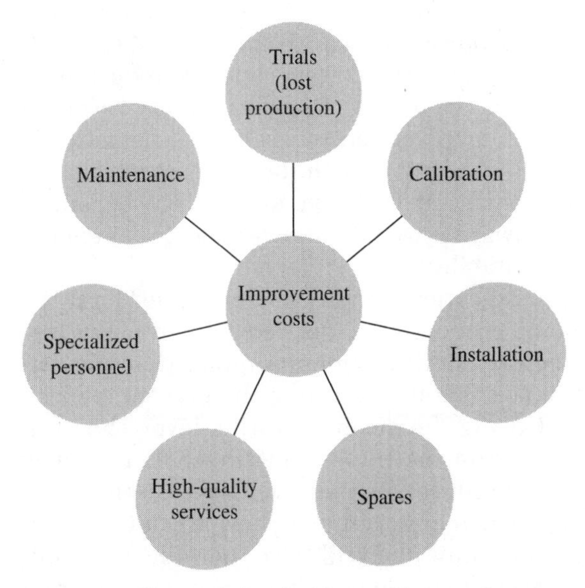

Figure 8.13 Costs associated with control system improvements

From discussions in the previous chapters it is apparent that improvements in control could result from improved instrumentation or from more sophisticated control schemes. The important message is that whatever the source, it is likely that the improvements bring with them difficulties in installation and long-term maintenance requirements.

Given the assumption that the variance of the process may be halved, the next step is to determine the opportunities to realize savings that result from improved control. These generally come from two related effects:

By reducing the variance of the operation, it is likely that the variance of the product quality will also be reduced. This should make the product more appealing from a customer perspective. However, the cost benefits that this brings are difficult to quantify.

In some cases, the best operating policy for a process is to closely approach constraints. Set points of process variables are specified as close as possible to constraints, given the level of variation experienced. If the variation can be reduced, then the set point can be moved closer toward the constraint and hence benefits gained. It is more straightforward to assess financial benefits arising from such situations.

8.6.3 Analyze

An example serves to highlight the procedure for calculating savings when constraints limit production. Consider the case study described in Chap. 3. It was found that in the process shown schematically in Fig. 8.11, poor operational policies on unit operation 2 were causing product quality to have significant variation. The degree of product quality variation prior to the implementation of improved control is shown schematically on the left-hand side of Fig. 8.14, together with the upper constraint on product quality. The upper constraint is shown, as, with the benefit of process knowledge, moving toward it would improve yield, lower energy costs, and hence increase profitability.

So how was the case for justification of improvement made? First the statistical variation of plant productivity was determined using historical process measurements. Making the assumption that the distribution is gaussian, it is possible to determine the percentage risk of violating this constraint using the mean operating level, the variance of the product quality, and the upper constraint. If it is now assumed that the variance is reduced by improved control, then it is possible with this new variance to determine by how much the mean can be moved toward the constraint and still have the same probability of violating the constraint. This idea is shown schematically on the right-hand side of Fig. 8.14, and the distribution of performance is shown in Fig. 8.15.

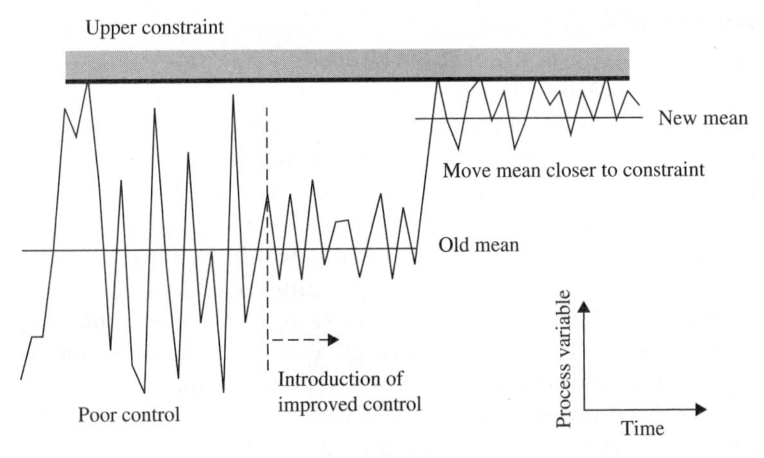

Figure 8.14 Improved control leads to change in mean operating level.

The dashed distribution in Fig. 8.15 shows the resulting distribution after the improvements have been made. Simple statistical analysis of the problem leads to the following movement of the mean (Δx):

$$\Delta x = (x_L - x) \times \left(1 - \frac{\sigma_{\text{new}}}{\sigma}\right) \tag{8.2}$$

where x_L = limit
x = current mean
σ = current operating variance of the plant
σ_{new} = assumed variance after improvement

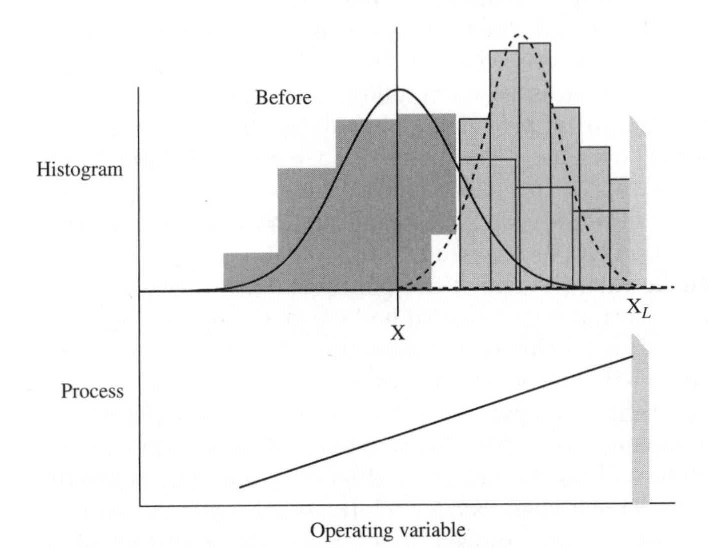

Figure 8.15 Statistical determination of the change in mean.

8.6.4 Improve/control

Process operating data were used to determine the likely change in variance and the resulting change in mean operating level that would be possible. An estimate of a 1 percent increase in yield was determined as a result. The question then follows, what is a 1 percent increase in yield worth? Since it was possible to sell any product that the plant made, it was relatively straightforward to measure the benefits from improved sales. In this case a modest benefit would be found but far more important were the implications on operating costs. Moving toward the constraint considerably reduced energy requirements and this exceeded any benefits from improved yield. Taking the hourly benefit and translating this to those achieved in a year is not just a matter of multiplying by the number of hours in a year. An allowance must be made for reduced process operating times due to plant maintenance and control system availability being less than 100 percent, resulting in a reduced benefit.

With the likely benefits determined, the next step was to consider the costs involved in obtaining these benefits. In this case, modifying the operating strategy involved changes in procedures, with little additional costs over current practice. In the general case, it would be necessary to account for the costs associated with the features shown in Fig. 8.15. With the cost-benefit analysis facts derived, a meeting with plant senior management took place to present the case for undertaking the project. With a strong case as a result of the cost-benefit analysis, permission was given to undertake the project. The technical details of the project are described in Chap. 3.

Identifying the potential savings that result from control improvements is only the first step in realizing the benefits. Implementation and commissioning of the system then follows. It is important at the completion of the project to assess whether the predicted benefits actually materialized. Performance assessment following system commissioning gives some indication of improvement, but determinations after the "honeymoon period" with long-term statistics are more realistic. In the early stages of use, extra concentration on the system can give misleading indications of improvement. Once normality resumes, the true benefits of the system can be assessed.

It is important toward the end of the improvement project to put in place the procedures that ensure continued usage of any system at its peak performance. It is for this very reason that the "champion" is part of the project team—if their vision for improvement is delivered then they will ensure that any system continues to function effectively. Without this champion, it is highly likely that following implementation, the control system will not be utilized sufficiently and will gradually become redundant. This is even true for relatively straightforward PID- (Chap. 3) based control loops. Indeed, in a large scale, multiunit plant one of the tests for scope of potential improvements is to determine the percentage of control loops in manual rather than automatic control. If a significant number are in manual mode, this indicates poorly tuned controllers or other equipment problems that, if rectified, could be a source of major benefit.

The system champion was apparent at the outset of the project and indeed in this case initiated the improvement study. Their role in implementing and maintaining a functioning control system was vital. The results presented in Chap. 3 proved that the variation determined over long-term performance was halved in line with the assumptions made. From a financial perspective the project payback time was around six months.

To summarize, the procedures to realize potential for improvement are as follows:

- Establish a team with intimate process knowledge.
- Establish process operating goals and constraints.
- List opportunities for improvement.
- Quantify the benefits associated with each opportunity.
- Produce an action plan to tackle appropriate opportunities.
- Estimate the cost of the action plan.
- Obtain financial approval.
- Implement the control system and measure the benefit.
- Establish a procedural framework to ensure that benefits are maintained.

Clearly, the cost-benefit analysis strategy presented is not just a method by which to determine the potential savings, it also constitutes a design and implementation philosophy to bring long-term achievement of the objectives. The procedure described has been tried and tested on control applications in the chemical industry. It is equally valid for application to other sectors, but a reasoned argument is not always the driving force for improvement. Experience suggests that occasionally when considering the implementation of some of the more advanced approaches, such as real-time knowledge-based systems, advances by competitors seem to be more influential than the pure financial argument. Clearly, the situation is extremely complex, as long-term financial gain may result from short-term loss for the purposes of the assessment of the performance characteristics and the determination of areas of potential application. This said, even when implementation comes under the banner of research and development, it is good practice to make some attempt to assess the financial implications prior to embarking on control improvements, whatever the driving force for the application.

Concluding Comments

A productivity-conscious organization has a distinct competitive advantage. Productivity should be viewed as an integration of knowledge, leadership, and siloless synergy. Successful productivity efforts require

- Transformational leadership that can lead the implementation of a comprehensive vision
- An improvement infrastructure that can support the vision and implementation
- Successful adoption of the Six Sigma methodology
- A learning organization

Productivity improvement has two logistical aspects: continuous and breakthrough. The continuous improvement concept has its origin in the Japanese manufacturing operating philosophy. This philosophy focuses on "ongoing operations introspection," which is closely monitoring, learning, and challenging the operations to improve the process. In this chapter the concept of continuous improvement was discussed in detail with industrial examples. These include the concept of design of experiment, database techniques for identifying continuous improvement, and so on. Breakthrough improvement could be characterized by quantum leap (step improvement) in the performance of the process. Breakthrough improvement can lead to reengineering the process.

References

Anderson, J. S. (1996), "Control for Profit," *Trans. of Inst. M.C.*, **18**(1):3–8.

Anderson, J. S., and M. L. Brisk (1992), "Estimating the Economic Benefits of Advanced Process Control," *I.Chem.E. Symposium Advances in Process Control III*, York, UK, 23–24 September.

Blanchard, B. S., and W. J. Fabrycky (1998), *Systems Engineering and Analysis*, Prentice-Hall, Upper Saddle River, NJ.

Bossidy, L., and R. Charan (2002), "The Discipline of Getting Things Done," *Execution*, Crown Business, New York.

Brealey, R. A., and S. C. Myers (1991), *Principles of Corporate Finance*, McGraw-Hill, New York.

Collins, J. (2001), *Good to Great*, HarperCollins, New York.

Covey, S. R. (1992), *Principled Centered Leadership*, Fireside, Rockefeller Center, New York.

Deming, W. E. (1986, 2000), *Out of Crisis*, MIT Press, Cambridge, MA.

Hambrick, D. C., and J. W. Fredrickson (2001), "Are You Sure You Have a Strategy," *Acad. Manage. Exec.*, **15**(4):48–59.

Hammer, M. (1996), *Beyond Reengineering*, Harper Business, New York.

Hammer, M., and J. Champy (2001), *Reengineering Corporation*, Harper Business, New York.

Hopp, W. J., and M. L. Spearman (2000), *Factory Physics*, 2d ed., McGraw-Hill, New York.

Imai, M. (1986), *Kaizen*, McGraw-Hill, New York.

Kaplan, R. S., and D. P. Norton (2001), *The Strategy Focused Organization*, Harvard Business School Press, Boston, MA.

Kossiakoff, A., and W. N. Sweet (2003), *Systems Engineering—Principles and Practices*, Wiley, Hoboken, NJ.

O'Hallaron, R., and D. O'Hallaron (1999), *The Mission Primer: Four Steps to an Effective Mission Statement*, Mission Incorporated, Richmond, VA.
Oakland, J. S. (1989), *Total Quality Management*, Butterworth-Heinemann, Oxford, UK.
Stern, J. M., J. S. Shiely, and I. Ross (1991), *The EVA Challenge-Implementating Value Added Change in an Organization*, Wiley, New York.
Welch, J., and J. Byrne (2003), *Straight from the Gut*, Headline Book, London.

Index

ABOUT THE AUTHORS

PANKAJ MOHAN, PHD, MBA, AMP, CENG, FICHEME, has over 16 years of diverse multinational (United States, United Kingdom, Germany, and India) experience in the pharmaceutical and biotechnology sector, which includes over 10 years of experience with Eli Lilly and Company in various leadership and technical assignments in the United Kingdom and the United States; 3 years of experience in academia with a center of excellence in biochemical engineering at University College London, UK; and about 2.5 years of experience in process utilities operations in India. Pankaj has a PhD in chemical engineering from the University of Birmingham, UK; master's in financial management from London, UK; an executive education (Advanced Management Program) from Fuqua School of Business, Duke University, U.S.; and a bachelor's degree in chemical engineering from the Indian Institute of Technology, India. Pankaj is a Chartered Engineer and Fellow of the Institution of Chemical Engineers, UK. He is recipient of the top leadership award within Lilly and has over 20 publications.

JARKA GLASSEY PHD graduated from the Slovak Technical University, Bratislava, Slovak Republic in 1990 with a First Class Honours in Biochemical Engineering. She was awarded a PhD in bioprocess modelling and control by the University of Newcastle in 1995. Dr Glassey joined the lecturing staff of the School of Chemical Engineering and Advanced Materials (formerly the Department of Chemical and Process Engineering), where she currently holds the position of senior lecturer. Her research areas concentrate on issues relating to bioprocesses, ranging from the development of novel postgenomic modeling tools to the investigations of artificial intelligence methods and their practical application mainly in biopharmaceutical industry. Her research and the results of close collaboration with a range of industrial partners are widely published.

GARY A. MONTAGUE, PHD, graduated from the Department of Chemical and Process Engineering, University of Newcastle in 1982 with a First Class Honours in Chemical Engineering. The following year he obtained an MEng in control systems from the University of Sheffield before returning to study at Newcastle for a PhD in bioprocess control. He joined the lecturing staff of the Department in Newcastle in 1985. Following research and teaching successes, he was appointed to a personal professorship in 1997. His research areas are quite diverse but include the development of artificial intelligence methods and their practical use in a variety of industrial sectors. Notable studies include an assessment for the Institution of Chemical Engineers in the United Kingdom on the industrial state of the art and requirements in the area of fermentation control. During the course of his research he was worked with many companies from a variety of industrial sectors. He has published the results of this work widely.